# NUTRACEUTICALS
# AND HEALTH
## Review of Human Evidence

# Nutraceuticals: Basic Research/Clinical Applications

Series Editor: Yashwant Pathak, PhD

*Nutraceuticals and Health: Review of Human Evidence*
Somdat Mahabir and Yashwant V. Pathak

*Marine Nutraceuticals: Prospects and Perspectives*
Se-Kwon Kim

NUTRACEUTICALS Basic Research/Clinical Applications

# NUTRACEUTICALS
# AND HEALTH

## Review of Human Evidence

Edited by
## Somdat Mahabir • Yashwant V. Pathak

CRC Press
Taylor & Francis Group
Boca Raton London New York

CRC Press is an imprint of the
Taylor & Francis Group, an **informa** business

CRC Press
Taylor & Francis Group
6000 Broken Sound Parkway NW, Suite 300
Boca Raton, FL 33487-2742

First issued in paperback 2016

© 2014 by Taylor & Francis Group, LLC
CRC Press is an imprint of Taylor & Francis Group, an Informa business

No claim to original U.S. Government works

Version Date: 20130524

ISBN 13: 978-1-138-19999-6 (pbk)
ISBN 13: 978-1-4665-1722-6 (hbk)

| Library of Congress Cataloging-in-Publication Data |
| --- |

Nutraceuticals and health : review of human evidence / editors, Somdat Mahabir, Yashwant V. Pathak.
  p. ; cm. -- (Nutraceuticals : basic research/clinical applications)
  Includes bibliographical references and index.
  ISBN 978-1-4665-1722-6 (hardcover : alk. paper)
  I. Mahabir, Somdat, editor of compilation. II. Pathak, Yashwant, editor of compilation.
III. Series: Nutraceuticals (Series)
  [DNLM: 1. Dietary Supplements. 2. Diet Therapy. QU 145.5]

RM216
615.8'54--dc23                                                                2013020082

**Visit the Taylor & Francis Web site at**
**http://www.taylorandfrancis.com**

**and the CRC Press Web site at**
**http://www.crcpress.com**

*Since ancient times, humans have tried to understand the concept of food as medicine. They have used food components and parts of plants as medicine and have made various concoctions out of these, not only to fight disease but also for well-being and good health. As we try to understand the scientific underpinnings of nutraceuticals in health and disease using modern tools and methodologies, we also acknowledge the fact that there is much to be learnt from the past, even the ancient past. Therefore, we dedicate this book to all the Rishis, sages, shamans, medicine men and women, and people who have followed ancient traditions and cultures and have contributed to the development of medicine and nutraceuticals, thereby keeping health sciences alive for the last several millennia.*

# Contents

## PART I    Introduction

### 1 Background to Nutraceuticals and Human Health. . . . . . . . . . . . . . . . . . . . . . . . 3
*Dipanwita Dutta, Somdat Mahabir,
and Yashwant V. Pathak*

## PART II    Cancer

### 2 Clinical Trials of Dietary Supplements and Cancer . . . . . . . . . . . . . . . . . . . . . . . . . . .15
*Shanti D. Mahabir, Renjit Thomas,
and Somdat Mahabir*

# Series Preface

## Nutraceuticals: Basic Research/Clinical Applications

The nutraceutical and functional food industries have been growing significantly in the last two decades. Foods that promote health beyond providing basic nutrition are termed "functional foods." The acceptance of these products by large segments of the population, especially in the West, is ever increasing, particularly in the United States.

As nutraceuticals have the potential to improve health and fight disease, their consumption and marketing have increased. This is also reflected by the incorporation of nutraceuticals and alternative medicines in the curriculum of several health-care professional education schemes.

The potential health benefits of nutraceuticals will contribute to an overall market growth of more than $90 billion by the end of 2015 in the United States and up to $180 billion worldwide. This has been attributed to the aging population worldwide, increased prevalence of serious diseases due to lifestyle changes, and enhanced focus on preventive medicines.

The challenges nutraceuticals, alternative medicines, dietary supplements, or functional foods face are proper characterization processes, reproducible activity, and clinical evidence to support the claims of their application in the prevention or treatment of disease. Looking ahead, it is likely that tougher FDA guidelines and GMPs will become a reality in the near term. In the future, FDA may ask for scientific evidence as a prerequisite to get nutraceutical and functional food products into the marketplace.

I feel it is most appropriate to address these issues, and CRC Press is giving utmost importance to them by starting a new book series "Nutraceuticals: Basic Research/Clinical Applications." The aim of this new series is to publish a range of books edited by distinguished scientists and researchers in the field. This series will address various aspects of nutraceutical products, including the historical perspective, traditional knowledge base, evidence-based approaches, and processing techniques and applications. The series will be very useful not only for researchers and academicians but will also be a valuable reference book for personnel in the nutraceutical and food industries.

This book will focus on the role of nutraceuticals in human health, disease prevention, health promotion, and as an adjunct to disease treatment. Although previous books have been published on nutraceuticals, none have focused on the epidemiological evidence in humans. The focus of previous work has been on cell culture experiments, and while these are usually considered the first step of investigations, the results of these experiments must be viewed with caution because of the probable lack of relevance to humans. The epidemiological focus is important because it will address whether and what kinds of evidence exist to support the role for nutraceuticals in disease prevention and treatment. Cutting-edge summaries will highlight both the biological and epidemiological findings of relevant studies of nutraceuticals in health and disease.

**Yashwant V. Pathak, PhD**
*Tampa, Florida*

# Preface

The term "nutraceutical" is derived from the words nutrition and pharmaceutical, and it refers to foods and food-derived substances with potential health benefits. Therefore, nutraceuticals collectively refer to a wide variety of products, including foods, herbs and spices, cholesterol-free foods, fat-free foods, beverages, energy drinks, specifically formulated diets, isolated food components, dietary supplements, vitamins, minerals, trace elements, and probiotics. These products have been marketed on the basis of potential health benefits, and there have been unsubstantiated claims made by certain products. Adequate scientific research on the benefits and harms of several types of nutraceuticals in humans has not been conducted. Limited population-based studies of humans have been conducted across different ethnic groups and subgroups such as the overweight and obese. Further, it is unknown whether nutraceuticals would be useful to populations that are well fed and well nourished such as found in Western societies. In addition, there is a lack of regulatory framework to govern nutraceuticals in most countries, and therefore marketing claims about the health benefits of these products can be inflated or harmful without proper scientific research. Even in the United States, nutraceuticals are widely available, and their use is largely unregulated. Therefore, information from well-conducted scientific studies on nutraceuticals in health and disease will be important for students, the general public, scientists, and health-care professionals to make informed decisions about the benefits and risks of these products.

The book opens with a general background to nutraceuticals and human health. It covers the following health and disease areas: cancer, lipidemia and cardiovascular disease, metabolic syndrome with obesity, diabetes and hypertension, respiratory health, the gut microbiome, and cognitive decline. It then concludes by addressing the methodological issues in conducting epidemiological research on nutraceuticals.

We hope that this book will be a good source of reference to students, researchers, public health professionals, and clinicians interested in the role of nutraceuticals on human health and disease.

**Somdat Mahabir**
**Yashwant V. Pathak**

# Acknowledgments

We would like to express our gratitude to all the authors and coauthors of the chapters in this book. We have been very fortunate to have had an outstanding group of international experts in the field of nutraceuticals related to health and disease.

We would like to express our sincere thanks to Steven Zollo, Kathryn Everett, and Rachael Panthier at Taylor & Francis Group/CRC Press for completing this book in a timely manner and for their professionalism.

Finally, we would like to thank our respective families for their cooperation and encouragement in completing this project even though the task of editing this book, at times, chipped away at personal family time.

# Editors

**Somdat Mahabir**, BS, MS, PhD, MPH, is a nutritional epidemiologist and program director in the Epidemiology and Genomics Research Program's (EGRP) Modifiable Risk Factors Branch (MRFB) at the National Cancer Institute (NCI), National Institutes of Health (NIH). Dr. Mahabir earned his PhD from New York University, MPH from New York Medical College, and MS from New York Institute of Technology. He also has a diploma in cancer prevention from the NCI, where he did his postdoctoral training in epidemiology.

Prior to joining EGRP in 2009, Dr. Mahabir was an assistant professor in the Department of Epidemiology, Division of Cancer Prevention and Population Sciences, at The University of Texas M.D. Anderson Cancer Center. He was an associate member of the Center for Research on Environmental Disease and an affiliated faculty member of the Center for Research on Minority Health at M.D. Anderson Cancer Center. He was also a member of the Psychosocial, Behavioral, and Health Services Research Committee Institutional Review Board (IRB) section at the same center. Dr. Mahabir was also a member of the American Cancer Society's (ACS) Institutional Research Grant (IRG) program and the Clinical, Translational, and Population-Based Research Projects review committees at M.D. Anderson Cancer Center. Before joining the M.D. Anderson Cancer Center, he was a cancer prevention fellow at NCI and a project director with Memorial Sloan-Kettering Cancer Center's Department of Epidemiology and Biostatistics.

Dr. Mahabir's current responsibilities include the scientific management of a research portfolio that focuses on modifiable cancer risk factors, such as diet and nutrition, alcohol, energy balance, and obesity. He is also involved with the advancement of new and innovative research ideas geared toward cancer prevention and control. In May 2011, Dr.

Mahabir organized the Expert Panel Workshop on Early-Life Events and Cancer at the NCI. In September 2012, he organized a meeting in Atlanta, Georgia, on "Birth Defects and Cancer: The Intersection of Genes and Prenatal Factors" in association with the Children's Oncology Group. Dr. Mahabir also holds an adjunct appointment in the Division of Cancer Epidemiology and Genetics at the NCI, where he pursues his research interests.

Dr. Mahabir has been the principal investigator and coinvestigator on grants funded by NIH, the ACS–M.D. Anderson Cancer Center IRG, and the Lance Armstrong Foundation. He is the recipient of the NCI Cancer Prevention Research Training Merit Award and academic awards from New York Medical College and New York Institute of Technology. Dr. Mahabir has published several peer-reviewed original papers, mostly first-authored papers, from prospective studies, case-control studies, and randomized intervention trials on cancer etiological factors.

**Yashwant V. Pathak**, MS, PhD, EMBA, is currently the associate dean for faculty affairs at the newly launched College of Pharmacy, University of South Florida, Tampa, Florida. Dr. Pathak earned his MS and PhD in pharmaceutical technology from Nagpur University, Nagpur, India and EMBA and MS in conflict management from Sullivan University, Louisville, Kentucky.

With extensive experience in academia and industry, Dr. Pathak has over 120 publications, research papers, abstracts, book chapters, and reviews to his credit. He has presented over 150 presentations, posters, and lectures worldwide in the field of pharmaceutics, drug delivery systems, and other related topics. He has also coedited six books on nanotechnology and drug delivery systems, two books on nutraceuticals, and several books on cultural studies. Dr. Pathak is the holder of two patents. He has travelled extensively to over 75 countries and is actively involved with many pharmacy colleges in different countries.

# Contributors

**Lina Badimon**
National Research Council (CSIC)
Cardiovascular Research Center
Hospital de la Santa Creu i Sant
    Pau, IIB-Sant Pau
and
CIBER$_{OBN}$-Pathophysiology of
    Obesity and Nutrition
Jesus Serra Foundation
Barcelona, Spain

**Fran T. Close**
College of Pharmacy and
    Pharmaceutical Sciences
Humphries Science Research
    Center
Institute of Public Health
Florida A&M University
Tallahassee, Florida

**Kylie Conroy**
School of Health Sciences
Queen Margaret University
Musselburgh, United Kingdom

**Malay K. Das**
Department of Pharmaceutical
    Sciences
Dibrugarh University
Dibrugarh, India

**Isobel Davidson**
School of Health Sciences
Queen Margaret University
Musselburgh, United Kingdom

**Dipanwita Dutta**
HerbaKraft Inc.
Piscataway, New Jersey

**Susan Ettinger**
Hunter College
The City University of New York
and
The New York Obesity Nutrition
    Research Center
New York, New York

**Zdenko Herceg**
Epigenetics Group
International Agency for Research
  on Cancer
Lyon, France

**Mark C. Houston**
School of Medicine
Vanderbilt University
and
Hypertension Institute
  and Vascular Biology
Saint Thomas Hospital
Nashville, Tennessee

**Young Sup Lee**
School of Life Sciences
College of Natural Sciences
Kyungpook National University
Daegu, South Korea

**Shanti D. Mahabir**
Department of Biology
School of Arts and Sciences
Howard University
Washington, District of Columbia

**Somdat Mahabir**
Epidemiology and Genomics
  Research Program
Division of Cancer Control
  and Population Sciences
National Cancer Institute
National Institutes of Health
Bethesda, Maryland

**Elizabeth Mazzio**
College of Pharmacy
  and Pharmaceutical Sciences
Florida A&M University
Tallahassee, Florida

**Prabhjot Singh Nijjar**
Division of Cardiology
University of Minnesota Medical
  School
Minneapolis, Minnesota

**Teresa Padro**
National Research Council (CSIC)
Cardiovascular Research Center
Hospital de la Santa Creu i Sant
  Pau, IIB-Sant Pau
and
$CIBER_{OBN}$-Pathophysiology of
  Obesity and Nutrition
Barcelona, Spain

**Awanish Kumar Pandey**
Department of Pharmacy
Institute of Technology and
  Management
Gorakhpur Industrial
  Development Authority
Gorakhpur, India

**Siva K. Panguluri**
Department of Pharmaceutical
  Sciences
College of Pharmacy
University of South Florida
Tampa, Florida

**Yashwant V. Pathak**
College of Pharmacy
University of South Florida
Tampa, Florida

**Angela Senders**
Department of Neurology
Oregon Health & Science
  University
Portland, Oregon

**Adeeb Shehzad**
School of Life Sciences
College of Natural Sciences
Kyungpook National University
Daegu, South Korea

**Karam F.A. Soliman**
College of Pharmacy and
  Pharmaceutical Sciences
Florida A&M University
Tallahassee, Florida

Rebecca I. Spain
Department of Neurology
Portland Veterans Affairs
    Medical Center
and
Oregon Health & Science
    University
Portland, Oregon

L. Joseph Su
Division of Cancer Control
    and Population Sciences
National Cancer Institute
National Institutes of Health
Bethesda, Maryland

Renjit Thomas
School of Medicine
Howard University
Washington, District of Columbia

Srinivas M. Tipparaju
Department of Pharmaceutical
    Sciences
College of Pharmacy
University of South Florida
Tampa, Florida

Poonam Tripathi
Department of Pharmacy
Institute of Technology and
    Management
Gorakhpur Industrial
    Development Authority
Gorakhpur, India

Jared Tur
Department of Pharmaceutical
    Sciences
College of Pharmacy
University of South Florida
Tampa, Florida

Mukesh Verma
Division of Cancer Control
    and Population Sciences
National Cancer Institute
National Institutes of Health
Bethesda, Maryland

Gemma Vilahur
National Research Council (CSIC)
Cardiovascular Research Center
Hospital de la Santa Creu i Sant
    Pau, IIB-Sant Pau
and
$CIBER_{OBN}$-Pathophysiology of
    Obesity and Nutrition
Barcelona, Spain

Francesco Visioli
Madrid Institute for Advanced
    Studies—Food
Madrid, Spain

Mary Warnock
School of Health Sciences
Queen Margaret University
Musselburgh, United Kingdom

Vijayshree Yadav
Department of Neurology
Multiple Sclerosis Center of
    Oregon
Oregon Health & Science
    University
Portland, Oregon

Zhenzhen Zhang
Department of Epidemiology and
    Biostatistics
Michigan State University
East Lansing, Michigan

# I

# Introduction

# Background to Nutraceuticals and Human Health

Dipanwita Dutta, Somdat Mahabir, and Yashwant V. Pathak

## Contents

## What are nutraceuticals?

The term nutraceutical is derived from the words nutrition and pharmaceutical and it refers to foods and food-derived substances with potential health benefits. Therefore, nutraceuticals collectively refer to a wide variety of products including foods, herbs, spices, cholesterol-free foods, fat-free foods, beverages, energy drinks, specifically formulated diets, isolated food components, dietary supplements, vitamins, minerals, trace elements, and probiotics.

With advancement of scientific research and health perception about nutraceutical products, the market has shown an affinity toward the usage of products obtained from natural sources [1]. The big wave for nutraceutical usage and marketing was predicted by Jim Wagnerthis, editor of the Nutritional Outlook, in June 2002 as "Nutraceuticals are in their formative years. But make no mistake, the nutraceutical boom is coming and it will be worth billions to the companies who define it." Since then, the market for nutraceuticals has been growing by leaps and bounds leading to a multibillion dollar industry worldwide [2]. Because there is no universal definition of "nutraceutical"

3

accepted by all the stakeholders, it is very difficult to forecast the size of the nutraceutical market.

From a historical perspective, looking at the traditional medicinal systems of African tribes, native American people, Australian Aboriginals, Chinese, Japanese, Hindus, Maoris, Tibetans, and many other indigenous tribes and cultures worldwide, it is obvious that all of them developed a functional, indigenous system of medicine that incorporated the extensive use of local herbs and natural minerals. Thus, they had a system of nutraceutical prod ucts, even in ancient times. Today in the allopathic medical world, it can be argued that many of the developed and marketed drugs have their origin based on the information acquired from plants used in ancient/traditional systems. In Indian traditional systems, such as Ayurvedic medicine, nutraceuticals are defined as extracts of foods claimed to have medical effects on human health and usually contained in a nonfood matrix such as capsules or tablets [3].

The term nutraceutical, as coined by Dr. Stephen DeFelice, founder of the Foundation for Innovation in Medicine in Cranford, New Jersey, covers a wide variety of products including dietary supplements, fortified foods that are enriched with nutrients not natural to the food (such as orange juice with added calcium), functional foods, and medical foods. Because of its content and formulation, nutraceuticals are also defined as parts of a food or a whole food that have a known or potential health benefit, including prevention and treatment of disease [4]. We also propose that nutraceuticals are products developed from whole foods, food components, or from traditional herbal or mineral substances, and their synthetic derivatives or forms, which can be delivered in pharmaceutical dosage forms such as pills, tablets, capsules, liquid orals, lotions, delivery systems, or other dermal preparations, and are manufactured under strict cGMP procedures. These are developed according to pharmaceutical principles and evaluations to ensure reproducibility and therapeutic efficacy of the products. Because of increasing scientific research and also purely on the basis of claims about health benefits, nutraceuticals are almost everywhere, including usage in clinical practice [5].

The concept of functional foods was first introduced in Japan and till 1998 it was the only country that legally defined foods for specified health use (FOSHU). A functional food is natural or formulated food that has enhanced physiological performance or prevents or treats a particular disease [6]. In Canada, functional food is defined as similar in appearance to conventional food, consumed as a regular part of the diet, while nutraceuticals are defined as a product produced from food but sold as pills, tablets, capsules, and other medicinal forms [7]. In the United Kingdom, the Department of Environment, Food and Rural Affairs (DEFRA) defines functional food as a food that has a component incorporated into it to give it a specific medical or physiological benefit other than purely nutritional benefits.

## Role of US Food and Drug Administration

The FDA is an agency within the US Department of Health and Human Services. The role of the US FDA is to protect "the public health by assuring the safety, effectiveness, and security of human and veterinary drugs, vaccines, and other biological products for human use, and medical devices. The agency also is responsible for the safety and security of our nation's food supply, cosmetics, dietary supplements, products that give off electronic radiation, and for regulating tobacco products" (FDA News Release, April 27, 2012).

Nutraceuticals are regulated by the FDA under the authority of the Federal Food Drug and Cosmetic Act, even though they are not specifically defined by law (http://www.fda.gov/Food/LabelingNutrition/default.htm). The FDA requires food labeling for most prepared foods and drinks, but nutrition labeling for raw products such as fruits, vegetables, and fish, for example, is voluntary. For certain categories of nutraceuticals such as dietary supplements, the FDA regulations fall under the Dietary Supplement Health and Education Act (DSHEA) of 1994 (www.fda.gov/food/dietarysupplements/default.htm). Under DSHEA, "the dietary supplement or dietary ingredient manufacturer is responsible for ensuring that a dietary supplement or ingredient is safe before it is marketed." In general, dietary supplement manufacturers are not required to register their products with FDA nor get FDA approval before producing or selling dietary supplements. "Manufacturers must make sure that product label information is truthful and not misleading. Under the FDA Final Rule 21 CFR 111, all domestic and foreign companies that manufacture, package, label, or hold dietary supplement, including those involved with testing, quality control, and dietary supplement distribution in the United States, must comply with the Dietary Supplement Current Good Manufacturing Practices (cGMPS) for quality control." The FDA also requires the manufacturer and other entities linked to the product marketing such as the packer and distributor whose names appear on the label of a dietary supplement marketed in the United States to submit reports of all serious adverse events associated with the particular dietary supplement to the FDA. If a dietary supplement product is deemed unsafe after it is sold, the FDA is responsible for taking action against it. And, indeed, the FDA does take action. For example, in April 2012 alone, the FDA issued warning letters to 10 manufacturers and distributors of dietary supplements containing dimethylamylamine, because they were marketing products for which evidence of the safety of the product had not been submitted to FDA (FDA News Release, April 27, 2012). Interestingly, according to DSHEA, the dietary supplement can be claimed to offer nutritional support in nutrient deficient diseases, but the companies are expected to write that this statement has not been evaluated by FDA. Further, they must state that this product is not intended to diagnose, treat, cure, or prevent any disease [8]. Looking at the trends worldwide and especially in the United States, it appears that the FDA will come up with more stringent regulations for nutraceuticals and functional food product claims and marketing.

# Historical perspectives

Ancient civilizations from India, China, Egypt, Japan, Africa, and Tibet have been using nutraceuticals for thousands of years against various diseases. *Ayurveda*, a 5000 year old Hindu medicinal system, has clearly stated that nutraceuticals are useful for therapeutic purposes. Around 2500 year ago, Hippocrates (a Greek physician) stated that "Let food be thy medicine and medicine be thy food." Hippocrates believed that the body should be treated as the sum of the whole and not just as a series of parts, and this is the philosophy behind a holistic medicine, focusing on all facets of an individual and putting attention on rebuilding and/or restoring one's physical health.

Ancient Hindus have made major contributions to the development of medical sciences by experimentation and use of nutraceuticals over 5000 years ago as evidenced by *Ayurvedic*, *Siddha*, and *Unani* systems of medical care and preventive health. The beneficial effects of nutraceuticals for maintaining a healthy life have been documented in the *Ayurveda* in the form of various *slokas* or verses. For example, "*Tat cha nityam prayunjeet svasthyam yen anuvartate. Ajaatanam vikaranam anuttpattikaram cha yat*" (Sutra Sthana: 5) and "*Pathye sati gadaartasya kim aushadh nishevane. Pathye asati gadaartasya kim aushadh nishevane*" (Vaidhya Jeevana: 1/10). These verses essentially state that one should consume a diet, which besides providing basic nutrition to the body, helps to maintain the body in a healthy state and prevents the occurrence of diseases.

*Ayurveda* has elaborately explained that plants contain active substances that have nutritive and health promotion properties. In *Ayurvedic* medicine, nutraceuticals like *Chyavanprash* (an immune booster that is also used in the management of respiratory disorder) was used, and is still being used today. Similarly, several other products like *Brahma Rasayana* (for protection from mental stress), *Phala Ghrita* (for reproductive health), *Arjuna Ksheerpaka* (for cardioprotection), *Shatavari Ghrita* (for general health of women during various physiological states), and *Rasona Ksheerpaka* (for cardioprotection) are also mentioned in *Ayurveda* [9].

Several lines of evidence are available indicating that ancient Hindu systems of medicine promoted the use of plant product formulations in disease treatment and health promotion. *Charaka*, an Ayurvedic physician, categorized all the food items into 12 classes: corns with bristles, pulses or legumes, meat, leafy vegetables, fruits, vegetables which are consumed raw, wines, water from different sources, milk and milk products, products of sugarcane, food preparations, and accessory food items such as oils and salts, and has further subcategorized these food groups. *Charaka* also provided details on *Jivaniya* (nutrients) in his text, the *Charaka Samhita* as ingredients for energy supplement, and to improve certain aspects of human metabolism.

According to Ayurveda, the concept of promoting health and preventing disease through nutrition is known as *Rasayana*. In *Rasayan-chikitsa* (science of rejuvenation), it has been stated that "from the *rasayan* treatment, one attains longevity, memory, intelligence, freedom from disorders, youthful age, excellence of luster, complexion and voice, oratory, optimum strength of physique and sense organs, respectability, and brilliance" [10].

*Rasayana* means acquisition, movement, or circulation, and nutrition is needed to provide nourishment to the organs, tissues, and tissue perfusion as mentioned in *Ashtanga Hridaya* dated about 300 AD (Chapter 29, *Sharangdhar Samhita*). In *Ayurvedic* medicine, the notion of *Rasayanas* is pathophysiology, a linking of health and disease with physiological balance/imbalance that leads to good health or disease depending on the physiology [11]. It is also reported in *Ayurveda* that *Rasayan* drugs and formulations provides longevity, memory, intelligence, freedom from disorders, youthful age, excellence of luster, complexion and voice, oratory, optimum strength of physique and sense organs, respectability, and brilliance.

The *sloka* (verse) indicating the benefits of *Rasayana* is as follows [12]:

दीर्घमायुः स्मृतिं मेधामारोग्यं तरुणं वयः । प्रभावर्णस्वरौदार्यं देहेन्द्रियबलं परम् ॥ ४९ ।
वाक्सिद्धिं प्रणतिं कान्तिं लभते ना रसायनात् । लाभोपायो हि शस्तानां रसायनमिति स्मृतम् ॥ ५० ।
न केवलं दीर्घमायुरश्नुते रसायनं यो विधिवद्‌बिषेवते ।
गतिं स देवर्षिनिषेवितां शुभां प्रपद्यते ब्रह्म तथैति चाक्षरम् ॥ ८० ॥

Several other physicians in various eras have described the usefulness of plant and plant products as medicines against various diseases. Some notable ones are as follows:

Theophrastus, in the fourth century BCE, who was a student of Aristotle, wrote *De Historia Plantarum*, describing the medicinal properties of 455 plants.

John Gerard's *The Herball* or *Generall Historie of Plantes* (1597) comprises 1392 pages and 2200 woodcut images of medicinal plants.

Nicholas Culpepper, an English botanist, herbalist, physician, and astrologer, wrote a book entitled *The Complete Herbal* (1653), which discussed the medicinal properties of herbal extract. The book focused on the medicinal uses of plants and influenced the *Doctrine of Signatures*—which held that the medicinal use of plants could be ascertained by recognizing distinct "signatures" visible on plants that correspond to human anatomy [13].

A Swiss-born physician, Theophrastus Bombastus von Hohenheim (1493–1541), popularly known as Paracelsus, argued that the medicinal value of plants came from the chemicals within them. The most desirable parts of plants were claimed not to be the leaves, bark, stems, or roots, but it is most beneficial when one extracts or distills its essence in the form of tinctures [14].

*Papyrus Ebers*, an ancient text written in 1500 BC, has mentioned plant medicine references to more than 700 herbal remedies and 800 compounds, and is thought to be a copy of the even more ancient *Book of Thoth* (3000 BC) [15].

## Changing lifestyles

Due to changing human lifestyles that encompass more sedentary behaviors and less physical activity energy expenditure, overweight, and obesity are major public health issues in most countries of the world today. Overweight (BMI between 25 and 29.9 $kg/m^2$), obesity (BMI of 30 or higher), and a sedentary lifestyle are escalating all over the world. In Europe, in the past decade, the prevalence of obesity has increased by about 10%–40% in the majority of European countries. Obesity is also becoming a serious problem throughout Asia, Latin America, and parts of Africa. There are also more widespread obesity in urban versus rural populations and some racial subgroups such as Mexican-Americans, African-Americans, American Indian communities, and Creoles in Mauritius. In the United States, there have been dramatic increases in overweight and obesity among all the races, but Mexicans and Blacks have the highest rates. In 1999–2002, ~65% of Americans aged ≥20 years were overweight and 30% were obese. During the same 1999–2002 period, ~71% Blacks and 72% Mexican-Americans were overweight and 39% Blacks and 33% Mexican-Americans were obese [4]. Because of this transition, in general, chronic disease rates are rising.

Trends in the pharmaceutical industry show that there are very potent drugs with severe side effects and toxic effects. Nutraceuticals are widely marketed and used as an alternative to perceived pharmaceuticals with toxic effects. However, some nutraceutical products may have similar toxicity problems. Obesity leads to many metabolic problems including nutrient metabolism. In addition, obesity is linked to complications such as type 2 diabetes, hyperlipidoma hypertension, and cardiovascular diseases. Individuals with diabetes or arthritis seem to be interested in preventive measures, which may delay the onset of these diseases or attenuate disease severity [16]. Several nutraceutical interventions are currently undergoing investigation as potential treatment options for obesity and its associated complications.

## Nutraceutical market

Recent progress in the scientific understanding of natural products and pressures from advertisements have led to an increased market demand for nutraceuticals. Proteins, fibers, and various specialized functional additives have become the top-selling group of nutraceutical ingredients. Due to the wide variety and complexity of nutraceutical products around the world, it is very difficult to accurately estimate the true market value and sales of these products. A report from the Freedonia Group noted that "nutraceutical application is forecasted to reach $6.0 billion in 2015, up 6.2% annually from 2010" (http://www.scribd.com/doc/77295623/World-Nutraceutical-Ingredients-Industry). Based on the increasing demand in the Western world for the nutraceutical, it is predicted that by 2020 countries like China, Brazil, Japan, India, Mexico, Poland, Russia, and South Korea will evolve as a largest producer of the nutraceutical [17]. Nutraceutical products in the form of formulated beverages, foods, and dietary supplements are becoming main stream in American households. The popular perception is that these products will prevent disease, enhance performance, and may delay the onset of aging, leading to a huge market share of nutraceutical products to over US $80 billion [18].

According to research from Freedonia global market, the nutraceutical and dietary supplement industry is yearly growing in excess of 7% to reach almost US $24 billion in 2015 (http://www.reportlinker.com/ci02038/Nutraceutical.html). According to "Global Industry Analysts," the global market for functional foods and drinks is expected to reach US $130 billion by 2015 (http://www.reportlinker.com/ci02036/Functional-Food.html).

Global Business Intelligence (GBI) researchers have estimated that the global nutraccutical markct worth is approximately US $128.6 billion, after increasing at a compound annual growth rate (CAGR) of 4.4% during 2002 through 2010. It has been projected that the market will swell to about US $180.1 billion by 2017, after growing at a CAGR of 4.9% from 2010 through 2017. GBI attributed the growth rates to an increase in the elderly population, the affluence of the working population, and increasing awareness of, and preference for, preventive medicine. These factors are expected to be important in stimulating market growth in the top seven markets of world including the United States, Japan, the United Kingdom, Spain, Italy, Germany, and France (http://www.nutraceuticalsworld.com/contents/view_online-exclusives/2012-01-23/all-signs-point-to-growth/).

According to Nutraceutical Product Market, Global Market Size, Segment and Country Analysis and Forecasts (2007–2017), the North America and Asia Pacific nutraceutical markets are expected to have market shares of 39.2% and 30.4% in 2017. According to a recent study, the number of people above 60 years will be >1 billion by the year 2020 of which 70% will be from developed nations. It is expected that the increasing elderly population will drive the global nutraceutical market upward to about US $250 billion by 2018 (http://www.nutraceuticalsworld.com/contents/view_breaking-news/2012-07-06/global-nutraceuticals-market-to-reach-250-billion-by-2018/).

According to a published report by the BBC, the global nutraceutical market is estimated to reach about US $207 billion in 2016. Functional beverages market is expected to experience the highest growth, at a CAGR of 8.8% during the 5 year period from 2011 to 2016 (http://www.marketresearch.com/BCC-Research-v374/Nutraceuticals-Global-Processing-Technologies-6481994/).

The other emerging markets worldwide include India, with US $540 million in sales in 2006, and an expected rise of almost 38% per year in the coming years crossing the billion mark by the end of 2009. China is another market that is growing significantly, followed by Brazil (US $881 million in 2006), Turkey (US $200 million in 2006), and Australia, New Zealand, and Middle East and African countries (US $300 million in 2006).

## Considerations in nutraceutical formulations

Nutraceuticals are also biologically active phytochemical substances obtained from plants and natural minerals and are major formulation products for industry. Nutraceutical manufacturers face some hurdles for formulation consideration since the idea behind nutraceuticals is to achieve a desirable therapeutic result, without the side effects often associated with pharmaceuticals. One problem is the drug content and drug uniformity in the dosage forms since nutraceuticals are a cluster of chemical entities and it will be comparatively difficult to identify and quantify all the ingredients in the products. In such situations, at least one major ingredient can be identified and quantified to ensure the uniform distribution of the product through the matrix. A second major hurdle nutraceutical products will face is defining and identifying impurities and ensuring that these impurities are not harmful to the consumer.

Other challenges with nutraceuticals include maintaining the correct balance of active compounds along with other ingredients during drug formulations. Most of the companies are trying to formulate nutraceuticals in the form of tablets, ointments, and capsules while maintaining the appropriate quantities of the active compounds. While doing this, the quantity of other associated ingredients might vary considerably. The scientific literature clearly indicates that active compounds present in botanical extracts, when consumed at a certain lower dose, show synergistic effects. However, the same compounds, at higher dosage, show antagonistic effects. The fact is each active compound that is present in the herbal extracts remain bounded by matrix of several other compounds, which at a particular dosage helps the active compounds reach the actual site of action/activity. The same matrix on some other dosage prevents the active compounds from showing the activity by exerting toxic effects. This problem might be offset if nutraceutical manufacturers first separate out the products and treat nutraceuticals separately from functional foods. Second, the formulation/processing of nutraceuticals needs to follow the norms of pharmaceutical formulations such as dosage forms, consistency of the product's therapeutic value, reproducibility, in vitro and in vivo evaluations, followed by clinical trials and epidemiological investigations. These dosage forms can be tablets, capsules, liquid orals, ointments, external products, and dermal products, pills, and also may explore the newer drug delivery systems like nanoparticulate drug delivery systems, microcapsules, and so on. Some of these evaluations can be easily performed and can contribute toward the quality of the products.

## Conclusion

Nutraceuticals are parts of a food or a whole food that have potential medical or health benefits, including the prevention and treatment of disease. Although the health benefits of specific nutraceuticals have been documented even during ancient times, today there is an explosion of new nutraceutical products competing in the marketplace and as a result there are several challenges including the possibility of implementation of stringent government regulations. While the nutraceutical industry is poised to grow on the basis of perceived health benefits on the basis of claims made by manufacturers, rigorous investigation of the usefulness, effectiveness, benefits, and harms of these products by epidemiological research will be a problem for the industry. For example, there is already strong human evidence that dietary supplements not only are ineffective in cancer prevention, but can also increase cancer risk (see Chapter 2, for example). We also know that in some cases nutraceutical formulation is a big challenge that needs to resolve because of the complex chemistry of the various active compounds involved and the lack of knowledge of their interactive effects.

## References

1. Kennedy ET, Luo H, and Ausman LMJ. 2012. Cost implications of alternative sources of (n-3) fatty acid consumption in the United States. *Nutrition* 142(3):605S–609S.
2. Nicoletti M. 2012. Nutraceuticals and botanicals: Overview and perspectives. *Int J Food Sci Nutr* 63(1):2–6.
3. Chaudhary A and Singh N. 2012. Intellectual property rights and patents in perspective of Ayurveda. *Ayurveda* 33(1):20–26.
4. Hedley A, Ogden C, Johnson C, Carroll M, Curtin L, Flegal K. 2004. Prevalence of overweight and obesity among US children, adolescents, and adults, 1999–2002. *JAMA* 291:2847–2850.
5. Pathak Y. 2011. *Handbook of Nutraceuticals, Volume II, Scale-Up, Processing and Automation*, CRC Press, Boca Raton, FL, pp. 1–14.
6. Pathak Y. 2009. *Handbook of Nutraceuticals, Volume I, Ingredients, Formulations, and Applications*, CRC Press, Boca Raton, FL, pp. 15–25.
7. Fitzpatrick KC. 2004. Regulatory issues related to functional foods and natural health products in Canada: Possible implications for manufacturers of conjugated linoleic acid 1–3. *Am J Clin Nutr* 79(Suppl):1217S–1220S.
8. Brent AB. 2000. Herbal Therapy: What a Clinician Needs to Know to Counsel Patients Effectively. *Mayo Clin Proc* 80(6):835–841.
9. Rani Y and Sharma NK. 2005. *Proceedings of WOCMAP III, Vol. 6: Traditional Medicine & Nutraceuticals*, UR Palaniswamy, LE Craker, and ZE Gardner, Eds. Acta Horticulture 680, ISHS 2005, pp. 131–136.
10. Samhita C. 1998. *Chaukhambha Orientalia, Varanasi, Section 6 Chikitsasthanam*, Chapter 1, Qtr 1, pp. 3–4, Shloka 7–8.
11. Samhita C. 700 BCE. *Chikitsa sthana*, Chapter 1, Parts 1–4 on rasayana, Edited and translated to English by PV Sharma, Choukhamba Prakashan, Varanasi, India.

12. Samhita C and Sharma PV. 1998. *Chaukhambha Orientalia, Varanasi, Section 6 Chikitsasthanam*, 4th Edn., Chapter 1, Qtr 1, pp. 12, Shloka 78–80.
13. Pearce JM. 2008. The Doctrine of Signatures. *Eur Neurol* 60(1):51–52.
14. Joseph FB. 2000. Paracelsus herald of modern toxicology. *Toxicol Sci* 53(1):2–4.
15. Filler AG. 2007. A historical hypothesis of the first recorded neurosurgical operation: Isis, Osiris, Thoth, and the origin of the djed cross. *Neurosurg Focus* 23(1):E6.
16. Rajasekaran A, Sivagnanam G, and Xavier R. 2008. Nutraceuticals as therapeutic agents: A review. *Res J Pharm Technol* 1(4):328–340.
17. Freedonia Group. 2011. *World Nutraceutical Ingredients: Forecasts for 2015&2020 in 40 Countries*. Report, FED00497, Clevland, OH, pp. 569.
18. Mark B, Ashley L, Carla O, and Mary EL. 2012. Herb supplement sales increase 4.5% in 2011. *HerbalGram* 95:60–64.

# Cancer

# 2

# Clinical Trials of Dietary Supplements and Cancer

Shanti D. Mahabir, Renjit Thomas, and Somdat Mahabir

## Contents

## Introduction

The term cancer refers to a group of diseases in which cells proliferate in an abnormal manner, reflecting loss of responsiveness to normal growth control signals. Once a cell has escaped normal controls, it can trespass aggressively from its origin and migrate to other tissues, where it forms a growth locus. This process, known as metastasis, is what makes cancer so lethal. If the spread of the cancer cells is not controlled, it can cause death. It was estimated that there were 12.7 million new cancer cases worldwide in 2008 (5.6 million in the developed countries and 7.1 million in developing countries). In 2008, 7.6 million people died of cancer (2.8 million in the developed countries and 4.8 million in developing countries). By 2030, it is estimated that there will be 21.4 million new cancer cases and 13.2 million cancer deaths [1]. In the United States, it was estimated that 11.4 million people alive in 2006 had a history of cancer and about 1.5 million new cancer cases will be diagnosed in 2010 [2].

Consumption of dietary supplements is now widespread in the United States and other countries. A systematic review has reported that over 50% of adults in the United States use vitamin and mineral dietary supplements and 33% use multivitamins [3]. Among cancer patients and longer-term cancer survivors, 64%–81% reported using vitamin and mineral supplements [3]. About 14%–32% of cancer survivors reported initiating the use of vitamin and mineral supplementation after the diagnosis of cancer [3]. However, the benefits and harms to individuals who take dietary supplements have not been fully resolved. The purpose of this review is to highlight the findings from the major clinical trials of dietary supplements and cancer risk conducted between 1982 and 2010.

## Trials of vitamin/mineral supplements

### physician's health study: Started 1982

The Physician's Health Study (PHS) was a randomized, double-blind, placebo-controlled trial that tested the effects of aspirin (325 mg on alternate days) and β-carotene (50 mg on alternate days), or a placebo for the prevention of cardiovascular disease (CVD). The study started in 1982 and enrolled 22,071 male physicians in the United States between the ages of 40 and 84 years [4]. The subjects had no history of cancer (except nonmelanoma skin cancer), myocardial infarction, stroke, or transient cerebral ischemia. The aspirin component of the study was terminated early, in 1988, on the advice of the data monitoring board because a 44% statistically significant reduction in the risk of myocardial infarction was found in the aspirin group [4]. The β-carotene component continued till the scheduled end of the study in 1995 (a duration of 12 years). β-Carotene recipients showed no benefit or harm for either CVD or cancer.

### Skin cancer prevention study: Started 1983

This study tested the effects of β-carotene (50 mg/day) versus a placebo in the prevention of recurrent nonmelanoma (basal cell carcinoma [BCC] and squamous cell carcinoma [SCC]) cancers of the skin [5]. The trial randomly assigned 1805 patients who had a recent nonmelanoma skin cancer to β-carotene or a placebo. The study started in 1983 and concluded in 1989 after 5 years of follow-up. The results showed no significant difference between the β-carotene and placebo groups on the rate of occurrence of the first new non-melanoma skin cancer in persons with previous nonmelanoma skin cancers. β-Carotene recipients showed no benefit or harm.

## Linxian nutrition intervention trials: Started 1983

Linxian is a rural county in Henan Province in north-central China. The Linxian trials were conducted based on epidemiological evidence that the people of Linxian suffered from low levels of several micronutrients and high incidence rates of esophageal and gastric cancers.

Linxian was selected for the conduct of two randomized, placebo-controlled trials. The first trial, known as the Linxian Dysplasia Trial (LDT), started in 1983 and recruited 3318 individuals between the ages of 40 and 69 years with a cyto-logical diagnosis of esophageal dysplasia [6]. The participants were randomly assigned to a daily supplement containing either 14 vitamins and 12 minerals (β-carotene, 18 mg; vitamin A, 10,000 IU; vitamin E, 60 IU, vitamin C, 180 mg; folic acid, 800 μg; vitamin B1, 5 mg; vitamin B2, 5.2 mg; niacinamide, 40 mg; vitamin B6, 6 mg; vitamin B12, 18 μg; vitamin D, 800 IU; biotin, 90 μg; pantothenic acid, 20 mg; calcium, 324 mg; phosphorous, 250 mg; iodine, 300 μg; iron, 54 mg, magnesium, 200 mg; copper, 6 mg; manganese, 15 mg; potassium, 15.4 mg; chloride, 14 mg, chromium, 30 μg; molybdenum, 30 μg; selenium, 50 μg; zinc, 45 mg) or a placebo. The duration of the trail was 6 years. The results showed that supplementation significantly reduced total deaths by 7%, but no significant differences were found for site-specific cancer deaths. In addition, compared to the placebo, the supplemented group had a nonsignificantly lower total cancer (4% lower), esophageal SCC (16% lower), and combined esophageal SCC/gastric cardiac cancer (8% lower). However, total stomach cancer mortality was significantly higher in the supplemented group (18% higher) versus the placebo [7].

The second trial, known as the Linxian General Population Trial (LGPT), started in 1986 and recruited a total of 29,584 subjects between the ages of 40 and 69 years with no history of cancer. The participants were randomly assigned to a daily supplement of either four combinations (*Supplement A*: retinol palmitate, 5000 IU and zinc oxide, 22.5 mg; *Supplement B*: riboflavin, 3.2 mg and niacin, 40 mg; *Supplement C*: ascorbic acid, 120 mg and molybdenum as Mo–yeast complex, 30 μg; *Supplement D*: β-carotene, 15 mg, selenium as Se–yeast and α-tocopherol, 30 mg) or a placebo. The duration of the trial was 6 years. This $2^4$ study design enabled the simultaneous testing for the effects of 16 combinations of nutrients, but used only 8 combinations (AB, AC, AD, BD, CD, ABCD, or placebo). For example, participants in the AB group received retinol, zinc, riboflavin, and niacin, while those in the ABCD group received all nine vitamins and minerals. A total of 2127 deaths occurred among trial participants during the intervention period. Cancer was the leading cause of death, with 32% of all deaths due to esophageal or stomach cancer, followed by cerebrovascular disease (25%). Results from this trial showed that those who received the β-carotene–vitamin E–selenium combination had a significant 13% reduction in cancer mortality, including a significant 21% reduction in stomach cancer mortality, and a significant 41% reduction in gastric noncardiac cancer mortality [8].

## Retinoid skin cancer prevention trial: Started 1984

The retinoid skin cancer prevention trial was a randomized, double-blind, placebo-controlled trial that was designed to test the effects of retinoids in the prevention of nonmelanoma skin cancer among subjects with a moderate degree of risk. The study trial started in 1984 and enrolled 2297 subjects between the ages of 21 and 84 years with actinic keratoses or SCC (not more than 2) or BCC (not more than 2). Subjects were randomly assigned to either a daily capsule of retinol (25,000 IU/day) or a placebo for up to 5 years [9]. The primary end points for the trial were the time to first new SCC or BCC. The trial started in 1984 and after 3.8 years of intervention, the subjects in the retinol group had a significant 26% fewer SCC compared to the placebo. There was no effect on BCC [10].

## Beta-Carotene and retinol efficacy trial: Started 1985

The β-Carotene and Retinol Efficacy Trial (CARET) was a multicenter, randomized, placebo-controlled study designed to test whether a daily combination of β-carotene (30 mg) plus retinol (25,000 IU in the form retinyl palmitate) was effective in preventing lung cancer in high risk subjects. The study recruited 18,314 smokers, former smokers, and subjects exposed to asbestos. The intervention was terminated early after ~4 years of follow-up. A total of 388 cases of lung cancer and 974 deaths occurred during the intervention phase of the study. Compared to the placebo, the β-carotene–retinol combination group had a significant 28% increased lung cancer risk and 17% increased total mortality risk [11]. Subgroup analyses showed that the increased risk from the β-carotene–retinol combination group was most pronounced in current smokers and heavy alcohol drinkers [12].

## Alpha-tocopherol and beta-carotene trial: Started 1985

The Alpha-Tocopherol and Beta-Carotene (ATBC) study was a randomized, placebo-controlled trial designed to test whether α-tocopherol (50 mg/day) alone, β-carotene (20 mg/day) alone, both α-tocopherol and β-carotene, or a placebo would reduce lung cancer incidence in high risk men. Between 1985 and 1993, a total of 29,133 Finnish male smokers aged 50–69 years were recruited into the study and were supplemented for 5–8 years. In general, the participants were heavy smokers and heavy drinkers. β-Carotene supplementation significantly increased lung cancer incidence by 16% and total mortality of 8% [13]. Subgroup analysis revealed that the higher incidence of lung cancer induced by β-carotene was more pronounced among those who smoked the most and drank the most. Several secondary cancers have also been assessed and reported. For example, there was a 32% reduction in the incidence of prostate cancer among men in the α-tocopherol group [14]. No clear benefits were observed for other cancers such as colorectal, urinary tract, or gastric cancer [15–17]. Follow-up after the intervention found no evidence of any lasting effect of the intervention [18].

## Physician health study II: Started 1987

The Physician Health Study II (PHS II), like PHS I, was a randomized, double-blind, placebo-controlled trial, but this study tested the effects of β-carotene (50 mg), vitamin E (400 IU α-tocopherol) on alternate days and daily vitamin C (500 mg) and a multivitamin pill (Centrum Silver) in preventing prostate cancer, CVD, cataracts, and macular degeneration [19]. Study subjects included 7641 participants, who completed PHS I, and 7000 physicians. The β-carotene intervention arm of the trial was stopped [20]. After 8 years of follow-up, neither vitamin E nor vitamin C had a significant effect on prostate cancer, or site-specific and total cancers combined [21].

## Calcium polyp prevention trial: Started 1988

The Calcium Polyp Prevention Trial was a randomized, double-blind, placebo-controlled study that tested the effects of supplemental calcium (3 g calcium carbonate, 1200 mg of elemental calcium) on the recurrence of colorectal adenomas. A total of 930 and women with an average age of 61 years were randomly assigned to receive a daily dose of calcium or a placebo with follow-up colonoscopies 1 and 4 years. The trial found that participants in the calcium group had a significantly lower 15% risk for adenoma recurrence compared to the placebo group [22].

## European cancer prevention trial: Started 1991

The European Cancer Prevention (ECP) trial was a multicenter study involving 10 European countries. It was a randomized, placebo-controlled trial. The study started in 1991 and enrolled 665 patients with a history of colorectal adenomas who were randomly assigned to receive either calcium gluconolactate and carbonate (2 g elemental calcium daily; $n=218$), fiber (3.5 g ispaghula husk; $n=226$), or a placebo ($n=221$). Participants had colonoscopy after 3 years of follow-up. Participants supplemented with ispaghula husk (fiber) had a significant increase risk for adenoma recurrence. Calcium supplementation was associated with a nonsignificant decreased risk for adenoma recurrence [23].

## Women's health study: Started 1992

The Women's Health Study (WHS) was a randomized, double-blind, placebo-controlled trial designed to test the effects of vitamin E (600 IU every other day) or a placebo in the primary prevention of CVD and cancer [24]. The study was conducted between 1992 and 2004 and a total of 39,876 apparently healthy women in the United States aged 45 years and higher participated in the trial. The follow-up period was 10.1 years. There were no significant effects on total cancer incidence, breast, lung, or colon cancer incidence. There were also no significant effects on cancer mortality [24].

## Calcium–Vitamin D women's health initiative trial. Started 1993

The Women's Health Initiative (WHI) was conducted between 1993 and 1998 and enrolled postmenopausal women between the age of 50 and 79 years at baseline to assess the risks and benefits of hormone replacement therapy and dietary modification. It was a randomized, double-blind, placebo-controlled trial of 36,282 postmenopausal women from 40 WHI centers and was conducted to test the effects of 500 mg elemental calcium as calcium carbonate plus 200 IU vitamin $D_3$ (18,176 women) twice daily (equivalent to 1000 mg elemental calcium and 400 IU vitamin $D_3$) or a placebo (18,106 women) on cancer incidence. After 7 years of follow-up, the incidence of colorectal cancer did not differ significantly between women who received the calcium plus vitamin D and those given the placebo [25]. Similarly, invasive breast cancer [26], benign proliferative breast disease [27], or nonmelanoma and melanoma skin cancer [28] were similar in the two groups. However, in a reanalysis investigating the effects of personal calcium or vitamin D supplementation in the WHI, among women who were not taking personal calcium or vitamin D supplements at randomization, twice daily elemental calcium and vitamin D decreased the risk of total, breast, and colorectal cancers [29].

## Health outcomes prevention evaluation trial: Started 1993

The Health Outcomes Prevention Evaluation (HOPE) study was an international, multicenter, double-blind, randomized trial designed to test the effect of ramipril (10 mg/day) or vitamin E (400 IU/day) versus a placebo on cardiovascular events and cancer in 9541 patients with vascular disease or diabetes mellitus. The initial HOPE trial was conducted between 1993 and 1999 and was extended as HOPE-The Ongoing Outcomes (HOPE-TOO) from 1999 to 2003. After 7 years of supplementation, there was no significant difference in cancer incidence in the vitamin E versus the placebo group [30].

## Supplementation en vitamines et mineraux antioxydants: Started 1994

The Supplementation en Vitamines et Mineraux Antioxydants (SUVIMAX) study was a population-based, double-blind, placebo-controlled, randomized trial designed to test the effects of a combination of daily supplementation of antioxidant vitamins and minerals in the prevention of cancer and heart disease. The trial started in 1994 and enrolled 5141 men aged

45–60 years and 7876 women aged 35–60 years. Participants were supplemented with a placebo or a daily capsule containing a combination of the following multivitamins and minerals: 120 mg vitamin C, 30 mg α-tocopherol, 6 mg β-carotene, 100 µg selenium, and 20 mg zinc daily. Total cancer incidence was not statistically different between the placebo and antioxidant vitamin/mineral group. However, among men only, the antioxidant vitamin/mineral supplement significantly protected against total cancer incidence [31]. Supplementation was also associated with a nonsignificant decrease in prostate cancer incidence [32]. The prostate-specific antigen (PSA) level of men entering the study was an apparent effect modifier because the antioxidant vitamin/mineral supplement significantly reduced prostate cancer incidence in the 94% of the men in the trial with PSA concentration <3 µg/L at baseline.

## Aspirin/Folate polyp prevention trial: Started 1994

The aspirin/folate polyp prevention study was a randomized, double-blind, placebo-controlled multicenter trial that was designed to test the effects of folic acid, with or without aspirin in the prevention of colorectal adenomas. The trial started in 1994 and enrolled 1021 men and women with a recent history of colorectal adenomas (removed within at least 3–16 months before recruitment) and no previous invasive large intestine carcinoma. The participants, aged 21–80 years, were randomly assigned to receive 1 mg/day folic acid ($n=516$) or a placebo ($n=505$), and were separately randomized to receive aspirin (81 or 325 mg/day) or placebo. The participants received two colonoscopic surveillance cycles, the first at 3 years and the second at 3 or 5 years later. The study found no benefits of folic acid in reducing the occurrence of at least one colorectal adenoma, but showed an increased risk for advanced lesions and adenoma multiplicity [33]. Further analysis revealed that baseline dietary folate intake, and plasma and RBC folate levels did not modify the association between folic acid treatment and risk of any adenoma or advanced lesions; but in the placebo group, there was a protective association of the highest tertile of dietary, plasma, and RBC folate with risk of adenoma [34].

## UK heart protection trial of antioxidant vitamins: Started 1994

The Heart Protection study was a randomized, placebo-controlled trial that was designed to test the effects of antioxidant vitamin supplementation on coronary events, cancer, and other morbidity. The trial started in 1994 and enrolled 20,536 adults in the United Kingdom between the ages of 40 and 80 years with coronary disease, other occlusive arterial disease, or diabetes. The participants were randomly assigned to receive antioxidant vitamin supplementation (600 mg vitamin E, 250 mg vitamin C, and 20 mg β-carotene daily) or a matching placebo. After the scheduled 5 year treatment period, there were no significant reductions in the 5 year mortality from, or incidence of any vascular disease or cancer [35].

## Selenium and vitamin E cancer prevention trial: Started 2001

The Selenium and vitamin E cancer prevention trial (SELECT) was a multi-center, randomized, double-blind, placebo-controlled, population-based trial designed to test the effects of selenium (200 μg/day of L-selenomethionine) and vitamin E (400 IU/day of *all-rac*-α-tocopherol acetate), either alone or in combination for the prevention of prostate cancer in healthy men [36]. The trial began in 2001 and enrolled 35,533 healthy men (78% white, 14% African American, 6% Hispanic, and 2% others). Initial results after a median follow-up of 5.5 years revealed no significant difference in prostate cancer incidence between any of the intervention groups (selenium, vitamin E, or placebo) [37]. The initial conclusion was that selenium or vitamin E, alone or in combination, at these specified doses and formula did not prevent prostate cancer in this population of relatively healthy men. However, with longer follow-up and more cancer cases, it was reported that supplementation with vitamin E significantly increased the risk of prostate cancer in the men [38].

## Vitamin D and omega-3 trial: Started 2010

The Vitamin D and Omega-3 trial (VITAL) is an ongoing randomized, double-blind, placebo-controlled study of vitamin D (in the form of vitamin D3 [cholecalciferol], 2000 IU/day) and marine ω-3 fatty acid (Omacor® fish oil, eicosapentaenoic acid [EPA] + docosahexaenoic acid [DHA], 1 g/day) for the primary prevention of cancer and CVD. The study started recruitment in 2010 and is expected to enroll a multi-ethnic population of 20,000 men in the United States aged 50 years or higher. The treatment period will be 5 years [39].

## Conclusion

This review has covered several large trials testing dietary supplements, especially antioxidants, in cancer prevention. From the completed trials, there is no evidence that single nutrients provided as pills/capsules were effective in preventing cancer. In fact, the results from the trials demonstrate that single nutrient supplements can be harmful, increasing risk for certain types of cancers. For examples, β-carotene supplements taken by smokers significantly increased risk for lung cancer as shown in the CARET, ATBC, and PHS II. However, the Dysplasia Trial in Linxian, China which tested a multivitamin–mineral (14 vitamins and 12 minerals) found that supplementation significantly reduced total deaths by 7%, nonsignificantly reduced total cancer incidence by 4%, and esophageal SCC by 16%. However, stomach cancer mortality increased by 18%. In the General Population Trial, also conducted in Linxian, supplementation with a combination of β-carotene–vitamin E–selenium significantly reduced total cancer mortality by 13%, including a significant 21% reduction in stomach cancer deaths and a significant 41% reduction in gastric noncardiac cancer mortality.

In general, the findings from the trials of dietary supplements and cancer incidence and mortality indicate that dietary supplementation with pills/capsule, in particular single antioxidant nutrients, to well-fed populations typical of Western countries is either useless or harmful. However, there is some evidence that multivitamin–mineral supplements or combination of nutrient supplement to populations that have compromised nutritional status might offer protection against cancer development. Future trials of dietary supplements should target populations and subgroups that suffer from compromised nutritional status.

## References

1. ACS. Global cancer facts and figures. Available from: http://www.cancer.org/acs/groups/content/@epidemiologysurveilance/documents/document/acspc-026238.pdf, 2010 (accessed on September 14, 2012).
2. ACS. American Cancer Society. *Cancer Facts & Figures*, 2010.
3. Velicer C, Ulrich C. Vitamin and mineral supplement use among US adults after cancer diagnosis: A systematic review. *J Clin Oncol* 2008;26:665–673.
4. Hennekens C, Buring J, Manson J et al. Lack of effect of long-term supplementation with beta carotene on the incidence of malignant neoplasms and cardiovascular disease. *N Engl J Med* 1996;334(18):1145–1149.
5. Greenberg E, Baron J, Stukel T et al. A clinical trial of beta carotene to prevent basal-cell and squamous cell cancers of the skin. *N Engl J Med* 1990;323:789–795.
6. Li B, Taylor P, Li J et al. Linxian nutrition intervention trials. Design, methods, participants characteristics, and compliance. *Ann Epidemiol* 1993;3(6):577–585.
7. Li J, Taylor P, Li B et al. Nutrition Intervention Trials in Linxian, China: Multiple vitamin/mineral supplementation, cancer incidence, and disease-specific mortality among adults with esophageal dysplasia. *J Natl Cancer Inst* 1993;85:1492–1498.
8. Blot W, Li Y, Taylor P et al. Nutrition intervention trials in Linxian, China: Supplementation with specific vitamin/mineral combinations, cancer incidence, and disease-specific mortality in the general population. *J Natl Cancer Inst* 1993;85:1483–1492.
9. Moon T, Levine N, Cartmel B et al. Design and recruitment for retinoid skin cancer prevention (SKICAP) trials. The Southwest Skin Cancer Prevention Study Group. *Cancer Epidemiol Biomarkers Prev* 1995;4(6):661–669.
10. Moon T, Levine N, Cartmel B et al. Effect of retinol in preventing squamous cell skin cancer in moderate-risk subjects: A randomized, double-blind, controlled trial. *Cancer Epidemiol Biomarkers Prev* 1997;6:949–956.
11. Omenn G, Goodman G, Thornquist M et al. Effects of a combination of beta carotene and vitamin A on lung cancer and cardiovascular disease. *N Engl J Med* 1996; 334:1150–1155.
12. Omenn G, Goodman G, Thornquist M et al. Risk factors for lung cancer and intervention effects in CARET, the beta-carotene and retinol efficacy trial. *N Engl J Med* 1996;88:1550–1559.
13. The Alpha-Tocopherol, Beta Carotene Cancer Prevention Study Group. The effect of vitamin E and beta carotene on the incidence of lung cancer and other cancers in male smokers. *N Engl J Med* 1994;330(15):1029–1035.
14. Heinonen O, Albanes D, Virtamo J et al. Prostate cancer and supplementation with alpha-tocopherol and beta-carotene: Incidence and mortality in a controlled trial. *J Natl Cancer Inst* 1998;90:440–446.
15. Albanes D, Malila N, Taylor P et al. Effects of supplemental alpha-tocopherol and beta-carotene on colorectal cancer: Results from a controlled trial (Finland). *Cancer Causes Control* 2000;11(3):197–205.

16. Virtamo J, Edwards B, Virtanen M et al. Effects of supplemental alpha-tocopherol and beta-carotene on urinary tract cancer: Incidence and mortality in a controlled trial (Finland). *Cancer Causes Control* 2000;11(10):933–939.

17. Malila N, Taylor P, Virtanen M et al. Effects of alpha-tocopherol and beta-carotene supplementation on gastric cancer incidence in male smokers (ATBC Study, Finland). *Cancer Causes Control* 2002;13(7):617–623.

18. Virtamo J, Pietinen P, Huttunen J et al. Incidence of cancer and mortality following alpha-tocopherol and beta-carotene supplementation: A postintervention follow-up. *JAMA* 2003;290(4):476–485.

19. Christen W, Gaziano J, Hennekens C. Design of Physicians' Health Study II—A randomized trial of beta-carotene, vitamins E and C, and multivitamins, in prevention of cancer, cardiovascular disease, and eye disease, and review of results of completed trials. *Ann Epidemiol* 2000;10(2):125–134.

20. Sesso H, Buring J, Christen W et al. Vitamins E and C in the prevention of cardiovascular disease in men: The Physicians' Health Study II randomized controlled trial. *JAMA* 2008;300(18):2123–2133.

21. Gaziano J, Glynn R, Christen W et al. Vitamins E and C in the prevention of prostate and total cancer in men: The Physicians' Health Study II randomized controlled trial. *JAMA* 2009;301(1):52–62.

22. Baron J, Beach M, Mandel J et al. Calcium supplements for the prevention of colorectal adenomas. *N Engl J Med* 1999;340:101–107.

23. Bonithon-Kopp C, Kronborg O, Giacosa A et al. Calcium and fibre supplementation in prevention of colorectal adenoma recurrence: A randomized intervention trial. *Lancet* 2000;356:1300–1306.

24. Lee I-M, Cook N, Gaziano J et al. Vitamin E in the primary prevention of cardiovascular disease and cancer. The Women's Health Study: A randomized controlled trial. *JAMA* 2005;294:56–65.

25. Wactawski-Wende J, Kotchen J, Anderson G et al. Calcium plus vitamin D supplementation and the risk of colorectal cancer. *N Engl J Med* 2006;354:684–696.

26. Chlebowski R, Johnson K, Kooperberg C et al. Calcium plus vitamin D supplementation and the risk of breast cancer. *J Natl Cancer Inst* 2007;100:1581–1591.

27. Rohan T, Negassa A, Chlebowski R et al. A randomized controlled trial of calcium plus vitamin D supplementation and risk of benign proliferative breast disease. *Breast Cancer Res Treat* 2009;116:339–350.

28. Tang J, Fu T, LeBlanc E et al. Calcium plus vitamin D supplementation and the risk of nonmelanoma and melanoma skin cancer: Post hoc analyses of the Women's Health Initiative Randomized Controlled Trial. *J Clin Oncol* 2011;29:3078–3084.

29. Bolland M, Grey A, Gamble G et al. Calcium and vitamin D supplements and health outcomes: A reanalysis of the Women's Health Initiative (WHI) limited-access data set. *Am J Clin Nutr* 2011;94:1144–1149.

30. The HOPE and HOPE-TOO Trial Investigators. Effects of long-term vitamin E supplementation on cardiovascular events and cancer. A randomized controlled trial. *JAMA* 2005;293:1338–1347.

31. Hercberg S, Galan P, Preziosi P et al. The SU.VI.MAX Study. A randomized, placebo-controlled trial of the health effects of antioxidant vitamins and minerals. *Arch Intern Med* 2004;164:2335–2342.

32. Meyer F, Galan P, Douville P et al. Antioxidant vitamin and mineral supplementation and prostate cancer prevention in the SU.VI.MAX trial. *Int J Cancer* 2005;116:182–188.

33. Cole B, Baron J, Sandler R et al. Folic acid for the prevention of colorectal adenomas: A randomized clinical trial. *JAMA* 2007;297(21):2351–2359.

34. Figueiredo J, Levine A, Grau M et al. Colorectal adenomas in a randomized folate trial: The role of baseline dietary and circulating folate levels. *Cancer Epidemiol Biomarkers Prev* 2008;17(10):2625–2631.

35. Heart Protection Study Collaborative Group. MRC/BHF heart protection study of antioxidant vitamin supplementation in 20536 high-risk individuals: A randomized placebo-controlled trial. *Lancet* 2002;360:23–33.
36. Lippman S, Goodman P, Klein E et al. Designing the selenium and vitamin E cancer prevention trial (SELECT). *J Natl Cancer Inst* 2005;97:94–102.
37. Lippman S, Klein E, Goodman P et al. Effect of selenium and vitamin E on risk of prostate cancer and other cancers: The Selenium and Vitamin E Cancer Prevention Trial (SELECT). *JAMA* 2009;301(1):39–51.
38. Klein E, Thompson IJ, Tangen C et al. Vitamin E and the risk of prostate cancer: The Selenium and Vitamin E Cancer Prevention Trial (SELECT). *JAMA* 2011; 306(14):1549–1556.
39. Manson J, Bassuk S, Lee I-M et al. The VITamin D and OmegA-3 TriaL (VITAL): Rationale and design of a large randomized controlled trial of vitamin D and marine omega-3 fatty acid supplements for the primary prevention of cancer and cardiovascular disease. *Contemporary Clin Trials* 2012;33:159–171.

# 3

# Curcumin in Human Cancer Prevention and Treatment

Adeeb Shehzad and Young Sup Lee

## Contents

## Introduction

Cancer is a fundamentally complex group of diseases characterized by abnormal regulation of tissue growth in which normal cells transform into cancer cells through altered genes, leading to unrestrained cellular proliferation. These genetic changes can occur at many levels, from expanding or deleting of entire chromosomes to mutation of a single DNA nucleotide [1]. Two categories of genes, oncogenes and tumor suppressor genes, mainly contribute to the development of cancer. Both types of genes exert their effects on tumor growth through their ability to control cell division or cell death (apoptosis).

Cancer arises from sequential mutations in oncogenes and truncations or deletions in the coding sequences of tumor suppressor genes [2].

Cancer is a major public health problem in many parts of the world, and it is estimated that it is the cause of one in four deaths in the United States. Indeed, it was recently suggested that a total of 1,638,910 new cancer cases and 577,190 deaths from cancer are expected to occur in the United States in 2012 [3]. Ineffective monotargeted drugs, which are very expensive and with many adverse effects, have highlighted the importance of multitargeted, nontoxic, inexpensive, and naturally occurring dietary agents or phytochemicals for the prevention and treatment of human diseases, including cancer.

Chemically, curcumin is bis-$\alpha,\beta$-unsaturated $\beta$-diketone, an extract of the dried ground rhizome of the perennial herb curcuma species, *Curcuma longa*, identified in 1910 by Lampe and Milobedzka [4]. Curcumin is a lipophilic agent that is nearly insoluble in water, yet quite stable in the acidic pH of the stomach. It exists in equilibrium with its enol tautomer and bis-keto form. The bis-keto form of curcumin predominates in acidic and neutral aqueous solutions, as well as in cell membranes [5,6]. Several clinical studies have used pure curcumin, but others have used either turmeric or a mixture of curcuminoids. It is known that commercially available turmeric contains 80% curcumin, 18% demethoxycurcumin, and 2% bisdemethoxycurcumin [6].

Recent research has demonstrated that curcumin is a highly antioxidative and anti inflammatory compound that possesses multifaceted therapeutic activities for the treatment of various chronic diseases. Curcumin mediates the modulation of enzymes, growth factors, and their receptors as well as cytokines and various kinase proteins that control cell proliferation and cell cycle progression [7]. Furthermore, it possesses potential therapeutic value for gastrointestinal cancers (colorectal and pancreatic cancers), genitourinary cancer (prostate cancer), reproductive cancer (cervical), hematological cancers (leukemia and multiple myeloma [MM]), pulmonary cancer, breast cancer, head and neck squamous cell carcinoma (HNSCC), and oral cancers [8].

The safety and efficacy of high doses of curcumin have been well documented in human clinical trials. The optimized daily intake of curcumin is 0–1 mg/kg body weight according to the Food and Agriculture Organization and the World Health Organization [8], and clinical studies have demonstrated that it is nontoxic and well tolerated, even at doses as high as 12 g/day [9]. The Food and Drug Administration has approved curcumin as generally recognized as safe (GRAS), and it is used as a dietary supplement in most parts of the world [10]. Curcumin has been marketed in the form of capsules, tablets, ointments, supplement drinks, and cosmetics. Clinical trials in which curcumin has been used either alone or in combination with other agents have indicated the therapeutic potential of curcumin against a wide range of human diseases.

This chapter focuses on published human clinical trials of curcumin covering multiple molecular pathways. The results presented herein will improve our understanding of the interconnected molecular pathways mediated by curcumin for prevention of different types of cancers.

## Anticancer effects of curcumin

Curcumin has long been used as a remedy in traditional Chinese and Indian medicine for the treatment of different ailments. The pharmacological activity and medicinal application of curcumin in human cancer types is due to its multitargeting activities. Specifically, curcumin disrupts the signaling pathways and molecular targets involved in the initiation and progression of various cancers (Figure 3.1). Following are the different forms of cancer for which curcumin has shown efficacy.

### Gastrointestinal cancers

#### Colorectal cancer

Cancers of the colon and rectum, together with colorectal cancer (CRC), comprise the third leading cause of cancer-related deaths in men and women. It is estimated that a total of 143,460 (73,420 male and 70,040 female) new cases and 51,690 (26,470 male and 25,220 female) deaths related to CRC will occur in the United States in 2012 [3]. Currently, there is no effective treatment except surgical resection; therefore, new drugs and strategies are needed to treat CRC cancer. In a dose escalation study, 15 patients with advanced CRC were treated for up to 4 months with curcumin extract at doses of 0.44 and 2.2 g/day, each containing 36–180 mg of curcumin. This dose did not induce any observed toxicity, but a 59% decrease in lymphocytic glutathione S-transferase (GST) activity was observed. A decline in cancer biomarker carcinoembryonic antigen (CEA) level from $310 \pm 15$ to $175 \pm 9$ was also observed after two months of curcumin administration [11]. In another study, curcumin doses between 0.45 and 3.6 g daily for up to four months in 15 patients with advanced CRC resulted in the dose-dependent inhibition of prostaglandin $E_2$ ($PGE_2$) production, illustrating the efficacy of curcumin against CRC. Curcumin and its metabolites (glucuronide and sulfate) were detected in plasma in the 10 nmol/L range and in urine [12]. Additionally, Garcea et al. [13] investigated the effects of administration of curcumin capsules containing three different doses (3600, 1800, or 450 mg/day) to 12 patients (5 female, 7 male, ages 47–72 years) for 7 days. Biopsy samples of normal and malignant colorectal tissue obtained at diagnosis and at 6–7 h after administration were then compared and found to contain $12.7 \pm 5.7$ and $7.7 \pm 1.8$ nmol/g curcumin, respectively. Furthermore, curcumin metabolites (sulfate and glucuronide) have been detected in the tissue of patients. This study also reported that trace levels of curcumin were found in the peripheral circulation. Curcumin treatment also decreased the levels of M(1)G in malignant colorectal tissue, but the levels of COX-2 were unaffected by the same dose of curcumin.

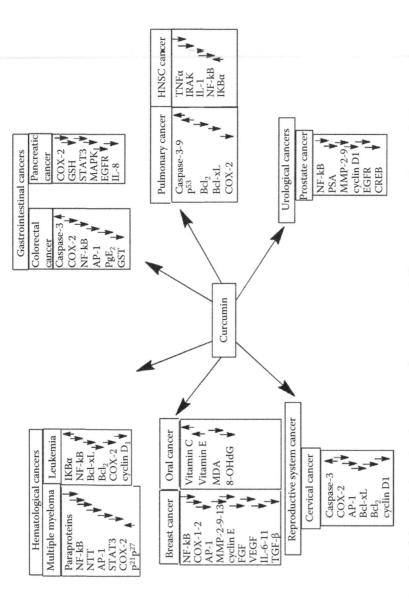

**Figure 3.1** A model of the multiple molecular targets of curcumin in different cancers is illustrated. The upward arrows represent up-regulation or activation of the target molecules, and the downward arrows represent down-regulation or inhibition of the specific molecules.

These findings indicate that curcumin has potential pharmacological effectiveness in CRC patients [13]. Additionally, curcumin has been shown to be effective for the prevention and treatment of CRC when administered concomitantly with other drugs. Curcumin 480 mg and quercetin 20 mg were administered orally to familial adenomatous polyposis (FAP) patients with prior colectomy (four with a retained rectum and one with an ileal anal pouch) three times a day. FAP is an autosomal-dominant disorder characterized by hundreds of colorectal adenomas that eventually develop into CRC. In all five patients, curcumin and quercetin decreased the polyp number and size from baseline after a mean of 6 months of treatment without any toxicity [14]. A nonrandomized, open-label clinical trial was conducted in 44 confirmed smokers in which curcumin was given orally at two different doses (2 or 4 g/day) for 30 days. The results revealed that curcumin treatment reduced the formation of aberrant crypt foci (ACF), the precursor of colorectal polyps [15]. Curcumin at both doses did not reduce the levels of $PGE_2$ and 5-hydroxyeicosatetraenoic acid in ACF or normal flat mucosa, but treatment at 4 g/day led to a significant reduction in ACF formation (40%). This reduction in ACF formation by curcumin was mediated by a significant fivefold increase in posttreatment plasma curcumin/conjugate levels. Overall, this study demonstrated that curcumin was well tolerated at both 2 and 4 g and effective against ACF formation [15]. Furthermore, He et al. determined the effects of curcumin in patients with CRC after diagnosis and presurgery. Curcumin capsule (360 mg thrice/day) administration for 10–30 days increased body weight, decreased serum TNF-α level, increased the number of apoptotic cells, and enhanced the expression of p53 in tumor tissue. Their study concluded that curcumin can improve the general health of CRC patients via the mechanism of increased p53 expression in tumor cells [16]. Further studies are warranted to confirm the safety and efficacy of curcumin in patients with CRC.

## Pancreatic cancer

Pancreatic cancer accounts for ~6% of all cancer-related deaths in men and women. It is estimated that a total of 43,920 (22,090 male and 21,830 female) new cases and 37,390 (18,850 male and 18,540 female) deaths related to pancreatic cancer will occur in the United States in 2012 [3]. Clinical trials dealing with the effects of orally administered curcumin (500 mg) and piperine (5 mg) in 20 patients with tropical pancreatitis resulted in a reduction in erythrocyte MDA levels with a significant increase in glutathione (GSH) levels. Oral administration of curcumin (8 g) to pancreatic cancer patients decreased the expression of NF-κB, COX-2, and phosphorylated signal transducer and activator of transcription 3 (STAT3); however, pain was not improved by this treatment [17,18]. This study concluded that oral curcumin with piperine may reverse lipid peroxidation in patients with tropical pancreatitis. Dhillon et al. [19] conducted a phase II clinical trial of patients with advanced pancreatic cancer that demonstrated the safety and efficacy of curcumin [19]. In their study, curcumin (8 g) was orally administered to 25 patients per day for 2 months. Circulating curcumin was detected as the glucuronide and sulfate conjugate forms, albeit at low steady-state levels, suggesting poor oral bioavailability. Despite this poor bioavailability, two patients showed clinical biological activity and one had ongoing stable disease for more than 18 months. In this study, one patient showed marked tumor regression accompanied by significant increase in levels of serum cytokines such as IL-6, IL-8, IL-10, and IL-1 receptor antagonists. Curcumin administration also down-regulated NF-κB, COX-2, and pSTAT3 in the peripheral blood mononuclear cells of patients. These findings demonstrate that orally administered curcumin is well tolerated and has biological activity in some patients with pancreatic cancer [19]. Curcumin efficacy was evaluated when administered concomitantly with gemcitabine in an open-label phase II trial against advanced pancreatic cancer [20]. Specifically, curcumin (8 g) was administered by mouth daily, with gemcitabine (1 g/m$^2$) intravenously three times a week to 17 patients. Five out of seventeen patients (29%) discontinued curcumin after a few days to 2 weeks due to intractable abdominal fullness or pain, and the dose of curcumin was reduced to 4000 mg/day because of abdominal complaints by two other patients. One of eleven evaluable patients (9%) showed a partial response, four (36%) had stable disease, and six (55%) had tumor progression. The time to tumor progression was 1–12 months (median, 2.5 months), and the overall survival was 1–24 months (median, 5 months). Taken together, this study suggested that 8 g/day of curcumin with gemcitabine seemed appropriate, but this should be confirmed in large number of patients [20]. Recently, the same therapeutic combinations of curcumin and gemcitabine were evaluated in the 21 patients with gemcitabine-resistant pancreatic cancer. Curcumin at 8 g/day in combination with gemcitabine-based chemotherapy was safe and well tolerated in patients with pancreatic cancer; therefore, it warrants further investigation into its efficacy [21].

## Genitourinary cancers

### Prostate cancer

Prostate cancer is the second greatest cause of cancer-related mortalities in men, and it has been reported that a total of 241,740 new cases will be diagnosed in the United States in 2012, leading to 28,170 deaths [3]. Prostate cancer is recognized by elevated prostate-specific antigen (PSA). A randomized clinical trial was conducted to evaluate the effects of soy isoflavones and curcumin on serum PSA levels in men who had undergone prostate biopsies, but were not found to have prostate cancer [22]. In this study, 85 subjects were administered isoflavones (40 mg) and curcumin (100 mg) and 42 subjects were administered placebo daily for 6 months, and PSA values measured before and 6 months after treatment. The level of PSA was apparently decreased in the group that received the isoflavones and curcumin as indicated by PSA values. These findings indicated that curcumin exerts synergistic effects with isoflavones to modulate serum PSA production [22]. Another clinical trial was conducted to determine the efficacy of the herbal preparation, Zyflamend (curcumin), against high-grade prostatic intraepithelial neoplasia (HGPIN) in 23 patients. In that study, Zyflamend was given to HGPIN patients three times a day for 18 months. A biopsy conducted at 18 months revealed no markers of HGPIN and a reduction in NF-κB and C-reactive protein [23].

## Reproductive cancers

### Cervical cancer

Cervical cancer typically develops in the cells of the cervix, which is the lower part of the uterus connected to the vagina. According to the American Cancer Society, a total of 12,170 new cases of cervical cancer will be diagnosed in 2012, leading to 4220 deaths [3]. Infection with human papilloma viruses (HPV) leads to a high risk for the development of cervical carcinoma, possibly through the actions of viral oncoproteins E6 and E7 [24]. In a phase I clinical trial, a daily 0.5–12 g oral dose of curcumin administered over 3 months resulted in the histological improvement of precancerous lesions in one of four patients suffering from cervical intraepithelial neoplasia [25].

## Hematological cancers

### Leukemia

Leukemia is the cancer of blood or bone marrow characterized by an abnormal proliferation of blood cells. It is estimated that a total of 47,150 (26,830 men and 20,320 women) new cases of leukemia will be diagnosed in the United States in 2012 leading to 23,540 (13,500 men and 10,040 women) deaths [3]. Curcumin has the ability to inhibit multiple drug resistance 1 (MDR1) and Wilms' tumor 1 (*WT1*) gene expression in acute myeloid leukemia (AML) patients [26]. The down-regulation of the *WT1* gene by curcumin may be achieved through the up-regulation of miR-15a/16-1 expression in leukemia [27].

### Multiple myeloma

MM is cancer of the plasma cells, a type of white blood cell present in bone marrow. Approximately 21,700 (12,190 men and 9510 women) patients in the United States will be diagnosed with MM in 2012, and approximately 10,710 (6020 men and 4690 women) will die of it within the same year [3]. Monoclonal gammopathy of undetermined significance (MGUS) and smoldering MM (SMM) were recently shown to be premalignant plasma cell proliferative disorders that can progress to MM [28]. It is also known that the early stages of this disease are characterized by a serum M-protein value of <30 g/L, but that an abnormal free light-chain ratio (rFLC) also increases the risk of initiation and progression of MM, regardless of the size and type of serum M-protein [28]. A randomized, blind, crossover pilot study was conducted to determine the effects of curcumin on plasma cells and osteoclasts in 26 patients with known MGUS. In this study, 17 patients were given 1 g (900 mg of curcumin, 80 mg of desmethoxycurcumin, and 20 mg of bis-desmethoxycurcumin) of C3 curcuminoid complex at the beginning and were then switched to placebo (1 g) after 3 months, while nine patients were initially given placebo and then switched to curcumin. Curcumin was found to decrease paraprotein load in 10 patients having a paraprotein level of >20 g/L, with five of these patients showing a 12%–30% reduction in their paraprotein levels after curcumin therapy. Moreover, 27% of patients on curcumin showed a >25% decrease in urinary N-telopeptide of type I collagen [29]. Recently, the same group also conducted a randomized, double-blind, placebo-controlled crossover study, followed by an open-label extension study using an 8 g dose to evaluate the outcome of curcumin administration on FLC response and bone turnover in patients with MGUS and SMM [28]. In that study, a total of 36 patients (19 MGUS and 17 SMM) were randomized to the curcumin (4 g)-treated group and the other 4 g-received placebo, with treatments being switched at 3 months, and open-label, 8 g dose extension study. Curcumin administration decreased the rFLC, reduced the difference between clonal and nonclonal light chain (dFLC), and involved FLC (iFLC). Curcumin also decreased urinary deoxypyridinoline (uDPYD), a marker of bone resorption, and serum creatinine levels when compared to the placebo group [28]. Furthermore, a clinical trial of curcumin was conducted in 29 MM patients in which the treated group was given curcumin at doses of 2, 4, 6, 8, or 12 g/day alone, or in combination with Bioperine (10 mg) for 12 weeks. The results indicated that curcumin down-regulated the activation of NF-κB and suppressed COX-2 expression [30,31].

## Pulmonary cancer

Pulmonary cancer, also known as lung cancer, is often revealed as a solitary pulmonary nodule. Pulmonary cancer is the leading cause of death by cancer in men in the United States. It is estimated that a total of 226,160 (116,470 male and 109,690 female) new cases of lung and bronchus cancer leading to 160,340 (87,750 male and 72,590 female) deaths will occur in 2012 [3]. Polasa et al. assessed very early antimutagenic effects of turmeric in 16 chronic smokers and six nonsmokers that were considered as a control [32]. They found that when turmeric was given at 1.5 g/day for 30 days, the urinary excretion of mutagens by smokers was significantly reduced but no changes were observed in that of nonsmokers. Their study also indicated that turmeric had no significant effect on serum aspartate aminotransferase and alanine aminotransferase, blood glucose, creatinine, or lipid profile [32], but that it was an effective antimutagen therapy in smokers to reduce the risk of lung cancer.

## Breast cancer

Breast cancer is the most frequently diagnosed cancer in pre- and postmenopausal women, but is very rarely observed in men. Approximately 229,060 (2,190 men and 226,870 women) patients in the United States will be diagnosed with breast cancer in 2012, approximately 39,920 (410 men and 39,510 women) of whom will die from this disease [3]. Several drug candidates are currently used for the treatment of breast cancer, among which docetaxel has commonly been used either alone or in combination with other chemotherapeutic agents for second-line treatment of advanced breast cancer [33]. An open-label phase I trial was conducted to assess the feasibility and tolerability of the combination of docetaxel and curcumin in 14 patients with advanced and metastatic breast cancer [34]. Docetaxel (100 mg/m$^2$) was administered as a 1 h intravenous infusion every 3 weeks on day 1 for six cycles, while curcumin was given orally from 500 mg/day for 7 days (from day 4 to day +2) and escalated until dose-limiting toxicity would occur. This study concluded that the feasibility and tolerability of curcumin dose was 8 g/day, whereas the recommended dose was 6 g/day for seven consecutive days every 3 weeks when administered in combination with a standard dose of docetaxel [34]. This combination warrants further clinical study with a greater number of patients to validate the results of their study.

## Head and neck cancer

HNSCC is the sixth most common non-skin cancer in the world, with an incidence of 600,000 cases per year, 50,000 of which occur in the United States [35]. Most HNSCCs arise in the hypopharynx, larynx, and trachea, as well as in the oral cavity and oropharynx. Despite medical advances, the 5 year survival rate for patients with HNSCC remains in the range of 40%–50%. Residents of industrialized areas are believed to be at high risk for this disease [36]. Aberrant signaling pathways have been identified in HNSCCs, and inhibition of NF-κB and inflammatory molecules such as IL-6, IL-8, and vascular endothelial growth factor (VEGF) and epidermal growth factor receptor (EGFR) has been shown to be a successful therapeutic strategy [35,37]. Kim et al. investigated the effects of curcumin in patients with HNSCC through inhibition of the activity of IκB kinase β (IKKβ), an enzyme involved in NF-κB activation that suppresses expression of inflammatory cytokines [38]. In this human study, 39 patients were enrolled in different groups (13 with dental caries, 21 with HNSCC, and 5 healthy volunteers). Saliva was collected before and after the subjects chewed two curcumin tablets for 5 min. Curcumin treatment led to a reduction in IKKβ kinase activity in the salivary cells of patients with HNSCC. Treatment of UM-SCC1 cell lines with curcumin as well as with post-curcumin salivary supernatant showed a reduction of IKKβ kinase activity. Additionally, a remarkable reduction in IL-8 levels was seen in post-curcumin samples from patients with dental caries. Although IL-8 expression was reduced in 8 of 21 post-curcumin samples of patients with HNSCC, the reduction was not significant. This study highlighted the potential role of IKKβ kinase as a biomarker for detection of the effects of curcumin in head and neck cancer [38].

## Oral cancer

Oral cancer is a subtype of head and neck cancer, and one of the 10 most common cancers. Indeed, there are 5000 new cases and 3000 deaths worldwide every year [39]. Oral cancer includes most common precancerous oral lesions, such as oral submucous fibrosis, oral leukoplakia, and oral lichen planus, which are strongly associated with smoking, chewing tobacco, and excessive consumption of alcohol. The number of micronucleated oral mucosal cells can be used to investigate preneoplastic effects by collecting cells directly from the affected tissues and assessing the potential of therapeutic agents [39,40]. One early study evaluated the effects of alcoholic extracts of turmeric oil, and those of turmeric oleoresin on the number of micronuclei in 58 clinically and histopathologically diagnosed male OSF patients from 15 to 35 years of age and compared these to 32 normal healthy subjects groups of the same age [41]. Both extracts protected against a benzo[a]pyrene-induced increase in micronuclei in circulating lymphocytes. Moreover, patients with submucous fibrosis were divided into three groups administered a daily oral dose of turmeric oil (600 mg) plus turmeric (3 g), turmeric oleoresin (600 mg) plus turmeric (3 g), or turmeric alone (3 g) for 3 months. All three turmeric extracts led to a decrease in the number of micronucleated cells in exfoliated oral mucosal cells and in circulating lymphocytes. However, turmeric oleoresin was more effective at reducing the number of micronuclei in oral mucosal cells [41]. Additionally, Cheng et al. conducted a phase I study of toxicology, pharmacokinetics, and biologically effective doses of curcumin in 25 patients with resected urinary bladder cancer, arsenic-associated Bowen's disease of the skin, uterine cervical intraepithelial neoplasm (CIN), oral leukoplakia, and intestinal metaplasia of the stomach [25]. In that study, curcumin was administered orally for 3 months to a total of 25 subjects, and biopsy of the lesion sites was conducted immediately before and 3 months after initiation of curcumin treatment. Curcumin did not induce any toxic effects at 8 g/day; however, at greater than 8 g/day the bulky volume of the drug was unacceptable to the patients. Additionally, one of four patients with CIN and one of seven patients with oral leukoplakia developed frank malignancies despite curcumin treatment. In contrast, histological improvement of precancerous lesions was observed in one of two patients with resected bladder cancer, two of seven patients with oral leukoplakia, one of six patients with intestinal metaplasia of the stomach, one of four patients with CIN, and two of six patients with Bowen's disease [25]. A randomized, double-blind, placebo-controlled trial was conducted in 100 patients with oral lichen planus to assess the effectiveness of curcuminoids [42].

The trial consisted of two interim analyses, the first of which was conducted in 33 subjects that were randomized to receive either placebo or curcuminoids at 2000 mg/day for 7 weeks. In addition, all participants received prednisone at 60 mg/day for the first week. The results of the first interim analysis did not reveal any significant difference between the placebo and curcuminoid groups. The conditional calculations suggested a <2% chance that the curcuminoid groups would have a significantly better outcome than the placebo group if the trial was continued to completion. The authors still suggested the use of a larger sample size and a higher dose for a longer duration of curcuminoids without an initial course of prednisone [42]. Rai et al. investigated the effects of a 1 g curcumin tablet (900 mg of curcumin, 80 mg of demethoxycurcumin, 20 mg of bisdemethoxycurcumin) in patients with oral leukoplakia, oral submucous fibrosis, or lichen planus, and healthy individuals ($n$ = 25 for each group) aged 17–50 years for 1 week. Curcumin administration was associated with decreased malonaldehyde and 8-hydroxydeoxyguanosine and an increase in vitamins C and E in the serum and saliva of patients with precancerous lesions [43]. These findings indicate that curcumin mediates its anti-precancer activities by increasing levels of vitamins C and E and preventing lipid peroxidation and DNA damage. Recently, Chainani-Wu et al. also confirmed the efficacy and the safety of curcuminoids at 6 g/day in a randomized, double-blind, placebo-controlled clinical trial of 20 patients with oral lichen planus. They concluded that curcuminoids at 6 g/day are well tolerated and may prove efficacious in controlling signs and symptoms of oral lichen planus [44].

## Curcumin bioavailability, safety, and toxicity

Available evidence suggests that curcumin diminishes the activities of undesirable cellular activities such as abnormal cell proliferation and uncontrolled growth in cancer. Although curcumin has shown potential against numerous human disorders, poor bioavailability due to poor absorption, rapid metabolism, and systemic elimination has been shown to limit its therapeutic efficacy [45]. Reduced bioavailability of curcumin involves the observation of extremely low serum levels after oral administration. As a result, several methods to improve curcumin's bioavailability and retention have been considered. The use of adjuvants that can block the metabolic pathway of curcumin is the most common strategy for enhancing its bioavailability [46]. In clinical trials, volunteers receiving a dose of 2 g of curcumin alone showed very low serum levels of curcumin. However, concomitant administration of 20 mg of piperine (a known inhibitor of hepatic and intestinal glucuronidation) with curcumin produced much higher concentrations within 30 min to 1 h of treatment, with an overall increase in bioavailability of 2000%. Other useful methods to increase the bioavailability of curcumin in humans include the use of liposomes and phytosomes.

Doses of 2, 3, and 4 g of solid lipid curcumin particle were evaluated in 11 patients with osteosarcoma and healthy subjects. In both healthy individuals and osteosarcoma patients, high interindividual variability in pharmacokinetics and nonlinear dose dependency was observed, suggesting potentially complex absorption kinetics [47]. Several studies have also shown that phospholipid complexes of curcumin also enhanced systemic bioavailability. The curcumin phytosome preparation Meriva® with soy phosphatidylcholine has better bioavailability than curcumin. Recently, Cuomo et al. examined the absorption of a curcuminoid mixture and Meriva in a randomized, double-blind, crossover human study [48]. They found that total curcuminoid absorption was about 29-fold higher for the Meriva mixture than the unformulated curcuminoid mixture. Furthermore, the phospholipid formulation increased the absorption of demethoxylated curcuminoids much more than that of curcumin [48]. The bioavailability of curcumin might also be enhanced through the preparation of nanoparticles [49], curcumin structural analogs, and nanocurcumin [45,50]. Antony et al. reported that reconstituted curcumin with the non-curcuminoid components of turmeric also increased the bioavailability in the trial of 11 healthy participants. The results of the study indicated that the relative bioavailability of BCM-95®CG (Biocurcumax™) was about 6.93-fold compared to normal curcumin without formulation and about 6.3-fold compared to curcumin–lecithin–piperine formula [50].

The development of therapeutic strategies for the treatment of various cancers with phytochemicals including curcumin has gained significant attention during the last decade. Curcumin is safe and well tolerated in human clinical trials of cancers. A very early study by the Food and Agriculture Organization and the World Health Organization demonstrated that the optimized daily intake of curcumin is 0–1 mg/kg body weight [51]. Cheng et al. investigated the toxicology, pharmacokinetics, pharmacodynamics, and physiologically effective doses of curcumin in different human inflammatory conditions. In their phase I clinical trial, curcumin was started at 500 mg/day and then escalated to 1, 2, 4, 8, and 12 g/day for 3 months in 25 subjects, out of which 24 subjects completed the study. No dose-limiting toxicity was observed in any subject under any condition [25]. In another study, curcumin (C3 Complex, Sabinsa Corporation) was administered to healthy volunteers at 500–12,000 mg and only 7 out of 24 subjects experienced minimal toxicity that was not dose related [52]. Recently, a randomized pilot study was conducted to assess the safety and effeminacy of curcumin in 45 patients (38 female, 7 male; mean age 47.88 years) with active rheumatoid arthritis. The results revealed that curcumin exhibits potent anti inflammatory effects against rheumatoid arthritis and that it was safe and not related to any adverse action [53]. It should also be noted that side effects including gastrointestinal upset, chest tightness, inflamed skin, and skin rashes have been reported; however, these are believed to occur in response to high doses.

Nevertheless of its proven efficacy over centuries of use and its demonstrated safety in several human studies, curcumin should be translated to the clinics for the treatment of cancers.

## Conclusion

The information provided in this chapter demonstrates that curcumin modulates multiple cellular signaling pathways and interacts with numerous molecular targets including transcription factors, growth factors and their receptors, cytokines, enzymes, and genes regulating cell proliferation and apoptosis. Human studies, mostly small pilot trials, demonstrate potentially important roles for curcumin in cancer treatment. To date, most of the studies suffer from methodological limitations such as small sample size and limited power to test effects. In the future, well-defined epidemiological studies with adequate sample size and power to test the effects of curcumin against human cancers will be needed.

## References

1. Morgan, D.O. 2007. *The Cell Cycle: Principles of Control.* New Science Press Ltd., London, UK.
2. Salk, J.J., Fox, E.J., and Loeb, L.A. 2010. Mutational heterogeneity in human cancers: Origin and consequences. *Annu Rev Pathol.* 5: 51–75.
3. Siegel, R., Naishadham, D., and Jemal, A. 2012. Cancer statistic. *CA Cancer J Clin.* 62: 10–29.
4. Shehzad, A., Ha, T., Subhan, F., and Lee, Y.S. 2011. New mechanisms and the anti-inflammatory role of curcumin in obesity and obesity-related metabolic diseases. *Eur J Nutr.* 50: 151–161.
5. Shehzad, A., Khan, S., and Lee, Y.S. 2012. Curcumin molecular targets in obesity and obesity-related cancers. *Future Oncol.* 8: 179–190.
6. Shehzad, A., Wahid, F., and Lee, Y.S. 2010. Curcumin in cancer chemoprevention: Molecular targets, pharmacokinetics, bioavailability, and clinical trials. *Arch Pharm (Weinheim).* 343: 489–499.
7. Shehzad, A., Khan, S., Shehzad, O., and Lee, Y.S. 2010. Curcumin therapeutic promises and bioavailability in colorectal cancer. *Drugs Today.* 46: 523–532.
8. Shehzad, A. and Lee, Y.S. 2010. Curcumin: Multiple molecular targets mediate multiple pharmacological actions: A review. *Drugs Future.* 35: 113–119.
9. Lao, C.D., Ruffin, M.T., Normolle, D. et al. 2006. Dose escalation of a curcuminoid formulation. *BMC Complement Altern Med.* 6: 10.
10. Goel, A., Jhurani, S., and Aggarwal, B.B. 2008. Multi-targeted therapy by curcumin: How spicy is it? *Mol Nutr Food Res.* 52: 1010–1030.
11. Sharma, R.A., McLelland, H.R., Hill, K.A. et al. 2001. Pharmacodynamic and pharmacokinetic study of oral curcuma extract in patients with colorectal cancer. *Clin Cancer Res.* 7: 1894–1900.
12. Sharma, R.A., Euden, S.A., and Platton, S.L. et al. 2004. Phase I clinical trial of oral curcumin: Biomarkers of systemic activity and compliance. *Clin Cancer Res.* 10: 6847–6854.

13. Garcea, G., Berry, D.P., and Jones, D.J. et al. 2005. Consumption of the putative chemopreventive agent curcumin by cancer patients: Assessment of curcumin levels in the colorectum and their pharmacodynamic consequences. *Cancer Epidemiol Biomarkers Prev.* 14: 120–125.
14. Cruz-Correa, M., Shoskes, D.A., and Sanchez, P. et al. 2006. Combination treatment with curcumin and quercetin of adenomas in familial adenomatous polyposis. *Clin Gastroenterol Hepatol.* 4: 1035–1038.
15. Carroll, R.E., Benya, R.V., Turgeon, D.K. et al. 2011. Phase IIa clinical trial of curcumin for the prevention of colorectal neoplasia. *Cancer Prev Res (Philadelphia).* 4: 354–364.
16. He, Z.Y., Shi, C.B., Wen, H., Li, F.L., Wang, B.L., and Wang, J. 2011. Upregulation of p53 expression in patients with colorectal cancer by administration of curcumin. *Cancer Invest.* 29: 208–213.
17. Shishodia, S., Chaturvedi, M.M., and Aggarwal, B.B. 2007. Role of curcumin in cancer therapy. *Curr Probl Cancer.* 31: 243–305.
18. Anand, P., Sundaram, C., Jhurani, S., Kunnumakkara, A.B., and Aggarwal, B.B. 2008. Curcumin and cancer: An "old-age" disease with an "age-old" solution. *Cancer Lett.* 267: 133–164.
19. Dhillon, N., Aggarwal, B.B., Newman, R.A. et al. 2008. Phase II trial of curcumin in patients with advanced pancreatic cancer. *Clin Cancer Res.* 14: 4491–4499.
20. Epelbaum, R., Schaffer, M., Vizel, B., Badmaev, V., and Bar-Sela, G. 2010. Curcumin and gemcitabine in patients with advanced pancreatic cancer. *Nutr Cancer.* 62: 1137–1141.
21. Kanai, M., Yoshimura, K., Asada, M. et al. 2011. A phase I/II study of gemcitabine-based chemotherapy plus curcumin for patients with gemcitabine resistant pancreatic cancer. *Cancer Chemother Pharmacol.* 68: 157–164.
22. Ide, H., Tokiwa, S., Sakamaki, K. et al. 2010. Combined inhibitory effects of soy isoflavones and curcumin on the production of prostate-specific antigen. *Prostate.* 70: 1127–1133.
23. Capodice, J.L., Gorroochurn, P., Cammack, A.S. et al. 2009. Zyflamend in men with high-grade prostatic intraepithelial neoplasia: Results of a phase I clinical trial. *J Soc Integr Oncol.* 7: 43–51.
24. Maher, D.M., Bell, M.C., O'Donnell, E.A., Gupta, B.K., Jaggi, M., and Chauhan, S.C. 2011. Curcumin suppresses human papillomavirus oncoproteins, restores p53, Rb, and PTPN13 proteins and inhibits benzo[*a*]pyrene-induced upregulation of HPV E7. *Mol Carcinog.* 50: 47–57.
25. Cheng, A.L., Hsu, C.H., Lin, J.K. et al. 2001. Phase I clinical trial of curcumin, a chemopreventive agent, in patients with high-risk or pre-malignant lesions. *Anticancer Res.* 21: 2895–2900.
26. Yang, C.W., Chang, C.L., Lee, H.C., Chi, C.W., Pan, J.P., and Yang, W.C. 2012. Curcumin induces the apoptosis of human monocytic leukemia THP-1 cells via the activation of JNK/ERK pathways. *BMC Complement Altern Med.* 12: 22.
27. Gao, S.M., Yang, J.J., Chen, C.Q. et al. 2012. Pure curcumin decreases the expression of WT1 by upregulation of miR-15a and miR-16-1 in leukemic cells. *J Exp Clin Cancer Res.* 31: 27.
28. Golombick, T., Diamond, T.H., Badmaev, V., Manoharan, A., and Ramakrishna, R. 2009. The potential role of curcumin in patients with monoclonal gammopathy of undefined significance—Its effect on paraproteinemia and the urinary N-telopeptide of type I collagen bone turnover marker. *Clin Cancer Res.* 15: 5917–5922.
29. Golombick, T., Diamond, T.H., Manoharan, A., and Ramakrishna, R. 2012. Monoclonal gammopathy of undetermined significance, smoldering multiple myeloma, and curcumin: A randomized, double-blind placebo-controlled cross-over 4 g study and an open-label 8 g extension study. *Am J Hematol.* 87: 455–460.
30. Sung, B., Kunnumakkara, A.B., Sethi, G., Anand, P., Guha, S., and Aggarwal, B.B. 2009. Curcumin circumvents chemoresistance in vitro and potentiates the effect of thalidomide and bortezomib against human multiple myeloma in nude mice model. *Mol Cancer Ther.* 8: 959–970.

31. Yang, C.L., Ma, Y.G., Xue, Y.X., Liu, Y.Y., Xie, H., and Qiu, G.R. 2012. Curcumin induces small cell lung cancer NCI-H446 cell apoptosis via the reactive oxygen species–mediated mitochondrial pathway and not the cell death receptor pathway. *DNA Cell Biol.* 31: 139–150.

32. Polasa, K., Raghuram, T.C., Krishna, T.P., and Krishnaswamy, K. 1992. Effect of turmeric on urinary mutagens in smokers. *Mutagenesis.* 7: 107–109.

33. Harvey, V., Mouridsen, H., Semiglazov, V. et al. 2006. Phase III trial comparing three doses of docetaxel for second-line treatment of advanced breast cancer. *J Clin Oncol.* 24: 4963–4970.

34. Bayet-Robert, M., Kwiatkowski, F., Leheurteur, M. et al. 2010. Phase I dose escalation trial of docetaxel plus curcumin in patients with advanced and metastatic breast cancer. *Cancer Biol Ther.* 9: 8–14.

35. Leemans, C.R., Braakhuis, B.J., and Brakenhoff, R.H. 2011. The molecular biology of head and neck cancer. *Nat Rev Cancer.* 11: 9–22.

36. Mignogna, M.D., Fedele, S., and Lo Russo, L. 2004. The World Cancer Report and the burden of oral cancer. *Eur J Cancer Prev.* 13: 139–142.

37. Aggarwal, B.B., Vijayalekshmi, R.V., and Sung, B. 2009. Targeting inflammatory pathways for prevention and therapy of cancer: Short-term friend, long-term foe. *Clin Cancer Res.* 15: 425–430.

38. Kim, S.G., Veena, M.S., Basak, S.K. et al. 2011. Curcumin treatment suppresses IKKbeta kinase activity of salivary cells of patients with head and neck cancer: A pilot study. *Clin Cancer Res.* 17: 5953–5961.

39. Halder, A., Chakraborty, T., Mandal, K. et al. 2004. Comparative study of exfoliated oral mucosal cell micronuclei frequency in normal, precancerous and malignant epithelium. *Int J Hum Genet.* 4: 257–260.

40. Jadhav, K., Gupta, N., and Ahmed, M.B. 2011. Micronuclei: An essential biomarker in oral exfoliated cells for grading of oral squamous cell carcinoma. *J Cytol.* 28: 7–12.

41. Hastak, K., Lubri, N., Jakhi, S.D. et al. 1997. Effect of turmeric oil and turmeric oleorisin on cytogenetic damage in patients suffering from oral submucous fibrosis. *Cancer Lett.* 116: 265–269.

42. Chainani-Wu, N., Silverman, S., Reingold, A. et al. 2007. A randomized, placebo-controlled, double-blind clinical trial of curcuminoids in oral lichen planus. *Phytomedicine.* 14: 437–446.

43. Rai, B., Kaur, J., Jacobs, R., and Singh, J. 2010. Possible action mechanism for curcumin in pre cancerous lesions based on serum and salivary markers of oxidative stress. *J Oral Sci.* 52: 251–256.

44. Chainani-Wu, N., Madden, E., Lozada-Nur, F., and Silverman, S. 2012. High-dose curcuminoids are efficacious in the reduction in symptoms and signs of oral lichen planus. *J Am Acad Dermatol.* 66: 752–760.

45. Anand, P., Kunnumakkara, A.B., Newman, R.A., and Aggarwal, B.B. 2007. Bioavailability of curcumin: Problems and promises. *Mol Pharm.* 4: 807–818.

46. Shoba, G., Joy, D., Joseph, T., Majeed, M., Rajendran, R., and Srinivas, P.S. 1998. Influence of piperine on the pharmacokinetics of curcumin in animals and human volunteers. *Planta Med.* 64: 353–356.

47. Gota, V.S., Maru, G.B., Soni, T.G., Gandhi, T.R., Kochar, N., and Agarwal, M.G. 2010. Safety and pharmacokinetics of a solid lipid curcumin particle formulation in osteosarcoma patients and healthy volunteers. *J Agric Food Chem.* 58: 2095–2099.

48. Cuomo, J., Appendino, G., Dern, A.S. et al. 2011. Comparative absorption of a standardized curcuminoid mixture and its lecithin formulation. *J Nat Prod.* 74: 664–669.

49. Sasaki, H., Sunagawa, Y., Takahashi, K. et al. 2011. Innovative preparation of curcumin for improved oral bioavailability. *Biol Pharm Bull.* 34: 660–665.

50. Antony, B., Merina, B., Iyer, V.S., Judy, N., Lennertz, K., and Joyal S. 2008. A pilot crossover study to evaluate human oral bioavailability of BCM-95CG (Biocurcumax), a novel bioenhanced preparation of curcumin. *Indian J Pharm Sci.* 70: 445–449.

51. WHO Geneva. 2000. Evaluation of certain food additives. WHO Technical Report Series 891.
52. Lao, C.D., Ruffin, M.T., Normolle, D. et al. 2006. Dose escalation of a curcuminoid formulation. *BMC Complement Altern Med.* 6: 10.
53. Chandran, B. and Goel, A. 2012. A randomized, pilot study to assess the efficacy and safety of curcumin in patients with active rheumatoid arthritis. *Phytother Res* 26: 1719–1725.

# Cancer Prevention Potential of Diet and Dietary Supplements in Epigenome Modulation

Zdenko Herceg

## Contents

# Introduction

The term "epigenetic" refers to all stable and mitotically heritable changes in gene expression and chromatin structure that occur without changes in underlying DNA code. Epigenome refers to the totality of epigenetic marks (DNA methylation, histone modifications, and noncoding RNAs [ncRNAs]) and it functions as an essential mechanism in setting up and governing tissue- and cell type–specific transcriptional programs that are essential for developing and maintaining cell identity (Suter et al. 2004; Tang and Ho 2007; Niemann et al. 2008; Chow and Heard 2009). Epigenome deregulation has been implicated in a wide range of human cancers. In contrast to changes in the genetic code (mutations), a wide range of epigenetic alterations occur in a gradual manner during cancer development and progression and they are in principle reversible. The epigenome is considered to be an interface between the genome and the environment and thus gene–environment interactions may be mediated by epigenetic mechanisms (Herceg 2007; Herceg and Vaissiere 2011). Many epidemiological and experimental studies provide evidence implicating dietary, environmental, and lifestyle factors, as well as endogenous stressors in epigenome deregulation resulting in impaired cellular functions associated with human diseases, most notably cancer. An important distinction between genetic and epigenetic deregulation is the gradual appearance and reversibility of the latter (Jones and Baylin 2002; Feinberg and Tycko 2004). These features make epigenetic alterations attractive targets for the development of novel strategies for cancer prevention, in addition to their exploitation in cancer therapy.

Several recent studies, including those with an intervention design, demonstrated that different chemical entities may have an important impact on epigenetic states when delivered as food supplements (Waterland and Jirtle 2004; Ulrich et al. 2008; Reuter et al. 2011). This highlighted the need for identifying critical epigenetic targets associated with cancer risk as well as windows of susceptibility to dietary modulation. Although several studies yielded discordant results, the data obtained may be exploited in the design of more efficient strategies for cancer control that are based on epigenome modulation. In this review, I summarize recent evidence demonstrating the impact of dietary supplementation on the epigenome and its potential for cancer prevention. For more comprehensive overviews of epigenetic mechanisms in regulation of cellular processes and epigenetic modifications in cancer cells induced by dietary, environmental, and lifestyle factors, readers are directed to excellent recent reviews (Herceg 2007; Perera and Herbstman 2011).

# Epigenetic mechanisms

Epigenetic inheritance includes three distinct self-reinforcing mechanisms: DNA methylation, histone modifications, and RNA-mediated gene silencing.

## DNA methylation

In multicellular eukaryotes, DNA methylation refers to the addition of a methyl group to cytosine bases that are located next to a guanosine base in the so-called CpG dinucleotide. This is the only known covalent modification of the DNA molecule and is carried out by several DNA methyltransferase (DNMT) enzymes. DNMT3A and DNMT3B are responsible for methylation of new CpG sites (de novo methylation), whereas DNMT1 is thought to be the main enzyme that maintains DNMTs as it shows a preference for hemimethylated DNA substrates. More recently, the molecules responsible for DNA demethylation have been identified. The human TET (10–11 translocation) gene products were shown to be capable of catalyzing the conversion of 5-methylcytosine (5mC) of DNA to 5-hydroxymethylcytosine (5hmC), 5-formylcytosine (5Fc), and 5-carboxylcytosine (5caC), which are involved in DNA demethylation (Tahiliani et al. 2009; He et al. 2011; Zhang et al. 2011). Importantly, TET2, a member of the TET family, was found to be frequently mutated in myeloid malignancies with low levels of 5hmC (Ko et al. 2010). Although the role of global and gene-specific DNA demethylation remains to be elucidated, these phenomena have been linked to development, cellular reprogramming, memory formation, neurogenesis, immune response, and tumorigenesis (Zhu 2009; Koh et al. 2011).

DNA methylation plays multiple roles in cellular processes including regulation of gene expression and cellular defense against parasitic DNA sequences, such as viruses. DNA methylation in gene promoter regions usually results in gene silencing, presumably due to steric inhibition of transcription complexes binding to regulatory DNA (Vaissiere et al. 2008). DNA methyl marks may also facilitate the recruitment of methyl CpG-binding domain proteins (proteins that bind methylated CpG sequences) that in turn recruit histone deacetylases (HDACs) and histone methyltransferases (HMTs), all of which may result in the creation of a transcriptionally silent state for surrounding chromatin.

## Histone modifications

Histone modifications include acetylation, phosphorylation, ubiquitination, and methylation of lysine residues on the N-terminal portions ("tails") of histones. Many studies provided evidence that different histone modifications may act in a combinatorial and consistent fashion in the regulation of cellular regulatory processes. This led to the concept known as the "histone code" that modulates genetic (DNA) code. Recently, with the discovery and characterization of a large number of histone modifying molecules and protein complexes, interest in histone modifications has grown. Different chromatin-modifying complexes act in physiological contexts to modulate DNA accessibility to the transcriptional and DNA repair machineries (Sawan and Herceg 2011). Deregulation of these chromatin-based processes inevitably leads to mutations, resulting in genomic instability, oncogenic transformation, and cancer. The histone-modifying complexes include histone acetyltransferases (HATs), enzymes responsible for acetylation of the tails of core histones, and HMTs, a group of enzymes that add methyl epitope on several histone residues and are responsible for diverse functions including gene silencing and generation of heterochromatin (Sawan and Herceg 2011).

Among all histone modifications, histone acetylation and methylation are the two main modifications that have been strongly associated with oncogenic transformation and phenotype of cancer cells. For example, several studies have identified changes in acetylation and methylation of specific residues in histone H3 and histone H4 as reliable markers of tumor cells (Martin and Zhang 2005).

Histone acetylation refers to the transfer of an acetyl group from acetyl CoA to the lysine ε-amino groups on the N-terminal tails of core histones. HATs and HDACs regulate the steady-state balance of histone acetylation in the cell. Recent interest in histone acetylation as a potential therapeutic target has further grown with the discovery of HDAC inhibitors that can induce cell cycle arrest and/or cell death (Herranz and Esteller 2007). Interestingly, HDAC inhibitors were shown capable of derepressing epigenetically silenced genes in cancer cells; therefore, many studies aimed at testing whether different chemical entities with properties of HDAC inhibitors can be used as dietary supplements. In this context, recent studies investigated the effects of resveratrol, curcumin, sulforafane, and butyrate (Marcu et al. 2006; Dashwood and Ho 2008; Nian et al. 2008; Dayangac-Erden et al. 2009).

## Noncoding RNAs (long and microRNAs)

ncRNAs in the form of small RNAs (microRNAs, miRNAs) represent the most recent class of epigenetic mechanisms that is important for the maintenance of the gene transcription state in a heritable manner. In addition to 20,000 protein-coding genes, the human genome can be transcribed in a way to produce either short or long ncRNAs (Krutovskikh and Herceg 2010). Among these, the most extensively studied ncRNAs are miRNAs, evolutionarily conserved 20–22 nt long RNAs that are located within the introns and exons of protein-coding genes or intergenic regions. Because they are involved in the regulation of almost all key cellular processes, including development, differentiation, proliferation, and apoptosis, miRNAs may prove to be powerful targets for cancer prevention and therapy (Niemann et al. 2008; Krutovskikh and Herceg 2010). Deregulation of miRNA expression patterns is believed to be involved in the development of human malignancies including metastatic progression. Importantly, the expression pattern of miRNAs was found to be specifically deregulated in a wide spectrum of human diseases including cancer (Suter et al. 2004; Krutovskikh and Herceg 2010). Curiously, a large fraction of 50 annotated human miRNAs (~50%) are mapped to fragile chromosomal regions, areas of the genome that are associated with a range of human malignancies (Calin and Croce 2006). Genomic instability that disrupts miRNA-target gene regulation has been associated with an increasing number of cancer types

(Selcuklu et al. 2009). Finally, deregulation of global miRNA processing has also been linked to deficient DNA repair, increased tumorigenic potential, and oncogenic transformation (Liu et al. 2011; Francia et al. 2012).

In addition to miRNA, recent studies revealed a network of long ncRNA (typically >200 nt). These studies revealed that changes in expression levels of long ncRNAs, as well as deregulation of their primary and/or secondary structure, may result in human disease, including cancer and neurodegeneration (Guttman et al. 2009; Mattick 2009; Qureshi et al. 2010).

Interestingly, a growing body of evidence suggests that dietary factors may promote tumor development and progression through deregulation of ncRNAs (Sun et al. 2008, 2009; Tsang and Kwok 2010). For example, miR168a (isolated from plants) was shown to regulate LDL levels in mammalian plasma, consistent with the silencing effect on LDLRAP1 mRNA and protein (Zhang et al. 2012). This finding is particularly important, considering that human food intake largely relies on plants; therefore, intake of ncRNAs from plants could be exploited in modulation of the epigenome and gene expression program.

## Epigenetic reprogramming in cell differentiation and embryonic development

Epigenetic mechanisms play a critical role in embryonic development and the epigenome is known to undergo a profound reconfiguration during lineage differentiation after fertilization. Notably, there are two waves of dramatic and genome-wide demethylation of the genome during embryonic development. These demethylations are followed by remethylation that in turn guides resetting of the gene expression program needed for lineage differentiation. Specifically, the parental genome first undergoes active DNA demethylation after fertilization, whereas passive demethylation (a replication-dependent event) is believed to occur in the cells of the preimplantation embryo. In addition, a discrete silencing of a specific set of genes, including the inactivation of the X chromosome, is established, an event that is accompanied by a set of changes in chromatin modifications. These events are followed by de novo methylation that takes place after implantation. This de novo methylation generates lineage-specific epigenetic patterns considered to be a lifelong memory system that, in some cases, may be passed to subsequent generations (Youngson and Whitelaw 2008; Skinner and Guerrero-Bosagna 2009). In this regard, a recent study identified permanent changes in the epigenome (methylome) of stem/ progenitor cell populations from neonates' cord blood (hematopoietic stem cells) following intrauterine growth restriction. These findings support the notion that epigenetic changes induced by dietary exposures during in utero life may be "remembered" in later life (Einstein et al. 2010). Therefore, it appears that a biological memory system exists which relies on epigenetic mechanisms and that epigenetic deregulation of developmental programming mediates the long-term consequences triggered by early-life events. These observations also argue that modulation of dietary and environmental exposures during critical windows during in utero life may be exploited in modulating phenotypic differences later in life.

Previous studies used several animal models for studying the effects of maternal supplementation on offspring. The best characterized model for studying the effect of dietary supplementation on the epigenome and phenotype of the offspring is the agouti viable yellow mouse (Avy). This model represents a sensitive indicator of diet-induced changes in epigenetic information and has proven to be very useful in studying the role of nutrition on epigenotype (Duhl et al. 1994; Wolff et al. 1998; Waterland and Jirtle 2003). For example, it was found that feeding pregnant female mice a diet supplemented with high doses of folic acid, choline, and vitamin B12 changes the coat color of their offspring (Wolff et al. 1998). In another study, feeding female mice with high levels of dietary methyl supplements before and during pregnancy resulted in significant phenotypic changes of the offspring. Importantly, a subsequent molecular characterization revealed that hypermethylation of the *agouti* locus is an underlying mechanism for this phenotypic change (Cooney et al. 2002; Waterland and Jirtle 2003).

A recent study using another model with murine metastable epiallele, *axin fused* (*Axin(Fu)*) (Rakyan et al. 2003), also suggested that mammalian cells exhibit epigenetic plasticity to maternal diet (Waterland 2006). Specifically, it was found that female mice fed diets containing high levels of methyl donors before and during pregnancy give rise to offspring with increased DNA methylation at *Axin(Fu)* locus and reduced incidence of tail kinking (Waterland 2006).

These findings suggest that early-life exposure to different nutritional and environmental agents may induce long-lasting changes in the epigenome and consequently the gene expression program associated with specific phenotype in later life. Although these studies were focused on two well-defined loci, it is believed that epigenome changes induced by dietary modulation are not limited to these loci but are likely to be a general feature of other metastable epialleles (Waterland and Jirtle 2004; Waterland 2006). The results obtained on animal models ultimately may lead to the development of epigenomic approaches for the identification of epialleles in humans. However, this raises the question of whether these findings can be extrapolated to humans and thus should be interpreted with caution (Rosenfeld 2010).

## One-carbon metabolism and modulation of the epigenome by diet

The methyl groups are required for establishing and maintaining DNA methylation. The addition of a methyl group to methylated cytosines utilizes the final methyl donor produced by one-carbon metabolism, $S$-adenosyl methionine (SAM) in the body. Methyl tetrahydrofolate ($CH_3$-THF) acts as a methyl group donor for the production of the cofactor tetrahydrofolate (THF) and is a precursor for homocysteine conversion to methionine. Methionine is then further activated to SAM by methionine adenosyltransferase. The addition of a methyl group to methylated cytosines utilizes the final methyl donor produced by SAM. This complex process, known as one-carbon metabolism, can be altered by different dietary factors. Folate, methionine, and B vitamins (including B2, B6, and B12) are considered as important nutrients involved in one-carbon metabolism (Selhub 2002), although other methyl donors (including choline and betaine) are capable of affecting this pathway (Ueland 2011). Because dietary methyl donors are an essential source of methyl groups, the epigenome is believed to be susceptible to modulation by diet (Ulrey et al. 2005). Thus, diet with sources of methyl groups may serve as powerful modulators of the epigenome in mammalian cells.

## Fundamental role of DNA methylation changes in cancer

Cancer epigenomes are characterized by both genome-wide DNA hypomethylation and promoter-specific hypermethylation, almost universal phenomenon that coexist in most human cancers (Sincic and Herceg 2011; Jones 2012). Global hypomethylation is believed to result in transcriptional activation of oncogenes, activation of latent retrotransposons, and chromosomal instability, whereas promoter-specific hypermethylation results in silencing tumor suppressors and other cancer-associated genes.

### Global DNA hypomethylation

Global hypomethylation is presented as the decrease of total methylcytosine content in tumor cells compared to normal tissues (Feinberg and Vogelstein 1983; Brena et al. 2006). This epigenetic phenomenon is detected as a loss of methylation of DNA repetitive sequences throughout the genome (subtelomeric and pericentric repeats). Global DNA hypomethylation is found in virtually all human cancers; however, the mechanism by which global loss of methyl marks promotes cancer development and progression is poorly understood. Global hypomethylation is believed to be an early event that precedes and promotes cancer development although different views (including one suggesting that global loss of methyl marks is a passive side effect of tumor development without functional consequences) have been expressed. Global hypomethylation may promote the neoplastic process through the induction of chromosomal instability and the activation of oncogenes (Robertson 2005). Genome-wide hypomethylation appears to affect to a larger extent the repeat elements that are normally heavily methylated (including long interspersed nuclear elements [LINE], moderately repeated DNA sequences of retrovirus K [HERV-K], and centromeric satellite repeats [Sat alpha and Sat2]), whereas other repetitive sequences (including short interspersed nuclear element [SINE] and Alu repeats) are less affected (Cordaux and Batzer 2009).

## Promoter-specific DNA hypermethylation and gene silencing

All human cancers express aberrant DNA hypermethylation at CpG islands of many tumor suppressor genes and other cancer-causing genes. A large number of tumor suppressor genes including *p16/CDKN2A*, *p15/CDKN2B*, *RASSF1A*, *MGMT*, *MLH1*, *VHL*, *RB*, *BRCA1*, *LKB1*, *CDH1*, and *GSTP1* have been found hypermethylated, although the list of genes targeted by aberrant hypermethylation is likely to grow rapidly (Jones and Martienssen 2005; Esteller 2007) with the application of next-generation sequencing and completion of large international sequencing initiatives (such as the Human Cancer Epigenome Project) (Eckhardt et al. 2006; Laird 2010). Genes that are targeted by aberrant hypermethylation are involved in a wide range of cellular processes including proliferation, cell cycle control, DNA repair, detoxification of carcinogens, and cellular migration. In contrast to global hypomethylation, the mechanism underlying unscheduled hypermethylation is relatively well understood. Several distinct mechanisms may operate during de novo hypermethylation. These include overexpression of DNMTs, loss of transcriptional activity, or stabilization of the DNMT on the promoter region (Tycko 2000). The mechanisms by which promoter-specific hypermethylation brings about gene silencing may include different pathways. These include the inhibition of the binding of transcription factors like CTCF (Bell and Felsenfeld 2000; Esteller et al. 2000), and creating binding sites for methyl-binding proteins and repressive complexes that establish a repressive chromatin configuration. These and other events are believed to establish a chromatin environment that ultimately silences the expression of the gene (Fuks 2005).

## Nutrition, DNA methylation, and cancer

An impact of diet on the development of a wide range of human cancers is supported by both epidemiological and laboratory-based studies. Studies investigating the role of dietary factors on epigenetic states have primarily been focused on alterations in the DNA methylome associated with dietary factors. Among a wide range of nutritional agents that have been investigated so far, many showed a potential of influencing DNA methylation states. In particular, the dietary factors related to one-carbon metabolism have been intensively studied. The impact of these agents has been examined using different approaches including animal models and human intervention studies.

## Folate

Folic acid, a water-soluble B vitamin (also known as folate, the form naturally occurring in the body), is a methyl donor that plays an essential role in DNA synthesis and biological methylation reactions, including DNA methylation.

Folate deficiency may be implicated in the development of genomic DNA hypomethylation, which is an early epigenetic event found in many cancers. Folate is a major component of one-carbon metabolism and for this reason it is among the most widely studied dietary factors. It is noteworthy that mammalian cells may do not synthesize folates making humans dependent on dietary folate intake. There is evidence that DNA methylation levels may be modulated by folate intake. The studies using agouti (Avy) mice revealed that dietary methyl supplementation of pregnant mothers with folic acid, vitamin B12, choline, and betaine may modulate the phenotype (coat color) of their offspring and that this is mediated through increased CpG methylation at the Avy locus (Waterland and Jirtle 2003). Importantly, it appears that modulation of methylation states by folate intake may have protective effects on cancer risk, including susceptibility to colorectal cancer (Duthie 2011). Modulation of dietary folate intake was also shown to alter miRNA expression patterns and consequently cancer risk (Ghoshal et al. 2006; Tryndyak et al. 2009).

## Choline

Choline is an important intermediate of methyl metabolism and is also required for normal development. In contrast to folate, choline can be produced by humans; however, the ability of cells to produce sufficient quantities of choline relies on the methyl exchange with folate. The studies in rodents showed that dietary supplementation of pregnant mothers with choline may result in profound and irreversible changes in the developing brain. Interestingly, low choline levels were shown to induce global changes in histone modifications (methylation) (Mehedint et al. 2010). Similarly, deficiency of dietary methyl group in adult mice and rats was shown to trigger changes in the levels of methylated histones (Pogribny et al. 2006; Dobosy et al. 2008). Finally, low levels of choline in pregnant dams was shown to result in DNA methylation changes in specific tissues of the fetus (Niculescu et al. 2006). The offspring of mice fed a choline-deficient diet had decreased memory and this is related to decreased cell proliferation in the hippocampus. Niculescu et al. (2006) showed that these effects are, at least in part, attributed to DNA methylation in brains.

## Alcohol

Alcohol consumption is considered as an important risk factor for several human cancers, including hepatocellular carcinoma, breast cancer, colorectal cancer, and head and neck cancer (Shukla et al. 2008). While alcohol may exert its carcinogenic effects through its main metabolite, acetaldehyde, it may also interfere with the intake of other nutrients and thus affect metabolism of different critical compounds such as vitamins. Alcohol may also promote the production of reactive oxygen species, and importantly it may interfere with several steps of methyl transfer and other one-carbon metabolism intermediates.

The evidence is accumulating that alcohol may contribute to cancer risk through disruption of epigenetic reprogramming events. This is supported by the impact of alcohol on one-carbon metabolism, the function of DNMT enzymes (Bielawski et al. 2002), DNA methylation patterns (Choi et al. 1999), and histone modification (Kim and Shukla 2006; Lee and Shukla 2007). Alcohol was also shown to alter the imprinting mechanism, an epigenetic phenomenon by which certain genes are expressed in a parent-of-origin-specific manner. For example, ethanol exposure at preimplantation stage was associated with changes in both fetal and placental weight. This coincided with reduced DNA methylation levels in specific imprinting loci (such as the *H19* gene) (Stouder et al. 2011). In mice, maternal exposure to alcohol before conception resulted in hypermethylation of Avy in offspring (Kaminen-Ahola et al. 2010). Other studies showed that in utero exposure to alcohol can result in neurodevelopmental disorders such as fetal alcohol spectrum disorders (FASD) (Liu et al. 2009a). For example, chronic ethanol treatment of mice resulted in CpG hypomethylation associated with up-regulation of the *N*-methyl-D-aspartate (*NMDA*) receptor 2B gene expression in fetal cortical neurons (Marutha Ravindran and Ticku 2004). In cancer cells, alcohol intake was associated with aberrant hypermethylation of specific genes, including *RASSF1A* in colorectal cancer (van Engeland et al. 2003), *E-cadherin* (Tao et al. 2011) in breast cancer, *MGMT* (Puri et al. 2005) and *SFRP* (Marsit et al. 2006) in head and neck cancer.

These observations are consistent with the notion that alcohol may act through disruption of epigenetic states, although it appears that adverse effects of alcohol may be locus-specific.

## Low-protein diet

Deficiency in maternal protein/caloric intake has been associated with an impairment in early fetal growth. Because early growth impairment is known to be linked with increased risk of chronic diseases in later life, including metabolic syndrome and cardiovascular disease, in utero and early infancy nutritional deficiency may result in metabolic adaptations that lead to persistent alterations of the phenotype and susceptibility to disease in later life (Green and Limesand 2010). The rodent model with low-protein diet is one of the most extensively exploited tools in studying nutritional programming in mammals. The studies using this model revealed that relatively moderate dietary restrictions may have a significant impact on fetal development through alteration of epigenetic states (Nijland et al. 2010). For example, in rats a low-protein diet during pregnancy was shown to induce a range of pathologies, including impaired glucose tolerance and insulin resistance, higher blood pressure, and alterations in liver structure and function, in the adult offspring. These pathologies were associated with changes in gene expression and promotor hypomethylation of specific genes such as the glucocorticoid receptor (*GR*), the peroxisome proliferator-activated receptor-α (*PPARα*), and the angiotensin receptor (Lillycrop et al. 2005; Bogdarina et al. 2007).

Bogdarina et al. (2010) reported that the offspring from the mothers fed with a low-protein diet during pregnancy exhibited hypomethylation of the *Agtr1b* gene promoter, possibly explaining higher blood pressure in these animals. Interestingly, down-regulation of DNMT1 expression was found to be associated with methylation changes induced by a protein-restricted diet, suggesting a potential mechanism by which dietary restriction may induce gene-specific changes in the methylome (Lillycrop et al. 2007).

While the transgenerational influence of maternal diet is well recognized, recent studies have suggested that changes in the epigenome may be an underlying mechanism. For example, in rats, low-protein diet during pregnancy was shown to induce methylation of the liver-specific genes transgenerationally (Burdge et al. 2007). This observation was supported by epidemiological studies in humans, where periconceptional dietary deprivation caused by severe famine during the Dutch hunger winter was found to be associated with hypomethylation of the maternally imprinted insulin-like growth factor 2 (*IGF2*) gene (Heijmans et al. 2008). Taken together, studies in animal and humans provide strong evidence that maternal nutritional insufficiency during pregnancy may influence the epigenome during critical windows of susceptibility during fetal development. This fact needs to be taken into account for the development of public health strategies aiming to reduce the burden of chronic disease in childhood and adulthood.

## Fat

An excessive consumption of fat is associated with obesity, a lifestyle factor associated with elevated risk for important human diseases including several malignancies. Although the underlying mechanism is poorly understood, maternal diet high in fat during pregnancy or lactation was found to be associated with lasting changes in the expression of the imprinted genes *IGF2* and *ncRNA* (*miR-122*) (Zhang et al. 2009; Cirera et al. 2010). High fat diet was also found to influence the methylation status of specific genes (such as the melanocortin-4 [*Mc4r*] gene) that are known to play an important role in body weight regulation (Widiker et al. 2010) and global hypomethylation in the placenta (Gallou-Kabani et al. 2010). In addition, Dudley et al. (2011) found neonates born to high fat fed mothers exhibit early hepatic dysfunction. Therefore, a high fat diet during pregnancy may predispose offspring to long-term health effects (Dudley et al. 2011). However, further studies are needed to define the precise epigenetic mechanisms through which a high fat diet contributes to the development of metabolic syndrome and obesity.

## Environmental chemicals

Dietary bisphenol A (BPA) is considered as an endocrine disruptor present in many commonly used products including food and beverage containers and baby bottles. BPA has been studied in the context of potential epigenetic deregulation using the agouti Avy model. Dolinoy et al. (2007) reported that

maternal exposure to BPA resulted in hypomethylation of the CDK5 activator protein IAP, a metastable epiallele, and suggested that BPA exposure may
promote hypomethylation of multiple metastable epialleles. In utero exposure to BPA also resulted in changes in DNA methylation at the promoter-
associated CpG islands in multiple loci in the fetal mouse brain (Yaoi et al.
2008). In addition, neonatal exposure to BPA was shown to alter the epigenome in prostate tissue (Ho et al. 2006). However, further studies are needed
to elucidate the effects of BPA on the epigenome and the precise molecular
mechanism that underlies the long-lasting effects of BPA on human health
(LeBaron et al. 2010).

## Impact of dietary supplementation on DNA methylome

Chemoprevention refers to the administration of natural or synthetic compounds with the aim of preventing or reducing the risk of human diseases.
These agents can be administered either as dietary supplements or as separate drugs. Chemopreventive agents are particularly attractive in the design
of strategies to prevent or delay cancer development. The development of
cancer prevention strategies is of particular interest for high-risk populations
(individuals with a family history of cancer or those that are frequently exposed
to carcinogens). Considering that epigenetic alterations have frequently been
identified among high-risk populations and that epigenetic changes are
reversible, targeting the epigenome of the target cells is particularly relevant
for chemoprevention. Previous studies provided evidence that administration
of individual chemopreventive agents or combinations thereof may prove to
be highly beneficial for epigenetics-based strategies for cancer prevention.

### Folic acid supplementation

The impact of folic acid supplementation on DNA methylation states and
cancer risk has been examined by many studies using different mice and
rats as the model systems. A recent study by Sie et al. (2011) revealed that
maternal supplementation with folic acid results in higher levels of global
methylcytosine content in the colon epithelium of the weaning offspring and
that this phenomenon is associated with decreased rectal epithelial proliferation and lower levels of DNA damage in the cells. However, the same group
reported that lower levels of global methylation in mammary glands resulting
from maternal and postweaning folic acid supplementation are associated
with increased risk of mammary tumors in the offspring (Ly et al. 2011).
These findings indicate that maternal folic acid supplementation may have
both protective and adverse effects in the offspring and that these effects are
tissue-specific. The effects on cancer risk may be dependent on the folic acid
dose with a high intrauterine and postweaning dietary exposure seeming to
increase the incidence of mammary tumors.

While the precise epigenome targets of folic acid remain to be identified, the imprinted genes may be particularly sensitive to dietary/environmental influences. This notion is supported by the study showing that postweaning diet folate level may alter the imprinting status of the *IGF2* locus (Waterland et al. 2006). The studies in rats showed that maternal supplementation with folic acid may also affect DNA methylation states in pubertal (Burdge et al. 2009) and aging animals (Choi et al. 2005; Keyes et al. 2007). Older animals exhibited particularly high sensitivity to folate and colon tissue was especially sensitive. These findings are consistent with the notion that older age and inadequate folate intake are associated with altered DNA methylation and are considered as important risk factors for colon cancer. This hypothesis is further supported by the study showing that global DNA methylation and promoter-specific methylation (such as the *p16* promoter methylation) in colon increased with increasing levels of dietary folate in old mice (18 months) but not in the younger animals (4 months) (Keyes et al. 2007). In rats, folate supplementation was found to increase global levels of DNA methylation in livers (Choi et al. 2005). Therefore, if these results are extrapolated to humans, availability of folate in the body may have a greater impact on cancer risk in the elderly.

The importance of the optimal levels of folate in the body is further emphasized by the association found between low-folate diet and changes in genomic methylation. For example, Kotsopoulos et al. (2008) found that in rats a low-folate diet induced a significant increase (34%–48%) in global DNA methylation in liver and that these changes persisted into adulthood. Other studies found that in both rats and humans low-folate diet during the postweaning period to puberty can influence global DNA methylation and hypomethylation of specific gene promoters in different tissues including colon and liver (Kim et al. 1996; Sohn et al. 2003; Kim et al. 2008). Interestingly, high-dose folate supplementation when applied in combination with low-protein diet was shown to prevent expression of the specific phenotypes and counteract changes in DNA methylation profiles (Lillycrop et al. 2005, 2007).

Developing human organisms are believed to have a window of susceptibility to epigenetic changes and that the periconceptional period is particularly important. Consistent with this notion, maternal supplementation by folic acid during the periconceptional period was found to influence methylation levels at the differentially methylated region (DMR) of the *IGF2* (Steegers-Theunissen et al. 2009), a growth-promoting hormone during gestation. In addition, methylation levels at DMR of *H19*, a long noncoding RNA that regulates the expression of *IGF2* gene, decreased in parallel with an increment of dietary folate levels before and during pregnancy (Hoyo et al. 2011). Therefore, the results obtained with folic acid supplementation in animal models and humans should be taken into consideration when chemopreventive strategies aiming to modulate folate status are designed.

The potential impact of folate supplementation on the epigenome and in preventing specific human disorders has also been examined. For example, the study investigating the effects of folate supplementation in patients with hyperhomocysteinemia revealed that folate may alter methylation pattern at imprinted genes (Ingrosso et al. 2003). Furthermore, a beneficial effect of folate has been inferred from the study showing that a greater intake of leafy greens was inversely correlated with aberrant hypermethylation of the specific genes involved in DNA repair (Stidley et al. 2010).

Although many studies in different model systems implied a beneficial effect of dietary folate supplementation, a randomized clinical trial has cast doubt on the chemoprotective utility of folate. Cole et al. (2007) reported that folic acid supplementation not only failed to prevent tumors but also increased the risk for advanced lesions in participants with a recent history of colorectal adenomas. While these findings need to be interpreted with caution, especially considering that a single dose of folic acid (1 mg/day) was used, these findings argue that folic acid supplementation may have, at least under certain circumstances, limited positive effect in chemoprevention.

A multicenter chemoprevention trial using a cross-sectional design and normal populations revealed that folate levels in red blood cells are positively associated with methylation levels in cancer-associated genes in normal colorectal mucosa. Together, these findings highlight the importance of dosage and timing of folate supplementation (Baron et al. 2010). They also underscore the need for further studies to assess the efficacy of folic acid supplementation in modulating colon cancer risk as well as the risk of other human malignancies.

## Methionine

Methionine is an essential amino acid, and its derivative SAM serves as the substrate involved in methyl group transfers. Previous studies provided evidence suggesting that methionine supplementation in mice results in DNA methylation changes in specific genes (Tremolizzo et al. 2002; Dong et al. 2005). In addition, a study in rats found that changes in gene expression program induced by vitamin B12 deficiency can be reversed by dietary methionine supplementation (Yamamoto et al. 2009). These findings suggest that dietary methionine supplementation may influence gene expression patterns although it remains unclear whether this methionine operates through the modulation of the epigenome.

## Phytochemical agents

Many phytochemical compounds that occur naturally in plants have the potential to modulate disease risk. These agents are referred to as chemopreventive agents and can be applied as an addition to conventional cancer therapies (Dorai and Aggarwal 2004). Phytochemicals may be efficient for specific stages of cancer development and progression and clinical trials assessed their efficiency in prevention and treatment of cancer (Surh 2003).

Among different phytochemicals, the family of flavonoids, a group of compounds that share a common polyphenol structure, is the most heavily studied. Among these, resveratrol (from grapes and berries), curcumin (from curry), and epigallocatechin-3-gallate (EGCG) from green tea have shown the most promising results as cancer chemopreventive agents. These compounds appear to exert their chemopreventive effects through their action on multiple levels of cellular regulation, including DNA repair carcinogen detoxification, cell proliferation, and cell-cycle progression. Interestingly, dietary polyphenols may act as DNMT inhibitors and their interaction with DNMT enzymes may result in DNA methylation changes and changes in gene expression program. The potential of phytochemicals in restoring the expression of tumor suppress genes has been demonstrated in different cellular and animal models (Link et al. 2010).

EGCG is a polyphenol with antioxidant properties found mostly in green tea. EGCG can be methylated by catechol-$O$-methyltransferases (COMT); therefore, SAM metabolite is used for this reaction. Interestingly, EGCG may regulate the function of DNMT1 enzyme through the formation of hydrogen bonds with different residues in the catalytic pocket of the enzyme (Link et al. 2010). This interaction is believed to have an inhibitory effect on DNMT1. Consistent with this notion, topical treatment with a hydrophilic cream containing EGCG was able to prevent UVB-induced global DNA hypomethylation in mice (Mittal et al. 2003). In addition, green tea polyphenols were found to lower methylation levels at the promoter of the retinoic X receptor α (RXR-α) in Apc(Min/+) mice treated with azoxymethane (Volate et al. 2009). Two studies using a retrospective analysis of lifestyle and dietary habits and gastric cancer also found a correlation between green tea phenols and DNA methylation levels in specific genes (*CDX2* and *BMP-2*) (Yuasa et al. 2005, 2009). Furthermore, in a phase 2 clinical trial, application of green tea extract exhibited a preventive effect in patients with high-risk oral premalignant lesions (Tsao et al. 2009). In contrast, Morey Kinney et al. (2009) found no effect of polyphenols on DNA methylation in specific genes and repetitive elements in a mouse prostate adenocarcinoma model (TRAMP), although polyphenols were able to inhibit tumor formation.

In contrast to EGCG, the impact of resveratrol and curcumin on DNA methylation patterns and cancer prevention was investigated in only few studies. Several studies investigating resveratrol were focused on its effects on histone-modifying activities and found that this agent is associated with activation of histone deacetylase SIRT1 and histone acetyltransferase p300 (Boily et al. 2009; Gracia-Sancho et al. 2010; Sulaiman et al. 2010). Other studies showed that resveratrol may also modulate DNA methylation levels although, compared to other polyphenols, resveratrol exhibited weaker inhibitory activity on DNMT enzymes (Howitz et al. 2003; Wang et al. 2008;

Gracia-Sancho et al. 2010; Stefanska et al. 2010). Curcumin and its metabolite can inhibit M.SssI as a DNMT analog and decrease global DNA methylation in leukemia cell lines to the levels comparable to decitabine-induced demethylation (Liu et al. 2009b).

In addition to polyphenols, phytoestrogens, naturally occurring plant-derived phytochemicals characterized by their estrogen-like effects (Rice and Whitehead 2006), have also been investigated as chemopreventive agents in different models. Among these, genistein (phytoestrogen isolated from soy) has been the subject of several studies. A study using the Avy mice found that prenatal exposure to genistein may alter both coat color and body weight of the Avy mouse offspring (Dolinoy et al. 2006). Neonatal exposure to genistein was also shown to result in gene-specific changes in DNA methylation (Tang et al. 2008). Furthermore, young mice (8 weeks old) fed high doses of genistein exhibited changes in DNA methylation (Day et al. 2002).

In newborn mice, coumestrol and equol, other phytoestrogens, were found to alter DNA methylation and induce silencing of the proto-oncogene *H-ras* (Lyn-Cook et al. 1995). High intake of genistein/daidzein in mice was found to result in natural gender-specific differences in DNA methylation of a developmentally regulated gene *Acta1* (Guerrero-Bosagna et al. 2008). Finally, a daily supplementation of healthy premenopausal women with isoflavones (genistein, daidzein, and glycitein) revealed methylation changes at specific genes (including dose-specific changes in *RARβ2* and *cyclin D2* genes) (Qin et al. 2009). These results suggest that phytoestrogens may interfere with gender-specific patterns of the epigenome, although their utility in cancer prevention remains to be carefully examined.

## ω-3 PUFA

Polyunsaturated fatty acids (PUFAs) represent various groups of fatty acids that contain more than one double bond in their backbone. PUFAs are an indispensable component of all cell membranes and their deficiency during the perinatal period is shown to be associated with developmental disorders. The ω-3 docosahexaenoic acid (DHA), enriched in fish oil, is a member of the PUFAs of the ω-3 family, which is believed to have protective effects against human malignancies (Spencer et al. 2009). A recent study found that at least in some tissues (placenta) DHA may be an important factor in one-carbon metabolism and that its levels may influence global methylation (Kulkarni et al. 2011). Although the underlying mechanism remains poorly understood, it was suggested that reduced DHA levels may result in diversion of methyl groups toward DNA in the one-carbon metabolic pathway ultimately resulting in DNA methylation. PUFAs may also modulate DNA methylation levels indirectly through their impact on folate and vitamin B12. For example, the enzyme phosphatidylethanolamine-*N*-methyltransferase (PEMT), an important player in DHA metabolism and choline metabolism, is a major user of SAM metabolite and its activity may have an impact on methyl transfer and DNA methylation.

The evidence is accumulating that ω-3 fatty acid intake may have an impact on cancer risk through the modulation of the epigenome, that is, DNA methylation and noncoding RNAs (Davidson et al. 2009). Future studies should provide information on the potential use of PUFA supplementation in decreasing cancer risk and whether the putative protective effects of PUFAs may be brought about through modulating the epigenome.

## Targeting DNA methylome with dietary supplementation in clinical trials

As discussed earlier, a wealth of experimental evidence demonstrates that dietary supplementation may have a profound effect on the epigenome and argues that "epigenetic drugs" may be exploited in cancer therapy and prevention. The effects of dietary supplementation on the epigenome have been investigated in several randomized trials. In these trials, the focus was on DNA methylation and to a lesser extent on other epigenetic marks (Table 4.1). While some of these studies found that in cancer patients dietary folate supplementation results in a moderate but significant reduction of total methylcytosine content (Cravo et al. 1994, 1998; Kim et al. 2001; Pufulete et al. 2005), others failed to detect consistent changes in methylation levels (Figueiredo et al. 2009; Jung et al. 2011; Scoccianti et al. 2011). Qin et al. (2009) also reported that in cancer-free population phytoestrogen, a plant-derived xenoestrogen, may induce significant changes in DNA methylation at specific cancer-related genes, whereas two studies reported that a daily supplementation with isoflavones (genistein, daidzein, and glycitein) in healthy premenopausal women may induce methylation alterations in specific genes (such as *RARβ2* and *cyclin D2* genes) in the intraductal tissues (Fenech et al. 1998; Basten et al. 2006). Curiously, these studies failed to find a noticeable effect of folate and polyphenols on methylation patterns. Together, these trials strongly argue that dietary supplementation may have an important impact on cancer development and progression and that it may be exploited in cancer prevention. However, important discrepancies between studies have been reported, likely due to the study design and methodologies used. Future studies should answer important questions on both beneficial and adverse effects of different dietary supplementation agents. For example, dietary supplementation regimes that could efficiently reduce the risk of malignant progression of mild-grade precancerous lesions are of particular interest.

## Concluding remarks and perspectives of cancer prevention

It is now widely recognized that nutritional, environmental, and lifestyle factors may induce changes in the epigenome and that epigenome deregulation is an important mechanism in the development of human cancer. The fact that the epigenome is susceptible to modulation by environmental factors offers an exciting opportunity for modulation of cancer risk. A large body of evidence supports the notion that epigenetic mark alterations in cancer and

**Table 4.1** Clinical Trials with Diet Supplementation that Modulates the Epigenome

| Dietary Agent | Subjects | Dose | Duration | End Point | Outcome | References |
|---|---|---|---|---|---|---|
| Folate | 22 patients (colorectal adenomas or carcinomas), 8 controls | 10 mg/day | 6 months | Global DNA methylation in colorectal tissues | 93% increase ($p < 0.002$) | Cravo et al. (1994) |
| | 20 patients with colorectal adenoma | 5 mg/day | 3 months | Global DNA methylation in colorectal tissues | 37% increase in patients with 1 adenoma ($p = 0.05$); no change in those with >1 adenomas | Cravo et al. (1998) |
| | 20 patients (colonic adenoma) | 5 mg/day | 6 months | Global DNA methylation in colorectal tissues | 57% increase ($p = 0.001$) | Kim et al. (2001) |
| | 31 patients (colorectal adenoma) | 0.4 mg/day | 10 weeks | Global DNA methylation in lymphocyte and colorectal tissues | Increased | Pufulete et al. (2005) |
| | 388 patients (colorectal adenoma) | 1 mg/day | 3 years | Global DNA methylation in colorectal tissues | No change | Figueiredo et al. (2009) |
| | 63 healthy volunteers | 2 mg/day | 12 weeks | Global DNA methylation in leukocytes | No change | Fenech et al. (1998) |
| | 61 volunteers | 1.2 mg/day | 12 weeks | Global DNA methylation in lymphocytes | No change | Basten et al. (2006) |
| | 216 healthy volunteers | 0.8 mg/day | 3 years | Global DNA methylation in leukocytes | No change | Jung et al. (2011) |
| Polyphenols | 88 healthy smokers | | 4 weeks | Global DNA and gene-specific genes methylation in PWBC | No change | Scaccianti et al. (2011) |
| Pyhtoestrogens | 34 healthy postmenopausal women | 40 mg/day; 140 mg/day | One menstrual cycle | Gene-specific methylation (cancer-specific genes) | Hypermethylation of RARβ2 and CCND2 | Qin et al. (2009) |

*Source:* See table for multiple data sources.

precancerous lesions constitute promising targets for the development of new strategies for cancer prevention (Cole et al. 2007). Because epigenetic changes are ubiquitous and appear early in tumor development, they are particularly attractive targets for epigenetics-based modulation. The quest is to identify and characterize the agents capable of modulating the epigenome and compatible with application in high-risk populations. Many studies on animals and human subjects provided the evidence that different naturally occurring and synthetic chemical agents delivered as dietary supplements may influence the epigenome.

The changes in the epigenome induced by different chemical entities may include a variety of cellular processes, including cell proliferation, cell cycle, DNA repair, differentiation, apoptosis, and carcinogen detoxification and altering these processes through dietary supplements may modulate the susceptibility to cancer in humans. Despite a wealth of evidence supporting the notion that dietary supplementation can in principle be used to prevent cancer development, many questions remain to be answered before these agents are applied in high-risk populations. This is exemplified by the fact that clinical randomized trials with folic acid supplementation not only failed to prevent the development of colorectal adenomas but also increased the incidence of advanced malignancy (Cole et al. 2007; Jung et al. 2011). The conflicting results from animal models and humans may be reconciled by the fact that different doses and administration regimens were used. It is also possible that different agents may have strong tissue-specific effects. Nevertheless, the findings from the randomized trials argue that dietary supplements may be a double-edged sword and that further studies are needed to fully assess the utility and potential side effects of different chemopreventive agents. In particular, future studies should focus on the compounds that are under serious consideration for dietary supplementation. These studies should incorporate genome-wide analysis of the epigenomic alterations in target tissues as well as of the impact of different timings, doses, and administration regimens on the epigenome and cancer risk. Although much remains unknown, supplementation strategies based on epigenomic modulation represent a great opportunity and may prove highly efficient in cancer prevention among high-risk populations. Studies aiming to improve our understanding of the mechanisms by which dietary supplements may modulate the epigenome should be instrumental in addressing the many challenges that lie ahead.

## Acknowledgments

The work in the Epigenetics Group at the International Agency for Research on Cancer (Lyon, France) is supported by grants from Institut National du Cancer (INCA, France), l'Agence Nationale de Recherche Contre le Sida et Hépatites Virales (ANRS, France), l'Association pour la Recherche sur le Cancer (ARC, France), and la Ligue Nationale (Française) Contre le Cancer,

France (to Z.H.). The founders had no role in study design, data collection and analysis, decision to publish, or preparation of the manuscript.

## Conflict of interest

The author declares that he has no conflict of interest.

## References

Baron, J.A., Wallace, K., Grau, M.V., Levine, A.J., Shen, L.L., Hamdan, R., Chen, X.L. et al. 2010. Association between folate levels and CpG island hypermethylation in normal colorectal mucosa. *Cancer Prev Res* 3(12): 1552–1564.

Basten, G.P., Duthie, S.J., Pirie, L., Vaughan, N., Hill, M.H., and Powers, H.J. 2006. Sensitivity of markers of DNA stability and DNA repair activity to folate supplementation in healthy volunteers. *Br J Cancer* 94(12): 1942–1947.

Bell, A.C. and Felsenfeld, G. 2000. Methylation of a CTCF-dependent boundary controls imprinted expression of the *Igf2* gene. *Nature* 405(6785): 482–485.

Bielawski, D.M., Zaher, F.M., Svinarich, D.M., and Abel, E.L. 2002. Paternal alcohol exposure affects sperm cytosine methyltransferase messenger RNA levels. *Alcohol Clin Exp Res* 26(3): 347–351.

Bogdarina, I., Haase, A., Langley-Evans, S., and Clark, A.J. 2010. Glucocorticoid effects on the programming of AT1b angiotensin receptor gene methylation and expression in the rat. *PLoS One* 5(2): e9237.

Bogdarina, I., Welham, S., King, P.J., Burns, S.P., and Clark, A.J. 2007. Epigenetic modification of the renin–angiotensin system in the fetal programming of hypertension. *Circ Res* 100(4): 520–526.

Boily, G., He, X.H., Pearce, B., Jardine, K., and McBurney, M.W. 2009. SirT1-null mice develop tumors at normal rates but are poorly protected by resveratrol. *Oncogene* 28(32): 2882–2893.

Brena, R.M., Auer, H., Kornacker, K., Hackanson, B., Raval, A., Byrd, J.C., and Plass, C. 2006. Accurate quantification of DNA methylation using combined bisulfite restriction analysis coupled with the Agilent 2100 Bioanalyzer platform. *Nucleic Acids Res* 34(3): e17.

Burdge, G.C., Lillycrop, K.A., Phillips, E.S., Slater-Jefferies, J.L., Jackson, A.A., and Hanson, M.A. 2009. Folic acid supplementation during the juvenile-pubertal period in rats modifies the phenotype and epigenotype induced by prenatal nutrition. *J Nutr* 139(6): 1054–1060.

Burdge, G.C., Slater-Jefferies, J., Torrens, C., Phillips, E.S., Hanson, M.A., and Lillycrop, K.A. 2007. Dietary protein restriction of pregnant rats in the F0 generation induces altered methylation of hepatic gene promoters in the adult male offspring in the F1 and F2 generations. *Br J Nutr* 97(3): 435–439.

Calin, G.A. and Croce, C.M. 2006. MicroRNA signatures in human cancers. *Nat Rev Cancer* 6(11): 857–866.

Choi, S.W., Friso, S., Keyes, M.K., and Mason, J.B. 2005. Folate supplementation increases genomic DNA methylation in the liver of elder rats. *Br J Nutr* 93(1): 31–35.

Choi, S.W., Stickel, F., Baik, H.W., Kim, Y.I., Seitz, H.K., and Mason, J.B. 1999. Chronic alcohol consumption induces genomic but not p53-specific DNA hypomethylation in rat colon. *J Nutr* 129(11): 1945–1950.

Chow, J. and Heard, E. 2009. X inactivation and the complexities of silencing a sex chromosome. *Curr Opin Cell Biol* 21(3): 359–366.

Cirera, S., Birck, M., Busk, P.K., and Fredholm, M. 2010. Expression profiles of miRNA-122 and its target CAT1 in minipigs (*Sus scrofa*) fed a high-cholesterol diet. *Comp Med* 60(2): 136–141.

Cole, B.F., Baron, J.A., Sandler, R.S., Haile, R.W., Ahnen, D.J., Bresalier, R.S., McKeown-Eyssen, G. et al. 2007. Folic acid for the prevention of colorectal adenomas: A random-ized clinical trial. *JAMA* 297(21): 2351–2359.

Cooney, C.A., Dave, A.A., and Wolff, G.L. 2002. Maternal methyl supplements in mice affect epigenetic variation and DNA methylation of offspring. *J Nutr* 132(8 Suppl): 2393S–2400S.

Cordaux, R. and Batzer, M.A. 2009. The impact of retrotransposons on human genome evolu-tion. *Nat Rev Genet* 10(10): 691–703.

Cravo, M., Fidalgo, P., Pereira, A.D., Gouveia-Oliveira, A., Chaves, P., Selhub, J., Mason, J.B., Mira, F.C., and Leitao, C.N. 1994. DNA methylation as an intermediate biomarker in colorectal cancer: Modulation by folic acid supplementation. *Eur J Cancer Prev* 3(6): 473–479.

Cravo, M.L., Pinto, A.G., Chaves, P., Cruz, J.A., Lage, P., Nobre Leitao, C., and Costa Mira, F. 1998. Effect of folate supplementation on DNA methylation of rectal mucosa in patients with colonic adenomas: Correlation with nutrient intake. *Clin Nutr* 17(2): 45–49.

Dashwood, R.H. and Ho, E. 2008. Dietary agents as histone deacetylase inhibitors: Sulforaphane and structurally related isothiocyanates. *Nutr Rev* 66(Suppl 1): S36–S38.

Davidson, L.A., Wang, N., Shah, M.S., Lupton, J.R., Ivanov, I., and Chapkin, R.S. 2009. n-3 Polyunsaturated fatty acids modulate carcinogen-directed non-coding microRNA sig-natures in rat colon. *Carcinogenesis* 30(12): 2077–2084.

Day, J.K., Bauer, A.M., DesBordes, C., Zhuang, Y., Kim, B.E., Newton, L.G., Nehra, V. et al. 2002. Genistein alters methylation patterns in mice. *J Nutr* 132(8 Suppl): 2419S–2423S.

Dayangac-Erden, D., Bora, G., Ayhan, P., Kocaefe, C., Dalkara, S., Yelekci, K., Demir, A.S., and Erdem-Yurter, H. 2009. Histone deacetylase inhibition activity and molecular docking of (e)-resveratrol: Its therapeutic potential in spinal muscular atrophy. *Chem Biol Drug Des* 73(3): 355–364.

Dobosy, J.R., Fu, V.X., Desotelle, J.A., Srinivasan, R., Kenowski, M.L., Almassi, N., Weindruch, R., Svaren, J., and Jarrard, D.F. 2008. A methyl-deficient diet modifies histone methylation and alters Igf2 and H19 repression in the prostate. *Prostate* 68(11): 1187–1195.

Dolinoy, D.C., Huang, D., and Jirtle, R.L. 2007. Maternal nutrient supplementation counter-acts bisphenol A-induced DNA hypomethylation in early development. *Proc Natl Acad Sci U S A* 104(32): 13056–13061.

Dolinoy, D.C., Weidman, J.R., Waterland, R.A., and Jirtle, R.L. 2006. Maternal genistein alters coat color and protects Avy mouse offspring from obesity by modifying the fetal epig-enome. *Environ Health Perspect* 114(4): 567–572.

Dong, E., Agis-Balboa, R.C., Simonini, M.V., Grayson, D.R., Costa, E., and Guidotti, A. 2005. Reelin and glutamic acid decarboxylase 67 promoter remodeling in an epigenetic methionine-induced mouse model of schizophrenia. *Proc Natl Acad Sci U S A* 102(35): 12578–12583.

Dorai, T. and Aggarwal, B.B. 2004. Role of chemopreventive agents in cancer therapy. *Cancer Lett* 215(2): 129–140.

Dudley, K.J., Sloboda, D.M., Connor, K.L., Beltrand, J., and Vickers, M.H. 2011. Offspring of mothers fed a high fat diet display hepatic cell cycle inhibition and associated changes in gene expression and DNA methylation. *PLoS One* 6(7): e21662.

Duhl, D.M., Vrieling, H., Miller, K.A., Wolff, G.L., and Barsh, G.S. 1994. Neomorphic agouti mutations in obese yellow mice. *Nat Genet* 8(1): 59–65.

Duthie, S.J. 2011. Folate and cancer: How DNA damage, repair and methylation impact on colon carcinogenesis. *J Inherit Metab Dis* 34(1): 101–109.

Eckhardt, F., Lewin, J., Cortese, R., Rakyan, V.K., Attwood, J., Burger, M., Burton, J. et al. 2006. DNA methylation profiling of human chromosomes 6, 20 and 22. *Nat Genet* 38(12): 1378–1385.

Einstein, F., Thompson, R.F., Bhagat, T.D., Fazzari, M.J., Verma, A., Barzilai, N., and Greally, J.M. 2010. Cytosine methylation dysregulation in neonates following intrauterine growth restriction. *PLoS One* 5(1): e8887.

Esteller, M. 2007. Cancer epigenomics: DNA methylomes and histone-modification maps. *Nat Rev Genet* 8(4): 286–298.

Esteller, M., Silva, J.M., Dominguez, G., Bonilla, F., Matias-Guiu, X., Lerma, E., Bussaglia, E. et al. 2000. Promoter hypermethylation and BRCA1 inactivation in sporadic breast and ovarian tumors. *J Natl Cancer Inst* 92(7): 564–569.

Feinberg, A.P. and Tycko, B. 2004. The history of cancer epigenetics. *Nat Rev Cancer* 4(2): 143–153.

Feinberg, A.P. and Vogelstein, B. 1983. Hypomethylation distinguishes genes of some human cancers from their normal counterparts. *Nature* 301(5895): 89–92.

Fenech, M., Aitken, C., and Rinaldi, J. 1998. Folate, vitamin B12, homocysteine status and DNA damage in young Australian adults. *Carcinogenesis* 19(7): 1163–1171.

Figueiredo, J.C., Grau, M.V., Wallace, K., Levine, A.J., Shen, L., Hamdan, R., Chen, X. et al. 2009. Global DNA hypomethylation (LINE-1) in the normal colon and lifestyle characteristics and dietary and genetic factors. *Cancer Epidemiol Biomarkers Prev* 18(4): 1041–1049.

Francia, S., Michelini, F., Saxena, A., Tang, D., de Hoon, M., Anelli, V., Mione, M., Carninci, P., and d'Adda di Fagagna, F. 2012. Site-specific DICER and DROSHA RNA products control the DNA-damage response. *Nature* 488(7410): 231–235.

Fuks, F. 2005. DNA methylation and histone modifications: Teaming up to silence genes. *Curr Opin Genet Dev* 15(5): 490–495.

Gallou-Kabani, C., Gabory, A., Tost, J., Karimi, M., Mayeur, S., Lesage, J., Boudadi, E. et al. 2010. Sex- and diet-specific changes of imprinted gene expression and DNA methylation in mouse placenta under a high-fat diet. *PLoS One* 5(12): e14398.

Ghoshal, K., Kutay, H., Bai, S.M., Datta, J., Motiwala, T., Pogribny, I., Frankel, W., and Jacob, S.T. 2006. Downregulation of miR-122 in the rodent and human hepatocellular carcinomas. *J Cell Biochem* 99(3): 671–678.

Gracia-Sancho, J., Villarreal, G., Jr., Zhang, Y., and Garcia-Cardena, G. 2010. Activation of SIRT1 by resveratrol induces KLF2 expression conferring an endothelial vasoprotective phenotype. *Cardiovasc Res* 85(3): 514–519.

Green, A.S. and Limesand, S.W. 2010. Remembering development—Epigenetic responses to fetal malnutrition. *J Physiol* 588(Pt 9): 1379–1380.

Guerrero-Bosagna, C.M., Sabat, P., Valdovinos, F.S., Valladares, L.E., and Clark, S.J. 2008. Epigenetic and phenotypic changes result from a continuous pre and post natal dietary exposure to phytoestrogens in an experimental population of mice. *BMC Physiol* 8: 17.

Guttman, M., Amit, I., Garber, M., French, C., Lin, M.F., Feldser, D., Huarte, M. et al. 2009. Chromatin signature reveals over a thousand highly conserved large non-coding RNAs in mammals. *Nature* 458(7235): 223–227.

He, Y.F., Li, B.Z., Li, Z., Liu, P., Wang, Y., Tang, Q., Ding, J. et al. 2011. Tet-mediated formation of 5-carboxylcytosine and its excision by TDG in mammalian DNA. *Science* 333(6047): 1303–1307.

Heijmans, B.T., Tobi, E.W., Stein, A.D., Putter, H., Blauw, G.J., Susser, E.S., Slagboom, P.E., and Lumey, L.H. 2008. Persistent epigenetic differences associated with prenatal exposure to famine in humans. *Proc Natl Acad Sci U S A* 105(44): 17046–17049.

Herceg, Z. 2007. Epigenetics and cancer: Towards an evaluation of the impact of environmental and dietary factors. *Mutagenesis* 22(2): 91–103.

Herceg, Z. and Vaissiere, T. 2011. Epigenetic mechanisms and cancer: An interface between the environment and the genome. *Epigenetics* 6(7): 804–819.

Herranz, M. and Esteller, M. 2007. DNA methylation and histone modifications in patients with cancer: Potential prognostic and therapeutic targets. *Methods Mol Biol* 361: 25–62.

Ho, S.M., Tang, W.Y., Belmonte de Frausto, J., and Prins, G.S. 2006. Developmental exposure to estradiol and bisphenol A increases susceptibility to prostate carcinogenesis and epigenetically regulates phosphodiesterase type 4 variant 4. *Cancer Res* 66(11): 5624–5632.

Howitz, K.T., Bitterman, K.J., Cohen, H.Y., Lamming, D.W., Lavu, S., Wood, J.G., Zipkin, R.E. et al. 2003. Small molecule activators of sirtuins extend *Saccharomyces cerevisiae* lifespan. *Nature* 425(6954): 191–196.

Hoyo, C., Murtha, A.P., Schildkraut, J.M., Jirtle, R.L., Demark-Wahnefried, W., Forman, M.R., Iversen, E.S. et al. 2011. Methylation variation at IGF2 differentially methylated regions and maternal folic acid use before and during pregnancy. *Epigenetics* 6(7): 928–936.

Ingrosso, D., Cimmino, A., Perna, A.F., Masella, L., De Santo, N.G., De Bonis, M.L., Vacca, M. et al. 2003. Folate treatment and unbalanced methylation and changes of allelic expression induced by hyperhomocysteinaemia in patients with uraemia. *Lancet* 361(9370): 1693–1699.

Jones, P.A. 2012. Functions of DNA methylation: Islands, start sites, gene bodies and beyond. *Nat Rev Genet* 13(7): 484–492.

Jones, P.A. and Baylin, S.B. 2002. The fundamental role of epigenetic events in cancer. *Nat Rev Genet* 3(6): 415–428.

Jones, P.A. and Martienssen, R. 2005. A blueprint for a Human Epigenome Project: The AACR Human Epigenome Workshop. *Cancer Res* 65(24): 11241–11246.

Jung, A.Y., Smulders, Y., Verhoef, P., Kok, F.J., Blom, H., Kok, R.M., Kampman, E., and Durga, J. 2011. No effect of folic acid supplementation on global DNA methylation in men and women with moderately elevated homocysteine. *PLoS One* 6(9): e24976.

Kaminen-Ahola, N., Ahola, A., Maga, M., Mallitt, K.A., Fahey, P., Cox, T.C., Whitelaw, E., and Chong, S. 2010. Maternal ethanol consumption alters the epigenotype and the phenotype of offspring in a mouse model. *PLoS Genet* 6(1): e1000811.

Keyes, M.K., Jang, H., Mason, J.B., Liu, Z., Crott, J.W., Smith, D.E., Friso, S., and Choi, S.W. 2007. Older age and dietary folate are determinants of genomic and p16-specific DNA methylation in mouse colon. *J Nutr* 137(7): 1713–1717.

Kim, J.S. and Shukla, S.D. 2006. Acute in vivo effect of ethanol (binge drinking) on histone H3 modifications in rat tissues. *Alcohol Alcohol* 41(2): 126–132.

Kim, Y.I., Baik, H.W., Fawaz, K., Knox, T., Lee, Y.M., Norton, R., Libby, E., and Mason, J.B. 2001. Effects of folate supplementation on two provisional molecular markers of colon cancer: A prospective, randomized trial. *Am J Gastroenterol* 96(1): 184–195.

Kim, Y.I., Kotsopoulos, J., and Sohn, K.J. 2008. Postweaning dietary folate deficiency provided through childhood to puberty permanently increases genomic DNA methylation in adult rat liver. *J Nutr* 138(4): 703–709.

Kim, Y.I., Pogribny, I.P., Salomon, R.N., Choi, S.W., Smith, D.E., James, S.J., and Mason, J.B. 1996. Exon-specific DNA hypomethylation of the *p53* gene of rat colon induced by dimethylhydrazine. Modulation by dietary folate. *Am J Pathol* 149(4): 1129–1137.

Ko, M., Huang, Y., Jankowska, A.M., Pape, U.J., Tahiliani, M., Bandukwala, H.S., An, J. et al. 2010. Impaired hydroxylation of 5-methylcytosine in myeloid cancers with mutant TET2. *Nature* 468(7325): 839–843.

Koh, K.P., Yabuuchi, A., Rao, S., Huang, Y., Cunniff, K., Nardone, J., Laiho, A. et al. 2011. Tet1 and Tet2 regulate 5-hydroxymethylcytosine production and cell lineage specification in mouse embryonic stem cells. *Cell Stem Cell* 8(2): 200–213.

Kotsopoulos, J., Sohn, K.J., and Kim, Y.I. 2008. Postweaning dietary folate deficiency provided through childhood to puberty permanently increases genomic DNA methylation in adult rat liver. *J Nutr* 138(4): 703–709.

Krutovskikh, V.A. and Herceg, Z. 2010. Oncogenic microRNAs (OncomiRs) as a new class of cancer biomarkers. *Bioessays* 32(10): 894–904.

Kulkarni, A., Dangat, K., Kale, A., Sable, P., Chavan-Gautam, P., and Joshi, S. 2011. Effects of altered maternal folic acid, vitamin B12 and docosahexaenoic acid on placental global DNA methylation patterns in Wistar rats. *PLoS One* 6(3): e17706.

Laird, P.W. 2010. Principles and challenges of genome-wide DNA methylation analysis. *Nat Rev Genet* 11(3): 191–203.

LeBaron, M.J., Rasoulpour, R.J., Klapacz, J., Ellis-Hutchings, R.G., Hollnagel, H.M., and Gollapudi, B.B. 2010. Epigenetics and chemical safety assessment. *Mutat Res* 705(2): 83–95.

Lee, Y.J. and Shukla, S.D. 2007. Histone H3 phosphorylation at serine 10 and serine 28 is mediated by p38 MAPK in rat hepatocytes exposed to ethanol and acetaldehyde. *Eur J Pharmacol* 573(1–3): 29–38.

Lillycrop, K.A., Phillips, E.S., Jackson, A.A., Hanson, M.A., and Burdge, G.C. 2005. Dietary protein restriction of pregnant rats induces and folic acid supplementation prevents epigenetic modification of hepatic gene expression in the offspring. *J Nutr* 135(6): 1382–1386.

Lillycrop, K.A., Slater-Jefferies, J.L., Hanson, M.A., Godfrey, K.M., Jackson, A.A., and Burdge, G.C. 2007. Induction of altered epigenetic regulation of the hepatic glucocorticoid receptor in the offspring of rats fed a protein-restricted diet during pregnancy suggests that reduced DNA methyltransferase-1 expression is involved in impaired DNA methylation and changes in histone modifications. *Br J Nutr* 97(6): 1064–1073.

Link, A., Balaguer, F., and Goel, A. 2010. Cancer chemoprevention by dietary polyphenols: Promising role for epigenetics. *Biochem Pharmacol* 80(12): 1771–1792.

Liu, A.M., Zhang, C., Burchard, J., Fan, S.T., Wong, K.F., Dai, H., Poon, R.T., and Luk, J.M. 2011. Global regulation on microRNA in hepatitis B virus-associated hepatocellular carcinoma. *OMICS* 15(3): 187–191.

Liu, Y., Balaraman, Y., Wang, G., Nephew, K.P., and Zhou, F.C. 2009a. Alcohol exposure alters DNA methylation profiles in mouse embryos at early neurulation. *Epigenetics* 4(7): 500–511.

Liu, Z., Xie, Z., Jones, W., Pavlovicz, R.E., Liu, S., Yu, J., Li, P.K., Lin, J., Fuchs, J.R., Marcucci, G., Li, C., and Chan, K.K. 2009b. Curcumin is a potent DNA hypomethylation agent. *Bioorg Med Chem Lett* 19(3): 706–709.

Ly, A., Lee, H., Chen, J., Sie, K.K., Renlund, R., Medline, A., Sohn, K.J., Croxford, R., Thompson, L.U., and Kim, Y.I. 2011. Effect of maternal and postweaning folic acid supplementation on mammary tumor risk in the offspring. *Cancer Res* 71(3): 988–997.

Lyn-Cook, B.D., Blann, E., Payne, P.W., Bo, J., Sheehan, D., and Medlock, K. 1995. Methylation profile and amplification of proto-oncogenes in rat pancreas induced with phytoestrogens. *Proc Soc Exp Biol Med* 208(1): 116–119.

Marcu, M.G., Jung, Y.J., Lee, S., Chung, E.J., Lee, M.J., Trepel, J., and Neckers, L. 2006. Curcumin is an inhibitor of p300 histone acetyltransferase. *Med Chem* 2(2): 169–174.

Marsit, C.J., McClean, M.D., Furniss, C.S., and Kelsey, K.T. 2006. Epigenetic inactivation of the *SFRP* genes is associated with drinking, smoking and HPV in head and neck squamous cell carcinoma. *Int J Cancer* 119(8): 1761–1766.

Martin, C. and Zhang, Y. 2005. The diverse functions of histone lysine methylation. *Nat Rev Mol Cell Biol* 6(11): 838–849.

Marutha Ravindran, C.R. and Ticku, M.K. 2004. Changes in methylation pattern of NMDA receptor *NR2B* gene in cortical neurons after chronic ethanol treatment in mice. *Brain Res Mol Brain Res* 121(1–2): 19–27.

Mattick, J.S. 2009. The genetic signatures of noncoding RNAs. *PLoS Genet* 5(4): e1000459.

Mehedint, M.G., Niculescu, M.D., Craciunescu, C.N., and Zeisel, S.H. 2010. Choline deficiency alters global histone methylation and epigenetic marking at the Re1 site of the *calbindin 1* gene. *FASEB J* 24(1): 184–195.

Mittal, A., Piyathilake, C., Hara, Y., and Katiyar, S.K. 2003. Exceptionally high protection of photocarcinogenesis by topical application of (–)-epigallocatechin-3-gallate in hydrophilic cream in SKH-1 hairless mouse model: Relationship to inhibition of UVB-induced global DNA hypomethylation. *Neoplasia* 5(6): 555–565.

Morey Kinney, S.R., Zhang, W., Pascual, M., Greally, J.M., Gillard, B.M., Karasik, E., Foster, B.A., and Karpf, A.R. 2009. Lack of evidence for green tea polyphenols as DNA methylation inhibitors in murine prostate. *Cancer Prev Res (Phila)* 2(12): 1065–1075.

Nian, H., Delage, B., Pinto, J.T., and Dashwood, R.H. 2008. Allyl mercaptan, a garlic-derived organosulfur compound, inhibits histone deacetylase and enhances Sp3 binding on the P21WAF1 promoter. *Carcinogenesis* 29(9): 1816–1824.

Niculescu, M.D., Craciunescu, C.N., and Zeisel, S.H. 2006. Dietary choline deficiency alters global and gene-specific DNA methylation in the developing hippocampus of mouse fetal brains. *FASEB J* 20(1): 43–49.

Niemann, H., Tian, X.C., King, W.A., and Lee, R.S. 2008. Epigenetic reprogramming in embryonic and foetal development upon somatic cell nuclear transfer cloning. *Reproduction* 135(2): 151–163.

Nijland, M.J., Mitsuya, K., Li, C., Ford, S., McDonald, T.J., Nathanielsz, P.W., and Cox, L.A. 2010. Epigenetic modification of fetal baboon hepatic phosphoenolpyruvate carboxykinase following exposure to moderately reduced nutrient availability. *J Physiol* 588(Pt 8): 1349–1359.

Perera, F. and Herbstman, J. 2011. Prenatal environmental exposures, epigenetics, and disease. *Reprod Toxicol* 31(3): 363–373.

Pogribny, I.P., Ross, S.A., Tryndyak, V.P., Pogribna, M., Poirier, L.A., and Karpinets, T.V. 2006. Histone H3 lysine 9 and H4 lysine 20 trimethylation and the expression of Suv4-20h2 and Suv-39h1 histone methyltransferases in hepatocarcinogenesis induced by methyl deficiency in rats. *Carcinogenesis* 27(6): 1180–1186.

Pufulete, M., Al-Ghnaniem, R., Khushal, A., Appleby, P., Harris, N., Gout, S., Emery, P.W., and Sanders, T.A. 2005. Effect of folic acid supplementation on genomic DNA methylation in patients with colorectal adenoma. *Gut* 54(5): 648–653.

Puri, S.K., Si, L., Fan, C.Y., and Hanna, E. 2005. Aberrant promoter hypermethylation of multiple genes in head and neck squamous cell carcinoma. *Am J Otolaryngol* 26(1): 12–17.

Qin, W., Zhu, W., Shi, H., Hewett, J.E., Ruhlen, R.L., MacDonald, R.S., Rottinghaus, G.E., Chen, Y.C., and Sauter, E.R. 2009. Soy isoflavones have an antiestrogenic effect and alter mammary promoter hypermethylation in healthy premenopausal women. *Nutr Cancer* 61(2): 238–244.

Qureshi, I.A., Mattick, J.S., and Mehler, M.F. 2010. Long non-coding RNAs in nervous system function and disease. *Brain Res* 1338: 20–35.

Rakyan, V.K., Chong, S., Champ, M.E., Cuthbert, P.C., Morgan, H.D., Luu, K.V., and Whitelaw, E. 2003. Transgenerational inheritance of epigenetic states at the murine Axin(Fu) allele occurs after maternal and paternal transmission. *Proc Natl Acad Sci U S A* 100(5): 2538–2543.

Reuter, S., Gupta, S.C., Park, B., Goel, A., and Aggarwal, B.B. 2011. Epigenetic changes induced by curcumin and other natural compounds. *Genes Nutr* 6(2): 93–108.

Rice, S. and Whitehead, S.A. 2006. Phytoestrogens and breast cancer—Promoters or protectors? *Endocr Relat Cancer* 13(4): 995–1015.

Robertson, K.D. 2005. DNA methylation and human disease. *Nat Rev Genet* 6(8): 597–610.

Rosenfeld, C.S. 2010. Animal models to study environmental epigenetics. *Biol Reprod* 82(3): 473–488.

Sawan, C. and Herceg, Z. 2011. Histone modifications and cancer. *Adv Genet* 70: 57–85.

Scoccianti, C., Ricceri, F., Ferrari, P., Cuenin, C., Sacerdote, C., Polidoro, S., Jenab, M., Hainaut, P., Vineis, P., and Herceg, Z. 2011. Methylation patterns in *sentinel* genes in peripheral blood cells of heavy smokers: Influence of cruciferous vegetables in an intervention study. *Epigenetics* 6(9): 1114–1119.

Selcuklu, S.D., Yakicier, M.C., and Erson, A.E. 2009. An investigation of microRNAs mapping to breast cancer related genomic gain and loss regions. *Cancer Genet Cytogenet* 189(1): 15–23.

Selhub, J. 2002. Folate, vitamin B12 and vitamin B6 and one carbon metabolism. *J Nutr Health Aging* 6(1): 39–42.

Shukla, S.D., Velazquez, J., French, S.W., Lu, S.C., Ticku, M.K., and Zakhari, S. 2008. Emerging role of epigenetics in the actions of alcohol. *Alcohol Clin Exp Res* 32(9): 1525–1534.

Sie, K.K., Medline, A., van Weel, J., Sohn, K.J., Choi, S.W., Croxford, R., and Kim, Y.I. 2011. Effect of maternal and postweaning folic acid supplementation on colorectal cancer risk in the offspring. *Gut* 60(12): 1687–1694.

Sincic, N. and Herceg, Z. 2011. DNA methylation and cancer: Ghosts and angels above the genes. *Curr Opin Oncol* 23(1): 69–76.

Skinner, M.K. and Guerrero-Bosagna, C. 2009. Environmental signals and transgenerational epigenetics. *Epigenomics* 1(1): 111–117.

Sohn, K.J., Stempak, J.M., Reid, S., Shirwadkar, S., Mason, J.B., and Kim, Y.I. 2003. The effect of dietary folate on genomic and p53-specific DNA methylation in rat colon. *Carcinogenesis* 24(1): 81–90.

Spencer, L., Mann, C., Metcalfe, M., Webb, M., Pollard, C., Spencer, D., Berry, D., Steward, W., and Dennison, A. 2009. The effect of omega-3 FAs on tumour angiogenesis and their therapeutic potential. *Eur J Cancer* 45(12): 2077–2086.

Steegers-Theunissen, R.P., Obermann-Borst, S.A., Kremer, D., Lindemans, J., Siebel, C., Steegers, E.A., Slagboom, P.E., and Heijmans, B.T. 2009. Periconceptional maternal folic acid use of 400 microg per day is related to increased methylation of the *IGF2* gene in the very young child. *PLoS One* 4(11): e7845.

Stefanska, B., Rudnicka, K., Bednarek, A., and Fabianowska-Majewska, K. 2010. Hypomethylation and induction of retinoic acid receptor beta 2 by concurrent action of adenosine analogues and natural compounds in breast cancer cells. *Eur J Pharmacol* 638(1–3): 47–53.

Stidley, C.A., Picchi, M.A., Leng, S., Willink, R., Crowell, R.E., Flores, K.G., Kang, H., Byers, T., Gilliland, F.D., and Belinsky, S.A. 2010. Multivitamins, folate, and green vegetables protect against gene promoter methylation in the aerodigestive tract of smokers. *Cancer Res* 70(2): 568–574.

Stouder, C., Somm, E., and Paoloni-Giacobino, A. 2011. Prenatal exposure to ethanol: A specific effect on the *H19* gene in sperm. *Reprod Toxicol* 31(4): 507–512.

Sulaiman, M., Matta, M.J., Sunderesan, N.R., Gupta, M.P., Periasamy, M., and Gupta, M. 2010. Resveratrol, an activator of SIRT1, upregulates sarcoplasmic calcium ATPase and improves cardiac function in diabetic cardiomyopathy. *Am J Physiol Heart Circ Physiol* 298(3): H833–H843.

Sun, M., Estrov, Z., Ji, Y., Coombes, K.R., Harris, D.H., and Kurzrock, R. 2008. Curcumin (diferuloylmethane) alters the expression profiles of microRNAs in human pancreatic cancer cells. *Mol Cancer Ther* 7(3): 464–473.

Sun, Q., Cong, R., Yan, H., Gu, H., Zeng, Y., Liu, N., Chen, J., and Wang, B. 2009. Genistein inhibits growth of human uveal melanoma cells and affects microRNA-27a and target gene expression. *Oncol Rep* 22(3): 563–567.

Surh, Y.J. 2003. Cancer chemoprevention with dietary phytochemicals. *Nat Rev Cancer* 3(10): 768–780.

Suter, C.M., Martin, D.I., and Ward, R.L. 2004. Germline epimutation of MLH1 in individuals with multiple cancers. *Nat Genet* 36(5): 497–501.

Tahiliani, M., Koh, K.P., Shen, Y., Pastor, W.A., Bandukwala, H., Brudno, Y., Agarwal, S., Iyer, L.M., Liu, D.R., Aravind, L., and Rao, A. 2009. Conversion of 5-methylcytosine to 5-hydroxymethylcytosine in mammalian DNA by MLL partner TET1. *Science* 324(5929): 930–935.

Tang, W.Y. and Ho, S.M. 2007. Epigenetic reprogramming and imprinting in origins of disease. *Rev Endocr Metab Disord* 8(2): 173–182.

Tang, W.Y., Newbold, R., Mardilovich, K., Jefferson, W., Cheng, R.Y., Medvedovic, M., and Ho, S.M. 2008. Persistent hypomethylation in the promoter of nucleosomal binding protein 1 (Nsbp1) correlates with overexpression of Nsbp1 in mouse uteri neonatally exposed to diethylstilbestrol or genistein. *Endocrinology* 149(12): 5922–5931.

Tao, M.H., Marian, C., Shields, P.G., Nie, J., McCann, S.E., Millen, A., Ambrosone, C. et al. 2011. Alcohol consumption in relation to aberrant DNA methylation in breast tumors. *Alcohol* 45(7): 689–699.

Tremolizzo, L., Carboni, G., Ruzicka, W.B., Mitchell, C.P., Sugaya, I., Tueting, P., Sharma, R., Grayson, D.R., Costa, E., and Guidotti, A. 2002. An epigenetic mouse model for molecular and behavioral neuropathologies related to schizophrenia vulnerability. *Proc Natl Acad Sci U S A* 99(26): 17095–17100.

Tryndyak, V.P., Ross, S.A., Beland, F.A., and Pogribny, I.P. 2009. Down-regulation of the microRNAs miR-34a, miR-127, and miR-200b in rat liver during hepatocarcinogenesis induced by a methyl-deficient diet. *Mol Carcinog* 48(6): 479–487.

Tsang, W.P. and Kwok, T.T. 2010. Epigallocatechin gallate up-regulation of miR-16 and induction of apoptosis in human cancer cells. *J Nutr Biochem* 21(2): 140–146.

Tsao, A.S., Liu, D., Martin, J., Tang, X.M., Lee, J.J., El-Naggar, A.K., Wistuba, I. et al. 2009. Phase II randomized, placebo-controlled trial of green tea extract in patients with high-risk oral premalignant lesions. *Cancer Prev Res (Phila)* 2(11): 931–941.

Tycko, B. 2000. Epigenetic gene silencing in cancer. *J Clin Invest* 105(4): 401–407.

Ueland, P.M. 2011. Choline and betaine in health and disease. *J Inherit Metab Dis* 34(1): 3–15.

Ulrey, C.L., Liu, L., Andrews, L.G., and Tollefsbol, T.O. 2005. The impact of metabolism on DNA methylation. *Hum Mol Genet* 14(Spec. No. 1): R139–R147.

Ulrich, C.M., Reed, M.C., and Nijhout, H.F. 2008. Modeling folate, one-carbon metabolism, and DNA methylation. *Nutr Rev* 66(Suppl 1): S27–S30.

Vaissiere, T., Sawan, C., and Herceg, Z. 2008. Epigenetic interplay between histone modifications and DNA methylation in gene silencing. *Mutat Res* 659(1–2): 40–48.

van Engeland, M., Weijenberg, M.P., Roemen, G.M., Brink, M., de Bruine, A.P., Goldbohm, R.A., van den Brandt, P.A., Baylin, S.B., de Goeij, A.F., and Herman, J.G. 2003. Effects of dietary folate and alcohol intake on promoter methylation in sporadic colorectal cancer: The Netherlands cohort study on diet and cancer. *Cancer Res* 63(12): 3133–3137.

Volate, S.R., Muga, S.J., Issa, A.Y., Nitcheva, D., Smith, T., and Wargovich, M.J. 2009. Epigenetic modulation of the retinoid X receptor alpha by green tea in the azoxymethane-Apc Min/+ mouse model of intestinal cancer. *Mol Carcinog* 48(10): 920–933.

Wang, R.H., Sengupta, K., Li, C., Kim, H.S., Cao, L., Xiao, C., Kim, S. et al. 2008. Impaired DNA damage response, genome instability, and tumorigenesis in SIRT1 mutant mice. *Cancer Cell* 14(4): 312–323.

Waterland, R.A. 2006. Assessing the effects of high methionine intake on DNA methylation. *J Nutr* 136(6 Suppl): 1706S–1710S.

Waterland, R.A. and Jirtle, R.L. 2004. Early nutrition, epigenetic changes at transposons and imprinted genes, and enhanced susceptibility to adult chronic diseases. *Nutrition* 20(1): 63–68.

Waterland, R.A. and Jirtle, R.L. 2003. Transposable elements: Targets for early nutritional effects on epigenetic gene regulation. *Mol Cell Biol* 23(15): 5293–5300.

Waterland, R.A., Lin, J.R., Smith, C.A., and Jirtle, R.L. 2006. Post-weaning diet affects genomic imprinting at the insulin-like growth factor 2 (Igf2) locus. *Hum Mol Genet* 15(5): 705–716.

Widiker, S., Karst, S., Wagener, A., and Brockmann, G.A. 2010. High-fat diet leads to a decreased methylation of the *Mc4r* gene in the obese BFMI and the lean B6 mouse lines. *J Appl Genet* 51(2): 193–197.

Wolff, G.L., Kodell, R.L., Moore, S.R., and Cooney, C.A. 1998. Maternal epigenetics and methyl supplements affect agouti gene expression in Avy/a mice. *FASEB J* 12(11): 949–957.

Yamamoto, Y., Uekawa, A., Katsushima, K., Ogata, A., Kawata, T., Maeda, N., Kobayashi, K.I., Maekawa, A., and Tadokoro, T. 2009. Change of epigenetic control of cystathionine beta-synthase gene expression through dietary vitamin B(12) is not recovered by methionine supplementation. *J Nutrigenet Nutrige* 2(1): 29–36.

Yaoi, T., Itoh, K., Nakamura, K., Ogi, H., Fujiwara, Y., and Fushiki, S. 2008. Genome-wide analysis of epigenomic alterations in fetal mouse forebrain after exposure to low doses of bisphenol A. *Biochem Biophys Res Commun* 376(3): 563–567.

Youngson, N.A. and Whitelaw, E. 2008. Transgenerational epigenetic effects. *Annu Rev Genomics Hum Genet* 9: 233–257.

Yuasa, Y., Nagasaki, H., Akiyama, Y., Hashimoto, Y., Takizawa, T., Kojima, K., Kawano, T., Sugihara, K., Imai, K., and Nakachi, K. 2009. DNA methylation status is inversely correlated with green tea intake and physical activity in gastric cancer patients. *Int J Cancer* 124(11): 2677–2682.

Yuasa, Y., Nagasaki, H., Akiyama, Y., Sakai, H., Nakajima, T., Ohkura, Y., Takizawa, T. et al. 2005. Relationship between *CDX2* gene methylation and dietary factors in gastric cancer patients. *Carcinogenesis* 26(1): 193–200.

Zhang, J., Zhang, F., Didelot, X., Bruce, K.D., Cagampang, F.R., Vatish, M., Hanson, M., Lehnert, H., Ceriello, A., and Byrne, C.D. 2009. Maternal high fat diet during pregnancy and lactation alters hepatic expression of insulin like growth factor-2 and key microRNAs in the adult offspring. *BMC Genomics* 10: 478.

Zhang, L., Hou, D., Chen, X., Li, D., Zhu, L., Zhang, Y., Li, J. et al. 2012. Exogenous plant MIR168a specifically targets mammalian LDLRAP1: Evidence of cross-kingdom regulation by microRNA. *Cell Res* 22(1): 107–126.

Zhang, Y., Ito, S., Shen, L., Dai, Q., Wu, S.C., Collins, L.B., Swenberg, J.A., and He, C. 2011. Tet proteins can convert 5-methylcytosine to 5-formylcytosine and 5-carboxylcytosine. *Science* 333(6047): 1300–1303.

Zhu, J.K. 2009. Active DNA demethylation mediated by DNA glycosylases. *Annu Rev Genet* 43: 143–166.

# III

# Lipidemia and Cardiovascular Diseases

# 5

# Role of Dietary Supplements in Lowering LDL-Cholesterol

Prabhjot Singh Nijjar

## Contents

## Introduction

Coronary heart disease (CHD) remains a major source of morbidity and mortality. With the growing epidemic of obesity among young adults and its antecedent complications of diabetes, hypertension, and dyslipidemia, the population at risk for atherosclerotic CHD is ever increasing. Aggressive risk reduction therapies are indicated for patients with CHD to reduce recurrent events and improve survival. More than a century of laboratory and human findings link cholesterol levels with a propensity to develop atherosclerosis [1]. Cholesterol travels in the blood in distinct particles called lipoproteins, the three major classes being low-density lipoproteins (LDL), very low-density lipoproteins (VLDL), and high-density lipoproteins (HDL). LDL is the major atherogenic lipoprotein and its levels correlate fairly well with atherosclerosis and CHD risk. It has been suggested that each percentage reduction in LDL-cholesterol (LDL-C) can reduce CHD risk by 2% [2]. Numerous clinical trials have shown the efficacy of lowering LDL-C for reducing CHD risk, and it is identified by the National Cholesterol Education Program Adult Treatment Panel III (NCEP ATP III) as the primary target for cholesterol-lowering therapy [3].

The Food and Drug Administration (FDA) defines dietary supplements (also known as nutraceuticals or functional foods) as products in capsule, tablet, or liquid form that provide essential nutrients. There was a sharp upward trend in the use of dietary supplements in the 1990s, and this resulted in widespread awareness and interest in these products [4]. As many as 17.7% of US adults used natural products in 2007 [5].

Many dietary supplements have found popular flavor in the lay community about their LDL-C-lowering effects, but only a few have withstood the rigors of controlled trials. Here, we review the evidence in support of specific dietary supplements (Table 5.1) and their LDL-C-lowering effects.

## Supplements with randomized controlled trial evidence to lower LDL-C
### Plant fiber

*Background and observational data*

Dietary plant fiber is a broad term for a variety of plant substances that are resistant to digestion in the human gut. They can be classified into two

**Table 5.1** Dietary Supplements Used to Lower LDL-C

| Supplements with RCT Evidence to Lower LDL-C | Supplements Lacking RCT Evidence to Lower LDL-C |
| --- | --- |
| • Dietary fiber | • Policosanol |
| • Phytosterols/stanols | • Garlic |
| • Soy protein | • Guggulipid |
| • Nuts | |
| • Red yeast rice | |

groups depending on their solubility in water. Common sources of soluble fiber include oats, psyllium, pectin, flaxseed, barley, and guar gum. The structural fibers such as cellulose, lignins, and wheat bran are insoluble. Large population-based observational studies found that diets high in total dietary fiber are associated with a reduced CHD risk [6,7]. A pooled analysis of 10 such prospective cohort studies observed a 12% reduction in risk for coronary events and 19% for coronary deaths, for each 10 g/day increment in dietary fiber [8]. Though risk reduction was observed for both types of fiber, it was more robust for soluble fiber.

## Proposed mechanism of action

Soluble fibers are thought to bind bile acids during the intraluminal formation of micelles [9], and this appears to be through physical entrapment rather than chemical binding [10]. This leads to increased bile acid synthesis, reduction in hepatic cholesterol content, up-regulation of LDL-receptors, and increased LDL clearance [11]. Other potential mechanisms include fibers increasing intraluminal viscosity and slowing macronutrient absorption [12], and increased satiety leading to lower energy intake [13]. In contrast, insoluble fiber does not seem to have any effect on LDL-C, unless it replaces foods supplying saturated fats and cholesterol [14]. The physical properties of soluble fiber, to increase intraluminal viscosity of the small intestine and bind bile acids, might explain why they are more effective than insoluble fiber in lowering LDL-C.

## Clinical trial evidence

Controlled studies have consistently demonstrated a modest but statistically significant reduction in LDL-C by approximately 5%–7% with ingestion of soluble fiber [3]. Investigators have used various sources with essentially similar results [15,16]. A meta-analysis of 67 trials showed that ingestion of soluble fiber, 2–10 g/day, was associated with a small but significant 7% LDL-C reduction [11]. The effect was independent of the type of soluble fiber, including oat, psyllium, pectin, and guar gum. A dose–response relationship was noted, with an absolute lowering of LDL-C by 1.12 mg/dL/g of soluble fiber [17]. Moreyra et al. reported similar LDL-C reductions in subjects taking 15 g/day of psyllium plus 10 mg simvastatin as compared to those taking 20 mg simvastatin plus placebo [18]. The observed percentage of LDL-C lowered by adding psyllium to statin was similar to that reported for doubling the dose of most statins (~6%).

## Clinical implications

The ATP-III panel recommends a daily intake of 5–10 g of soluble fiber as a therapeutic option to enhance LDL-C lowering [3]. However, palatability of foods containing fiber can be limiting. A possible increase in flatulence may result from colonic fermentation. Symptoms are usually mild in the recommended doses of 5–10 g/day, and less fermented fibers such as psyllium may be helpful [10]. In addition, gradually increasing amount of fiber in the diet and drinking adequate amount of fluids are recommended to limit symptoms.

# Plant sterols and stanols

## Background and observational data

Plant sterols, or phytosterols, are naturally occurring sterols of plant origin, with sitosterol being the most abundant. The only difference between cholesterol and sitosterol consists of an additional ethyl group at position C-24 in sitosterol, which is responsible for its poor absorption [19]. In the late 1950s, Eli Lilly Company introduced the first plant sterol product, "Cytellin," as a cholesterol-lowering agent [20]. The daily supply of plant sterols in a typical western diet amounts to an average of 150–400 mg [21]. Vegetable oils are the main natural sources of plant sterols; the one used most commercially being tall pine-tree and soybean oils [22]. Significant quantities are also found in various nuts [23].

## Proposed mechanism of action

Phytosterols compete with cholesterol for space within bile salt micelles in the intestinal lumen, thereby reducing cholesterol absorption [24]. The presence of increased quantities of phytosterols in the gut lowers the micellar solubility of cholesterol, lowering the amount available for absorption [25]. Phytosterols can be esterified to increase solubility and facilitate incorporation into fat-based food products. This allows for much lower doses to be given as compared to the earlier used insoluble material while achieving similar cholesterol reductions [26]. Phytosterols may also be hydrogenated to stanols, such as the conversion of sitosterol to sitostanol. Phytostanols are absorbed negligibly (0.02%–0.3%) compared to phytosterols (0.4%–5%), resulting in lower serum levels for stanols than sterols [27]. There is conflicting data whether plant stanols are more effective than sterols in reducing cholesterol [28,29]. Most likely, both plant sterols and stanols, either esterified or not, lower LDL-C similarly.

## Clinical trial evidence

There is now a large body of evidence showing that plant sterols/stanols at dosages of 2–3 g/day lower LDL-C by 6%–15% [30–36]. Table 5.2 lists some of the most important randomized controlled trials. In summary, studies are in agreement on the effectiveness of plant sterols in reducing LDL-C, independent of the food matrix used as a vehicle. A recent meta-analysis of 41 trials showed that intake of 2 g/day of sterols reduced LDL-C by 10%, and higher intake did not significantly improve on this [37]. Five randomized controlled studies evaluating the combination of use of statins and plant sterols have shown an additive effect of 4.5% ± 2.4% per gram of sterols [38–42]. There have been no reported nutrient–drug

**Table 5.2** Phytosterols and LDL-C: Randomized Controlled Studies

| Authors | Study Design | Intervention | Effect on LDL-C |
|---|---|---|---|
| Miettinen et al. [30] | Mild hypercholesterolemia, $N = 153$, 1 year (RCT, double-blind design) | Margarine with sitostanol ester, 2.6 g/day | ↓ 14% ($P < 0.001$) |
| Hallikainen et al. [31] | Hypercholesterolemia, $N = 22$, 4 weeks (RCT, single blind design) | Margarine with stanol ester, 1.6 g/day | ↓ 5.6% ($P < 0.05$) |
| Volpe et al. [32] | Moderate hypercholesterolemia, $N = 30$, 8 weeks (RCT, crossover parallel-arm design) | Yogurt-based drink, 2 g/day | ↓ 15.6% ($P < 0.001$) |
| Davidson et al. [33] | Mild hypercholesterolemia, $N = 84$, 8 weeks (RCT, double blind) | Reduced fat salad dressing, 9 g/day | ↓ 9% ($P = $ NS) |
| Vanstone et al. [34] | Familial hypercholesterolemia, $N = 15$, 3 weeks (RCT, crossover) | Plant sterols, 1.8 g/day | ↓ 11.3% ($P < 0.03$) |
| Devaraj et al. [35] | Mild hypercholesterolemia, $N = 72$, 8 weeks (RCT, placebo-controlled) | Orange juice with sterols, 2 g/day | ↓ 12.4% ($P < 0.01$) |
| Goldberg et al. [36] | Hypercholesterolemia on statin therapy, $N = 26$, 6 weeks (RCT, double-blind, placebo-controlled, parallel-arm design) | Stanol tablets, 1.8 g/day | ↓ 9.1% ($P < 0.007$) |

*Source:* See table for multiple data sources.
*Abbreviations:* *N*, Number; NS, not significant; RCT, randomized controlled trial.

interactions. So far, no studies have been conducted to assess the effect of plant sterols/stanols on CHD risk [27].

## Clinical implications

The ATP-III panel recommends a daily intake of 2 g of plant sterol/stanol esters as a therapeutic option to enhance LDL-C lowering [3]. The vehicles used have varied from fat-based food products such as spreads, milk, and yogurt to low-fat food products such as cereal, bread, and orange juice [43]. These fortified foods should be isocalorically substituted for other foods of similar nutritional value to prevent excess energy intake and weight gain [44].

There have been concerns regarding the potential adverse effects of phytosterols in certain individuals that are hyper-absorbers [27]. Sitosterolemia is one such rare autosomal recessive disorder, characterized by increased intestinal absorption and decreased biliary excretion of dietary sterols [45]. These patients are homozygous for mutations in the ATP-binding cassette half transporters ABCG5 or ABCG8, which regulate the export of sterols from the hepatocyte and enterocyte [46]. Patients absorb between 15% and 60% of ingested sitosterol as compared to <5% absorbed in normal subjects. This very rare condition leads to elevated plasma levels of plant sterols and cholesterol, manifesting clinically with premature atherosclerosis [47]. However, the more common heterozygotes do not appear to have an increased rate of absorption of sterols [48]. Overall, the role of dietary plant sterols in the development of atherosclerotic plaque is unclear, with conflicting studies [27,49,50].

Consumption of plant sterols can interfere with the intestinal absorption of other lipid-soluble compounds such as fat-soluble vitamins and carotenoids. In fact, they have been shown to reduce plasma levels of $\beta$-carotene, $\alpha$-carotene, and vitamin E, with the effect being more pronounced at higher levels of intake [51]. However, existing studies indicate this to be of small magnitude and clinically insignificant, as long as a nutritionally balanced diet is consumed with the recommended daily intake of 3–5 servings of fruits and vegetables [52]. In summary, phytosterols are well tolerated, and no significant adverse effects have been noted at the recommended doses [33].

# Soy protein

## Background and observational data

Substitution of animal protein with vegetable protein appears to be associated with a lower risk of CHD [53]. Epidemiological studies from Asian countries, where large amounts of soy products are consumed, showed a lower incidence of hypercholesterolemia and CHD [54] sparking interest in the cholesterol-lowering properties of soy protein. The beneficial effects of soy protein on cholesterol lowering were shown in animals more than a century ago by Ignatowski, and in humans by Sirtori et al. in the 1970s [55]. Since then, extensive research has been done to confirm the earlier findings. This culminated in an FDA-approved health claim in 1999 stating that 25 g/day of soy protein, as part of a diet low in saturated fat and cholesterol, may reduce the risk of heart disease [59].

## Proposed mechanism of action

The exact mechanism by which soy lowers cholesterol is not clear. Carroll did much of the earlier work reviewing the various hypotheses and concluded that it was the lack of essential amino acids in soy, as their excess in animal protein enhanced hepatic cholesterol biosynthesis [56,57]. Soy protein products also contain bioactive products called phytoestrogens or isoflavones. The three main isoflavones found in soybeans are genistin, glycitin, and daidzin. These remain in soy preparations that are not extracted with alcohol, and have both weak estrogenic and antiestrogenic properties [58]. There has been debate on whether it is these isoflavones [59] or the protein moiety, especially the 7S globulin fraction [60] that is responsible for the cholesterol-lowering effect. A recent meta-analysis concluded that isolated isoflavones in capsules do not affect blood cholesterol and are not recommended [61]. According to the American Heart Association (AHA) advisory committee on soy, current evidence favors soy protein rather than soy isoflavones as the responsible nutrient, with the caveat that another yet unknown component could be the active factor [54].

## Clinical trial evidence

Interpretation of studies on soy protein has been complicated by the differences in the amount and form of soy used, baseline lipid levels, and nonstandardized methodology. A meta-analysis by Anderson et al. in 1995 showed that soy ingestion decreased LDL-C significantly by 21.7 mg/dL or 12.9% [62]. The response was determined proportionally by the degree of hypercholesterolemia, with nonsignificant reductions in subjects with normal baseline cholesterol levels. There was no relation to the amount of soy eaten, which ranged widely from 18 to 124 g/day (average 47 g/day). Since then, further meta-analyses have found more modest reductions, and a 2006 AHA advisory committee estimated the average effect to be about 3% [54]. This might be due to the quality of studies included, with a broader inclusion criteria used earlier, and the later analyses giving more weight to well-controlled trials. A recent meta-analysis showed that ingestion of moderate amounts of soy (about 25 g) by adults with normal or mild hypercholesterolemia resulted in a small but significant 6% reduction in LDL-C [63]. The reduction was substantially lower than the initial meta-analysis, but is in keeping with more recent trial evidence. There was no relationship to soy dose in the range of 15–40 g, or to baseline cholesterol levels. A dose–response relationship has been reported in the range of 20–106 g [64]. At these very high protein intakes, it is likely that fat or carbohydrate intake is also reduced and it becomes difficult to separate the effect directly due to soy from that related to different macronutrient intake. Contrary to earlier thoughts, effect on LDL-C does not seem to be modulated by the saturated fat and cholesterol content of the baseline diet [54]. In summary, current evidence points to a small reduction in LDL-C, but soy protein is recommended to replace foods high in animal protein that contain saturated fat.

## Clinical implications

The ATP-III panel recommends food sources containing soy protein as replacements for animal food products as a therapeutic option to enhance LDL-C lowering [3]. Soy protein contains calories and should be used in the diet to replace animal or other vegetable protein. It could also replace other sources of calories, namely carbohydrates or fat [54]. The potential impact of high-protein diets on CHD risk is not fully known.

# Nuts

## Background and observational data

Epidemiological studies have consistently shown an association between nut consumption and CHD mortality. Large cohort studies showed that increased nut consumption was associated with reduced risk of heart disease [65–68]. This was followed by attempts to elucidate the mechanism of this effect.

## Proposed mechanism of action

Nuts are a good source of mono- and polyunsaturated fatty acids, and they also contain dietary fiber, phytosterols, and polyphenols. These components likely combine to reduce LDL-C beyond the effects predicted by equations based solely on fatty acid profiles [69]. Walnuts in particular are unique since they are composed largely of polyunsaturated fatty acids, especially α-linolenic acid and linoleic acid, which may benefit CHD risk by non-lipid mechanisms [70]. The ratio of polyunsaturated to saturated fats in walnuts is 7:1, one of the highest among naturally occurring foods.

## Clinical trial evidence

Although most randomized controlled studies have been on almonds and walnuts, several clinical studies on other individual nuts, including hazelnuts, pecans, pistachios, and macadamias, have shown reductions in LDL-C [71–78]. Table 5.3 summarizes some of the most important randomized controlled trials. In one of the earlier such studies, Sabate et al. showed in 1993 that a diet

**Table 5.3** Nuts and LDL-C: Randomized Controlled Studies

| Authors | Study Design | Nuts Consumed | Effect on LDL-C |
|---|---|---|---|
| Sabate et al. [71] | Healthy men/4 weeks (RCT) | Walnuts | ↓ 16.3% (P<0.001) |
| Morgan and Clayshulte [75] | Healthy men and women/8 weeks (RCT) | Pecans | ↓ 19% (P<0.05) |
| Iwamoto et al. [143] | Healthy men and women/2 weeks (RCT, crossover design) | Walnuts | ↓11% (P=0.0008) in women, NS in men |
| Lovejoy et al. [144] | Healthy men and women/4 weeks (RCT, crossover design) | Almonds | ↓ 29% (P<0.001) |
| Jenkins et al. [79] | Hyperlipidemic men and women/1 month (RCT, crossover design) | Almonds | ↓ 9.4% (P<0.001) |
| Sabate et al. [145] | Healthy men and women/4 weeks (RCT, crossover design) | Almonds | ↓ 7% (P<0.001) |
| Tapsell et al. [146] | Men and women with DM—2/6 months (randomized case–control parallel-arm study) | Walnuts | ↓10% (P<0.03) |
| Griel et al. [147] | Mildly hypercholesterolemic men and women per 5 weeks (RCT, crossover design) | Macadamia nuts | ↓8.9% (P<0.0001) |

*Source:* See table for multiple data sources.
*Abbreviations:* NS, not significant; RCT, randomized controlled trial.

that includes moderate quantities of walnuts decreased LDL-C by 18.2 mg/dL (P<0.001) lower than a NCEP step 1 diet (<10% saturated fat) over 4 weeks in healthy men [71]. Jenkins et al. showed almonds significantly reduced LDL-C as compared to a NCEP step 2 diet (<7% saturated fat) (4.4%±1.7%, P=0.018 for half dose almonds and 9.4%±1.9%, P<0.001 for full dose almonds) [79]. This suggests a dose response in which ~7 g of almonds per day reduce LDL-C by ~1%. One percent reductions in LDL-C for walnuts, pecans, peanuts, macadamias, and pistachios would be achieved with daily intakes of 4, 11, 4, 10, 4 g, respectively [71,74–78,80,81].

In a recent meta-analysis of 13 trials, diets supplemented with walnuts resulted in a 6.7% significantly greater decrease in LDL-C compared with the control diet (weighted mean difference=−9.23 mg/dL, P<0.001) [82]. However, the authors note that the amount of walnuts consumed in these trials was relatively large, representing 5%–25% of total calories. Trials incorporating smaller amounts of walnuts showed more modest lipid improvements.

## Clinical implications

In the past, nuts were considered a rich fat source and therefore calorically dense and hence not recommended in CHD patients. However, this concern has not been borne out, with some studies even suggesting that they may be part of a successful low carbohydrate weight loss program [83,84]. Consumption of 1 oz or 28 g of unsalted nuts daily or up to 5 oz weekly, isocalorically substituted for other foods, is recommended to enhance LDL-C lowering and decrease CVD risk [85]. The relatively large amount of nuts required for LDL-C lowering might be difficult to incorporate into a diet. Nut allergies are also common in our society.

# Red yeast rice

## Background and observational data

Red yeast rice (RYR) is the fermented product of rice on which red yeast (*Monascus purpureus*) has been grown. Its use was first documented in China during the Tang dynasty around 800 AD. It has been used as a food preservative, to make rice wine, for its medicinal properties including lipid lowering, and continues to be a dietary staple in many Asian countries [86]. American consumers spent US $17 million on RYR in 2006 [87].

## Proposed mechanism of action

In 1979, Endo discovered that a strain of *Monascus* yeast naturally produced a substance, which he named monacolin K, that inhibits cholesterol synthesis [88]. Monacolin K is also known as lovastatin. RYR also contains a family of eight monacolin-related substances with the ability to inhibit 3-hydroxy-3-methylglutaryl coenzyme A (HMG-CoA) reductase. In addition, RYR has been found to contain sterols, isoflavones, and monounsaturated fatty acids. Little is known about their pharmacodynamics, but these monacolins may either have lipid-lowering effects or potentiate the effects of monacolin K.

## Clinical trial evidence

Extensive safety and efficacy studies of RYR extracts have been conducted. Heber et al. conducted a double-blind, placebo-controlled randomized trial in an American population in which 83 healthy subjects with hyperlipidemia were treated with RYR (2.4 g/day) or placebo for 12 weeks [86]. Compared to baseline, RYR significantly reduced LDL-C ($39 \pm 19$ mg/dL) and this differed significantly from placebo ($P<0.001$). Other studies have shown similar results in different populations [89,90]. In a meta-analysis of 93 randomized trials (9625 participants) by Liu et al., three RYR preparations showed significant reductions in LDL-C ($-28.22$ mg/dL, $P<0.00001$) [91].

A trial evaluating simvastatin (40 mg) vs. an alternate regimen (lifestyle changes, RYR, fish oil) for 12 weeks showed significant reductions in LDL-C in both the groups, and no difference was noted between the groups [92]. Becker et al. evaluated the efficacy and tolerability of RYR in 62 patients with dyslipidemia and a history of statin-associated myalgias [87]. They were randomized to receive RYR (1800 mg) or placebo for 24 weeks, in addition to a 12 week therapeutic lifestyle change program. In the RYR group, LDL-C decreased by 35 mg/dL compared to baseline at week 24, and this was significantly lower compared to placebo ($P=0.011$). Levels of liver enzymes, creatine phosphokinase (CPK), and pain severity scores did not differ between the groups. The dose of RYR in this study was equivalent to a daily lovastatin dose of only 6 mg. RYR may be less likely to deplete mevalonate metabolites distal to HMG-CoA reductase, such as ubiquinone and GTP-binding regulatory proteins, which are believed to mediate statin-induced muscle injury [93]. A recent trial by the same group compared the tolerability of RYR (2400 mg twice daily) vs. pravastatin (20 mg twice daily) in patients with previous statin intolerance. Both were well tolerated and showed comparable LDL-C lowering (30% and 27%, respectively), with no difference in discontinuation rates due to myalgias [94]. This shows that RYR may be a treatment option for dyslipidemic patients who cannot tolerate statins.

Lu et al. conducted a double-blind placebo-controlled randomized trial of Xuenzhikang (XZK), an extract of RYR, on coronary events in a Chinese population with previous myocardial infarction [95]. The study medication consisted of a 300 mg capsule of XZK, each containing 2.5–3.2 mg/capsule of monacolin K or lovastatin. Treatment with XZK reduced LDL-C significantly by 20% compared to baseline ($P<0.001$), as opposed to 3.5% in the placebo group. XZK group also showed a highly significant ($P<0.001$) decrease in frequency of major coronary events (10.4% in placebo vs. 5.7% in XZK group), a relative and absolute decrease of 45% and 4.7%, respectively. No significant treatment-related side effects were reported from this study. This makes RYR the only supplement shown to reduce CV events in a randomized controlled trial.

## Clinical implications

In 2001, the FDA ruled that the RYR product Cholestin was a drug and not a dietary supplement because it contained lovastatin [96]. Companies were asked to reformulate products to remove RYR. Other RYR dietary supplements continue to be sold in the United States; however, any supplement containing more than trace amounts of lovastatin may technically be in violation of FDA policy. The lack of consistency of RYR between different manufacturers is a major deterrent to its widespread use [97]. It is not clear what the recommended dose is, as studies have used different formulations. RYR has been reported to cause myopathy [98–100], rhabdomyolysis [101], and hepatotoxicity [102]. Citrinin, a potential toxin, is a by-product of the fermentation process during RYR preparation. It has been shown to be nephrotoxic in animal models; similar effects in humans are not yet proven [103]. A review by ConsumerLab.com found citrinin in 4 of the 10 supplements that were tested [104]. Due to lack of human data, it is not known if the levels detected

**Table 5.4** RYR Formulations Tested by ConsumerLab.com

| Product Name | Suggested Daily Serving (600 mg Extract/Capsule) (per day) |
| --- | --- |
| Twenty-first century 100% vegetarian RYR extract | 1–2 |
| Cholestene™ RYR | 4 |
| Cholesterin RYR | 4 |
| Healthy America RYR | 2–4 |
| Natural balance RYR concentrated extract | 2 |
| Nature's Plus herbal actives RYR | 1 |
| Schiff® New RYR | 2 |
| Solaray RYR | 2 |
| VegLife 100% vegan RYR | 2 |
| Walgreens Finest Natural RYR | 2 |

*Abbreviation:* RYR, Red yeast rice.

are toxic. Table 5.4 lists the different formulations tested by ConsumerLab. com, several of which are considered to be high quality preparations. Despite some of these concerns, evidence is mounting that RYR is a safe and effective supplement for LDL-C lowering, especially in statin-intolerant patients. It is also the only supplement shown to reduce CV events in a randomized controlled trial.

## Supplements lacking randomized controlled trial evidence for lowering LDL-C

### Policosanol

Policosanol is a natural mixture of long-chain aliphatic alcohols isolated and purified from sugarcane wax, the main component being octocosanol. It can also be obtained from a variety of other plant sources as well as beeswax. It was originally developed by Dalmer Laboratories in Havana, Cuba, and is used mainly in the Caribbean and Latin America. It has been suggested that policosanol acts through some effect on the enzyme activity of HMG-CoA reductase [105], but this remains to be conclusively proven [106].

In the 1990s, numerous well-designed, randomized, controlled studies showed lipid-lowering effects of policosanol, sparking interest in this compound. The response was shown across a wide range of patient subsets, including healthy volunteers, hypercholesterolemics [107], diabetics [108], and postmenopausal women [109]. The reported effect on LDL-C was generally large, in the range of 25%. In comparative trials with statins, policosanol showed similar or even greater improvement in lipid profiles [110,111]. There were also reports of improvement in clinical outcomes, including intermittent claudication [112] and cardiac ischemia [113]. Policosanol seems to be well tolerated.

Most of these clinical studies were performed by a small group of investigators in limited centers, mainly in Cuba, and exclusively with policosanol from Cuban sugarcane. Attempts to replicate these results outside of Cuba have failed to confirm the cholesterol-lowering effects of policosanol [106,114–116]. A randomized placebo-controlled double-blind study by Greyling et al. showed no significant effect on serum lipid levels with an intake of policosanol supplement (differed slightly in composition from the policosanol supplement used in earlier studies) for 12 weeks [106]. Berthold et al. conducted a randomized, placebo-controlled, double-blind parallel-group trial using Cuban sugarcane-derived policosanol provided by Dalmer Laboratories in Cuba (same as the supplement used in earlier studies). No statistically significant difference between policosanol and placebo was observed [115]. The general consensus now seems to be that policosanol does not work to lower LDL-C.

## Garlic

Garlic has been used medicinally since ancient times to enhance physical and mental health. Garlic supplements are among the top-selling herbal supplements [4], mainly for its purported lipid-lowering properties. Allicin is the active ingredient and is formed when the bulb is crushed, through action of alliinase enzymes on the stable precursor alliin. Because of its instability, allicin instantly decomposes to other compounds, some of which are responsible for the characteristic odor.

The proposed mechanisms for lipid lowering include decreasing cholesterol absorption in the intestine [117], inhibiting enzymes involved in cholesterol synthesis [118], and deactivation of HMG-CoA reductase [119]. Extensive in vitro and animal studies supported a strong plausibility of effect [120]. Earlier trials conducted before 1995 reported significant reductions in total cholesterol levels, and two meta-analysis reported reductions in the range of 9%–12% [121,122]. However, they were criticized for serious design and conduct limitations [121].

Subsequent well-designed, randomized, controlled studies have shown either a more modest effect [123] or no effect [124–128]. There was also concern about the differences in bioavailability of important sulfur-containing constituents between raw garlic and specific garlic supplement formulations that were tested, possibly confounding results. Gardner et al. conducted a parallel-design trial in which 192 adults with moderate hypercholesterolemia were randomly assigned to four treatment arms: raw garlic, powdered garlic supplement, aged garlic extract supplement, or placebo [124]. Extensive chemical characterization of different products was done before and throughout the study to ensure similar composition. There were no statistically significant effects of the three forms of garlic on LDL-C. No side effects of garlic have been reported, apart from the offensive odor. Based on recent evidence, it appears unlikely that garlic has any reasonable effect on the lipid profile. Garlic may have other protective effects with regard to cardiovascular disease (CVD) such as reduced blood pressure and platelet inhibition [129], but these also need to be scrutinized in large, carefully designed trials.

## Guggulipid

Guggul has been used medicinally in India as far back as 600 BC [130], and is an extract from the resin of the mukul myrrh tree (*Commiphora mukul*). The plant sterols E- and Z-guggulsterone are thought to be the active components [131], reportedly through antagonism of two nuclear hormone receptors involved in bile acid metabolism [132]. Trials, mainly uncontrolled, from India reported a large LDL-C-lowering effect, some up to 60%–80% [133,134]. Szapary et al. conducted a randomized, placebo-controlled trial to study the safety and efficacy of a standardized guggul extract (guggulipid) in healthy adults with hyperlipidemia eating a western diet [135]. Surprisingly, guggulipid did not reduce and in fact modestly increased levels of LDL-C by 4%–5%. In addition, it seemed to cause a hypersensitivity drug reaction in a few patients. Not only was guggulipid ineffective in lowering cholesterol, but this was a reminder that supplements require rigorous clinical testing before being recommended and that they cannot be assumed to be safe.

## Discussion

There are many over-the-counter dietary supplements that promote cholesterol-lowering benefits. Current evidence indicates that soluble plant fiber, plant sterols, soy protein, nuts, and RYR definitely seem to have modest but significant LDL-C-lowering properties. There seems to be consensus that garlic, guggulipid, and policosanol do not have any significant effects on LDL-C. The history of supplements such as garlic and policosanol highlights the importance of rigorous testing and cautious skepticism before adopting a new product into clinical practice. Concerns remain about compliance and effective ways to incorporate supplements into the diets of free-living individuals. Though seemingly safe from toxic side effects, they should be tested under strict conditions to ensure safety, just as pharmaceuticals are. Since dietary supplements are not under the regulatory preview of the FDA, there are also problems with consistency among different products available in the market. The FDA permits a health claim to be made for CVD risk reduction through the lowering of LDL-C for soluble fiber, plant sterols, soy protein, and nuts [136,137]. The NCEP ATP III recommends soluble fibers, plant sterols, and soy protein as therapeutic options to enhance LDL-C lowering [3] (Table 5.5). Table 5.6 lists different formulations of these supplements.

**Table 5.5** NCEP ATP III Guidelines

1. 5–10 g of soluble fiber per day reduces LDL-C by ~5%
2. 2–3 g of plant stanol/sterol esters per day reduces LDL-C by 6%–15%
3. High intake of soy protein can cause small reductions in LDL-C, especially when it replaces animal protein

**Table 5.6** Dietary Supplement Formulations with Content

| Dietary Supplement (Recommended Amount) | Example of Supplement | Content |
|---|---|---|
| Soluble fiber (10 g/day) | Oatmeal, 1 cup cooked | 2.0 g |
| | Whole wheat bread, 1 slice | 0.5 g |
| | Apple, 1 medium | 1.0 g |
| | Orange, 1 medium | 2.0 g |
| | Kidney beans, ½ cup | 3.0 g |
| | Broccoli, cooked ½ cup | 1.0 g |
| | Metamucil (psyllium husk), 1 tsp | 2.0 g |
| | Benefiber (partially hydrolyzed guar gum), 1 tbsp | 3.0 g |
| Phytosterols/stanols (2 g/day) | Promise Activ buttery spread, 1 tbsp | 1.0 g plant sterol |
| | Smart Balance Heart Right | 1.0 g plant sterol |
| | 1% low-fat milk, 8 oz | |
| | Benecol margarine spread, 1 tbsp | 0.5 g plant stanol |
| | Heart Health Advantage from Bayer, 2 tablets | 0.8 g plant sterols |
| | Centrum Cardio Tablets, 2 tablets | 0.8 g plant sterols |
| Soy protein (25 g/day minimum) | Tofu, firm, 1 slice (84 g) | 6.0 g protein |
| | Soy veggie burger, 1 patty (70 g) | 11.0 g protein |
| | Edamame, ½ cup (90 g) | 11.0 g protein |
| Nuts (1 oz, 28 g/day) | Almonds (22 pieces) | |
| | Walnuts (14 halves) | |
| | Pecans (20 halves) | |
| Red yeast rice supplements (2–4 tabs/day) | Schiff New Red Yeast Rice 600 mg | 2.7 mg lovastatin/tablet |
| | Distributed by Schiff Nutrition Group | |
| | Cholestene Red Yeast Rice 600 mg | 2.5 mg lovastatin/tablet |
| | Distributed by HPF, LLC | |

*Sources:* Adapted from www.nutritiondata.com, USDA National Nutrient Database for Standard Reference, www.nal. usda.gov; Executive summary of the third report of The National Cholesterol Education Program (NCEP) expert panel on detection, evaluation, and treatment of high blood cholesterol in adults (adult treatment panel III), *JAMA*, 285(19), 2486, 2001; Sacks, F.M. et al., *Circulation*, 113(7), 1034, 2006; Lin, Y.L. et al., *Appl. Microbiol. Biotechnol.*, 77(5), 965, 2008.

A series of studies have evaluated the effects of combining different dietary supplements under metabolically controlled conditions. Gardner et al. showed a significant reduction in LDL-C with a combination of soy, nuts, and viscous fiber [138]. Combination therapy with plant sterols and soy resulted in a 14.8% reduction in LDL-C [139]. The "portfolio" dietary approach consisted of a combination of plant sterols, viscous fiber, soy protein, and nuts in addition to a low-saturated fat diet. After 1 month, subjects on this diet had a mean LDL-C reduction of 28.6% (P<0.001) compared with 30.9% (P<0.001) for those on a low-saturated fat diet taking lovastatin (20 mg/day), with no significant difference between the groups [140]. This provides evidence that a wholesome approach incorporating different dietary components works well to reduce LDL-C, and is possibly as effective as low-dose statin therapy. A follow-up 1 year study in which subjects were not fed the "portfolio" diet in a metabolic ward showed a slightly more modest 13.6% LDL-C reduction [141]. There was a large variation seen in the group, and this was thought to be related to compliance with the diet under free-living conditions.

Rapid advent of biotechnology has provided us with various pharmaceuticals, including statins, which work very well at lowering cholesterol levels. With new trial data being available, we have seen LDL-C goals being lowered over time. This will lead to a large cohort of the population that would be candidates for lipid-lowering therapy to decrease their lifetime risk. Even though statins are relatively safe and well tolerated, there are still a significant number of patients who cannot tolerate them leading to discontinuation of therapy, and many others who have only mildly elevated LDL-C levels and prefer nonprescription alternatives to statin therapy. It is difficult to estimate the number of patients who seek alternative treatments to statins [142].

Dietary supplements cannot replace the use of prescription pharmaceuticals for LDL-C reduction, but they can complement them to increase the number of patients whose lipids are well controlled. They may also allow drug therapy to be used at a lower dose, thereby decreasing the risk of dose-related side effects. Young adults with borderline high cholesterol may benefit from the modest cholesterol-lowering property of nutraceuticals, and hence avoid or delay being on drug therapy. New roles might emerge for different supplements, such as RYR for statin intolerance. In addition to a diet that is low in saturated fat and cholesterol, the combination of different supplements offers an exciting possibility of LDL-C lowering that is comparable to low-dose statin, albeit by natural means.

# References

1. Libby, P., M. Aikawa, and U. Schonbeck, Cholesterol and atherosclerosis. *Biochim Biophys Acta*, 2000, 1529: 299–309.
2. Law, M.R., N.J. Wald, and S.G. Thompson, By how much and how quickly does reduction in serum cholesterol concentration lower risk of ischaemic heart disease? *BMJ*, 1994, 308(6925): 367–372.

3. Executive summary of the third report of The National Cholesterol Education Program (NCEP) expert panel on detection, evaluation, and treatment of high blood cholesterol in adults (adult treatment panel III). *JAMA*, 2001, 285(19): 2486–2497.

4. Kelly, J.P. et al., Recent trends in use of herbal and other natural products. *Arch Intern Med*, 2005, 165(3): 281–286.

5. Barnes, P.M., B. Bloom, and R.L. Nahin, Complementary and alternative medicine use among adults and children: United States, 2007. *Natl Health Stat Report*, 2008, (12): 1–23.

6. Liu, S. et al., Whole-grain consumption and risk of coronary heart disease: Results from the Nurses' Health Study. *Am J Clin Nutr*, 1999, 70(3): 412–419.

7. Bazzano, L.A. et al., Dietary fiber intake and reduced risk of coronary heart disease in US men and women: The National Health and Nutrition Examination Survey I Epidemiologic Follow-up Study. *Arch Intern Med*, 2003, 163(16): 1897–1904.

8. Pereira, M.A. et al., Dietary fiber and risk of coronary heart disease: A pooled analysis of cohort studies. *Arch Intern Med*, 2004, 164(4): 370–376.

9. Anderson, J.W. and J. Tietyen-Clark, Dietary fiber: Hyperlipidemia, hypertension, and coronary heart disease. *Am J Gastroenterol*, 1986, 81(10): 907–919.

10. Jenkins, D.A., Nutraceuticals and functional foods for cholesterol reduction. *Clinical Lipidology*, Ballantyne, C.M., ed., Saunders Elsevier, PA, 2009.

11. Brown, L. et al., Cholesterol-lowering effects of dietary fiber: A meta-analysis. *Am J Clin Nutr*, 1999, 69(1): 30–42.

12. Leinonen, K.S., K.S. Poutanen, and H.M. Mykkanen, Rye bread decreases serum total and LDL cholesterol in men with moderately elevated serum cholesterol. *J Nutr*, 2000, 130(2): 164–170.

13. Blundell, J.E. and V.J. Burley, Satiation, satiety and the action of fibre on food intake. *Int J Obes*, 1987, 11(Suppl 1): 9–25.

14. Kris-Etherton, P.M. et al., The effect of diet on plasma lipids, lipoproteins, and coronary heart disease. *J Am Diet Assoc*, 1988, 88(11): 1373–1400.

15. Ripsin, C.M. et al., Oat products and lipid lowering. A meta-analysis. *JAMA*, 1992, 267(24): 3317–3325.

16. Arvill, A. and L. Bodin, Effect of short-term ingestion of konjac glucomannan on serum cholesterol in healthy men. *Am J Clin Nutr*, 1995, 61(3): 585–589.

17. Anderson, J.W. et al., Cholesterol-lowering effects of psyllium intake adjunctive to diet therapy in men and women with hypercholesterolemia: Meta-analysis of 8 controlled trials. *Am J Clin Nutr*, 2000, 71(2): 472–479.

18. Moreyra, A.E., A.C. Wilson, and A. Koraym, Effect of combining psyllium fiber with simvastatin in lowering cholesterol. *Arch Intern Med*, 2005, 165(10): 1161–1166.

19. Fernandez, M.L. and S. Vega-Lopez, Efficacy and safety of sitosterol in the management of blood cholesterol levels. *Cardiovasc Drug Rev*, 2005, 23(1): 57–70.

20. Pollak, O.J. and D. Kritchevsky, eds., *Sitosterol Monographs Atherosclerosis*, Vol. 10, 1981. Basel, Switzerland: Karger.

21. European Commission. *General view of the Scientific Committee on Food on the Long Term Effects of the Intake of Elevated Levels of Phytosterols from Multiple Dietary Sources.* Scientific committee on food, 2002. Brussels, Belgium. http://ec.europa.eu/food/fs/sc/scf/out143_en.pdf (accessed on July 7, 2010).

22. Ito, T., T. Tamura, and T. Matsumoto, Sterol composition of 19 vegetable oils. *J Am Oil Chem Soc*, 1973, 50(4): 122–125.

23. Amaral, J.S. et al., Determination of sterol and fatty acid compositions, oxidative stability, and nutritional value of six walnut (*Juglans regia* L.) cultivars grown in Portugal. *J Agric Food Chem*, 2003, 51(26): 7698–7702.

24. Patch, C.S. et al., Plant sterols as dietary adjuvants in the reduction of cardiovascular risk: Theory and evidence. *Vasc Health Risk Manag*, 2006, 2(2): 157–162.

25. Jones, P.J. et al., Modulation of plasma lipid levels and cholesterol kinetics by phytosterol versus phytostanol esters. *J Lipid Res*, 2000, 41(5): 697–705.

26. Hallikainen, M.A. et al., Comparison of the effects of plant sterol ester and plant stanol ester-enriched margarines in lowering serum cholesterol concentrations in hypercholesterolaemic subjects on a low-fat diet. *Eur J Clin Nutr*, 2000, 54(9): 715–725.
27. Weingartner, O., M. Bohm, and U. Laufs, Controversial role of plant sterol esters in the management of hypercholesterolaemia. *Eur Heart J*, 2009, 30(4): 404–409.
28. Miettinen, T.A. and H. Gylling, Plant stanol and sterol esters in prevention of cardiovascular diseases. *Ann Med*, 2004, 36(2): 126–134.
29. Fransen, H.P. et al., Customary use of plant sterol and plant stanol enriched margarine is associated with changes in serum plant sterol and stanol concentrations in humans. *J Nutr*, 2007, 137(5): 1301–1306.
30. Miettinen, T.A. et al., Reduction of serum cholesterol with sitostanol-ester margarine in a mildly hypercholesterolemic population. *N Engl J Med*, 1995, 333(20): 1308–1312.
31. Hallikainen, M.A., E.S. Sarkkinen, and M.I. Uusitupa, Plant stanol esters affect serum cholesterol concentrations of hypercholesterolemic men and women in a dose-dependent manner. *J Nutr*, 2000, 130(4): 767–776.
32. Volpe, R. et al., Effects of yoghurt enriched with plant sterols on serum lipids in patients with moderate hypercholesterolaemia. *Br J Nutr*, 2001, 86(2): 233–239.
33. Davidson, M.H. et al., Safety and tolerability of esterified phytosterols administered in reduced-fat spread and salad dressing to healthy adult men and women. *J Am Coll Nutr*, 2001, 20(4): 307–319.
34. Vanstone, C.A. et al., Unesterified plant sterols and stanols lower LDL-cholesterol concentrations equivalently in hypercholesterolemic persons. *Am J Clin Nutr*, 2002, 76(6): 1272–1278.
35. Devaraj, S., I. Jialal, and S. Vega-Lopez, Plant sterol-fortified orange juice effectively lowers cholesterol levels in mildly hypercholesterolemic healthy individuals. *Arterioscler Thromb Vasc Biol*, 2004, 24(3): e25–e28.
36. Goldberg, A.C. et al., Effect of plant stanol tablets on low-density lipoprotein cholesterol lowering in patients on statin drugs. *Am J Cardiol*, 2006, 97(3): 376–379.
37. Katan, M.B. et al., Efficacy and safety of plant stanols and sterols in the management of blood cholesterol levels. *Mayo Clin Proc*, 2003, 78(8): 965–978.
38. Miettinen, T.A., T.E. Strandberg, and H. Gylling, Noncholesterol sterols and cholesterol lowering by long-term simvastatin treatment in coronary patients: Relation to basal serum cholestanol. *Arterioscler Thromb Vasc Biol*, 2000, 20(5): 1340–1346.
39. Vanhanen, H., Cholesterol malabsorption caused by sitostanol ester feeding and neomycin in pravastatin-treated hypercholesterolaemic patients. *Eur J Clin Pharmacol*, 1994, 47(2): 169–176.
40. Gylling, H., R. Radhakrishnan, and T.A. Miettinen, Reduction of serum cholesterol in postmenopausal women with previous myocardial infarction and cholesterol malabsorption induced by dietary sitostanol ester margarine: Women and dietary sitostanol. *Circulation*, 1997, 96(12): 4226–4231.
41. Neil, H.A., G.W. Meijer, and L.S. Roe, Randomised controlled trial of use by hypercholesterolaemic patients of a vegetable oil sterol-enriched fat spread. *Atherosclerosis*, 2001, 156(2): 329–337.
42. Simons, L.A., Additive effect of plant sterol-ester margarine and cerivastatin in lowering low-density lipoprotein cholesterol in primary hypercholesterolemia. *Am J Cardiol*, 2002, 90(7): 737–740.
43. Clifton, P.M. et al., Cholesterol-lowering effects of plant sterol esters differ in milk, yoghurt, bread and cereal. *Eur J Clin Nutr*, 2004, 58(3): 503–509.
44. Van Horn, L. et al., The evidence for dietary prevention and treatment of cardiovascular disease. *J Am Diet Assoc*, 2008, 108(2): 287–331.
45. Bhattacharyya, A.K. and W.E. Connor, Beta-sitosterolemia and xanthomatosis. A newly described lipid storage disease in two sisters. *J Clin Invest*, 1974, 53(4): 1033–1043.
46. Hubacek, J.A. et al., Mutations in ATP-cassette binding proteins G5 (ABCG5) and G8 (ABCG8) causing sitosterolemia. *Hum Mutat*, 2001, 18(4): 359–360.
47. Salen, G. et al., Sitosterolemia. *J Lipid Res*, 1992, 33(7): 945–955.

48. Kwiterovich, P.O., Jr. et al., Response of obligate heterozygotes for phytosterolemia to a low-fat diet and to a plant sterol ester dietary challenge. *J Lipid Res*, 2003, 44(6): 1143–1155.

49. Wilund, K.R. et al., No association between plasma levels of plant sterols and atherosclerosis in mice and men. *Arterioscler Thromb Vasc Biol*, 2004, 24(12): 2326–2332.

50. Miettinen, T.A. et al., Plant sterols in serum and in atherosclerotic plaques of patients undergoing carotid endarterectomy. *J Am Coll Cardiol*, 2005, 45(11): 1794–1801.

51. Clifton, P.M. et al., High dietary intake of phytosterol esters decreases carotenoids and increases plasma plant sterol levels with no additional cholesterol lowering. *J Lipid Res*, 2004, 45(8): 1493–1499.

52. Noakes, M. et al., An increase in dietary carotenoids when consuming plant sterols or stanols is effective in maintaining plasma carotenoid concentrations. *Am J Clin Nutr*, 2002, 75(1): 79–86.

53. Hilleboe, H.E., Some epidemiologic aspects of coronary artery disease. *J Chronic Dis*, 1957, 6(3): 210–228.

54. Sacks, F.M. et al., Soy protein, isoflavones, and cardiovascular health: An American Heart Association Science Advisory for professionals from the Nutrition Committee. *Circulation*, 2006, 113(7): 1034–1044.

55. Sirtori, C.R. et al., Soybean–protein diet in the treatment of type-II hyperlipoproteinaemia. *Lancet*, 1977, 1(8006): 275–277.

56. Carroll, K.K., Hypercholesterolemia and atherosclerosis: Effects of dietary protein. *Fed Proc*, 1982, 41(11): 2792–2796.

57. Carroll, K.K., Review of clinical studies on cholesterol-lowering response to soy protein. *J Am Diet Assoc*, 1991, 91(7): 820–827.

58. Cassidy, A., S. Bingham, and K. Setchell, Biological effects of isoflavones in young women: Importance of the chemical composition of soyabean products. *Br J Nutr*, 1995, 74(4): 587–601.

59. Crouse, J.R., 3rd et al., A randomized trial comparing the effect of casein with that of soy protein containing varying amounts of isoflavones on plasma concentrations of lipids and lipoproteins. *Arch Intern Med*, 1999, 159(17): 2070–2076.

60. Lovati, M.R. et al., Soy protein peptides regulate cholesterol homeostasis in Hep G2 cells. *J Nutr*, 2000, 130(10): 2543–2549.

61. Weggemans, R.M. and E.A. Trautwein, Relation between soy-associated isoflavones and LDL and HDL cholesterol concentrations in humans: A meta-analysis. *Eur J Clin Nutr*, 2003, 57(8): 940–946.

62. Anderson, J.W., B.M. Johnstone, and M.E. Cook-Newell, Meta-analysis of the effects of soy protein intake on serum lipids. *N Engl J Med*, 1995, 333(5): 276–282.

63. Harland, J.I. and T.A. Haffner, Systematic review, meta-analysis and regression of randomised controlled trials reporting an association between an intake of circa 25 g soya protein per day and blood cholesterol. *Atherosclerosis*, 2008, 200(1): 13–27.

64. Reynolds, K. et al., A meta-analysis of the effect of soy protein supplementation on serum lipids. *Am J Cardiol*, 2006, 98(5): 633–640.

65. Albert, C.M. et al., Nut consumption and decreased risk of sudden cardiac death in the Physicians' Health Study. *Arch Intern Med*, 2002, 162(12): 1382–1387.

66. Ellsworth, J.L., L.H. Kushi, and A.R. Folsom, Frequent nut intake and risk of death from coronary heart disease and all causes in postmenopausal women: The Iowa Women's Health Study. *Nutr Metab Cardiovasc Dis*, 2001, 11(6): 372–377.

67. Fraser, G.E. et al., A possible protective effect of nut consumption on risk of coronary heart disease. The Adventist Health Study. *Arch Intern Med*, 1992, 152(7): 1416–1424.

68. Hu, F.B. et al., Frequent nut consumption and risk of coronary heart disease in women: Prospective cohort study. *BMJ*, 1998, 317(7169): 1341–1345.

69. Mensink, R.P. and M.B. Katan, Effect of dietary fatty acids on serum lipids and lipoproteins. A meta-analysis of 27 trials. *Arterioscler Thromb*, 1992, 12(8): 911–919.

70. Li, L. et al., Fatty acid profiles, tocopherol contents, and antioxidant activities of heartnut (*Juglans ailanthifolia* var. *cordiformis*) and Persian walnut (*Juglans regia* L.). *J Agric Food Chem*, 2007, 55(4): 1164–1169.

71. Sabate, J. et al., Effects of walnuts on serum lipid levels and blood pressure in normal men. *N Engl J Med*, 1993, 328(9): 603–607.

72. Spiller, G.A. et al., Nuts and plasma lipids: An almond-based diet lowers LDL-C while preserving HDL-C. *J Am Coll Nutr*, 1998, 17(3): 285–290.

73. Chisholm, A. et al., A diet rich in walnuts favourably influences plasma fatty acid profile in moderately hyperlipidaemic subjects. *Eur J Clin Nutr*, 1998, 52(1): 12–16.

74. Edwards, K. et al., Effect of pistachio nuts on serum lipid levels in patients with moderate hypercholesterolemia. *J Am Coll Nutr*, 1999, 18(3): 229–232.

75. Morgan, W.A. and B.J. Clayshulte, Pecans lower low-density lipoprotein cholesterol in people with normal lipid levels. *J Am Diet Assoc*, 2000, 100(3): 312–318.

76. Curb, J.D. et al., Serum lipid effects of a high-monounsaturated fat diet based on macadamia nuts. *Arch Intern Med*, 2000, 160(8): 1154–1158.

77. Zambon, D. et al., Substituting walnuts for monounsaturated fat improves the serum lipid profile of hypercholesterolemic men and women. A randomized crossover trial. *Ann Intern Med*, 2000, 132(7): 538–546.

78. Rajaram, S. et al., A monounsaturated fatty acid-rich pecan-enriched diet favorably alters the serum lipid profile of healthy men and women. *J Nutr*, 2001, 131(9): 2275–2279.

79. Jenkins, D.J. et al., Dose response of almonds on coronary heart disease risk factors: Blood lipids, oxidized low-density lipoproteins, lipoprotein (a), homocysteine, and pulmonary nitric oxide: A randomized, controlled, crossover trial. *Circulation*, 2002, 106(11): 1327–1332.

80. Almario, R.U. et al., Effects of walnut consumption on plasma fatty acids and lipoproteins in combined hyperlipidemia. *Am J Clin Nutr*, 2001, 74(1): 72–79.

81. Kris-Etherton, P.M. et al., High-monounsaturated fatty acid diets lower both plasma cholesterol and triacylglycerol concentrations. *Am J Clin Nutr*, 1999, 70(6): 1009–1015.

82. Banel, D.K. and F.B. Hu, Effects of walnut consumption on blood lipids and other cardiovascular risk factors: A meta-analysis and systematic review. *Am J Clin Nutr*, 2009, 90(1): 56–63.

83. Sabate, J. et al., Does regular walnut consumption lead to weight gain? *Br J Nutr*, 2005, 94(5): 859–864.

84. Foster, G.D. et al., A randomized trial of a low-carbohydrate diet for obesity. *N Engl J Med*, 2003, 348(21): 2082–2090.

85. Kris-Etherton, P.M. et al., The effects of nuts on coronary heart disease risk. *Nutr Rev*, 2001, 59(4): 103–111.

86. Heber, D. et al., Cholesterol-lowering effects of a proprietary Chinese red-yeast-rice dietary supplement. *Am J Clin Nutr*, 1999, 69(2): 231–236.

87. Becker, D.J. et al., Red yeast rice for dyslipidemia in statin-intolerant patients: A randomized trial. *Ann Intern Med*, 2009, 150(12): 830–839, W147–W149.

88. Endo, A. and Monacolin, K., A new hypocholesterolemic agent produced by a *Monascus* species. *J Antibiot (Tokyo)*, 1979, 32(8): 852–854.

89. Lin, C.C., T.C. Li, and M.M. Lai, Efficacy and safety of *Monascus purpureus* Went rice in subjects with hyperlipidemia. *Eur J Endocrinol*, 2005, 153(5): 679–686.

90. Huang, C.F. et al., Efficacy of *Monascus purpureus* Went rice on lowering lipid ratios in hypercholesterolemic patients. *Eur J Cardiovasc Prev Rehabil*, 2007, 14(3): 438–440.

91. Liu, J. et al., Chinese red yeast rice (*Monascus purpureus*) for primary hyperlipidemia: A meta-analysis of randomized controlled trials. *Chin Med*, 2006, 1: 4.

92. Becker, D.J. et al., Simvastatin vs therapeutic lifestyle changes and supplements: Randomized primary prevention trial. *Mayo Clin Proc*, 2008, 83(7): 758–764.

93. Thompson, P.D., P. Clarkson, and R.H. Karas, Statin-associated myopathy. *JAMA*, 2003, 289(13): 1681–1690.

94. Halbert, S.C. et al., Tolerability of red yeast rice (2,400 mg twice daily) versus pravastatin (20 mg twice daily) in patients with previous statin intolerance. *Am J Cardiol*, 105(2): 198–204.

95. Lu, Z. et al., Effect of Xuezhikang, an extract from red yeast Chinese rice, on coronary events in a Chinese population with previous myocardial infarction. *Am J Cardiol*, 2008, 101(12): 1689–1693.

96. Moore, R., Letter to Sonia Rodriguez, Mason Vitamins. May 5, 2001. US Department of Health and Human Services. http://www.fda.gov/ohrms/dockets/dailys/01/Jun01/061101/let0494.pdf (accessed October 12, 2009).

97. Heber, D. et al., An analysis of nine proprietary Chinese red yeast rice dietary supplements: Implications of variability in chemical profile and contents. *J Altern Complement Med*, 2001, 7(2): 133–139.

98. Cartin-Ceba, R., L.B. Lu, and A. Kolpakchi, A 'natural' threat. *Am J Med*, 2007, 120(11): e3–e4.

99. Lapi, F. et al., Myopathies associated with red yeast rice and liquorice: Spontaneous reports from the Italian Surveillance System of Natural Health Products. *Br J Clin Pharmacol*, 2008, 66(4): 572–574.

100. Mueller, P.S., Symptomatic myopathy due to red yeast rice. *Ann Intern Med*, 2006, 145(6): 474–475.

101. Prasad, G.V. et al., Rhabdomyolysis due to red yeast rice (*Monascus purpureus*) in a renal transplant recipient. *Transplantation*, 2002, 74(8): 1200–1201.

102. Roselle, H. et al., Symptomatic hepatitis associated with the use of herbal red yeast rice. *Ann Intern Med*, 2008, 149(7): 516–517.

103. Lin, Y.L. et al., Biologically active components and nutraceuticals in the *Monascus*-fermented rice: A review. *Appl Microbiol Biotechnol*, 2008, 77(5): 965–973.

104. http://www.consumerlab.com/reviews/Red_Yeast_Rice_Supplements-Lovastatin_Monacolin/Red_Yeast_Rice/ (accessed November 1, 2009).

105. Janikula, M., Policosanol: A new treatment for cardiovascular disease? *Altern Med Rev*, 2002, 7(3): 203–217.

106. Greyling, A. et al., Effects of a policosanol supplement on serum lipid concentrations in hypercholesterolaemic and heterozygous familial hypercholesterolaemic subjects. *Br J Nutr*, 2006, 95(5): 968–975.

107. Pons, P. et al., Effects of successive dose increases of policosanol on the lipid profile of patients with type II hypercholesterolaemia and tolerability to treatment. *Int J Clin Pharmacol Res*, 1994, 14(1): 27–33.

108. Torres, O. et al., Treatment of hypercholesterolemia in NIDDM with policosanol. *Diabetes Care*, 1995, 18(3): 393–397.

109. Castano, G. et al., Effects of policosanol on postmenopausal women with type II hypercholesterolemia. *Gynecol Endocrinol*, 2000, 14(3): 187–195.

110. Crespo, N. et al., Comparative study of the efficacy and tolerability of policosanol and lovastatin in patients with hypercholesterolemia and noninsulin dependent diabetes mellitus. *Int J Clin Pharmacol Res*, 1999, 19(4): 117–127.

111. Castano, G. et al., Comparison of the efficacy and tolerability of policosanol with atorvastatin in elderly patients with type II hypercholesterolaemia. *Drugs Aging*, 2003, 20(2): 153–163.

112. Castano, G. et al., A long-term study of policosanol in the treatment of intermittent claudication. *Angiology*, 2001, 52(2): 115–125.

113. Batista, J. et al., Effect of policosanol on hyperlipidemia and coronary heart disease in middle-aged patients. A 14-month pilot study. *Int J Clin Pharmacol Ther*, 1996, 34(3): 134–137.

114. Lin, Y. et al., Wheat germ policosanol failed to lower plasma cholesterol in subjects with normal to mildly elevated cholesterol concentrations. *Metabolism*, 2004, 53(10): 1309–1314.

115. Berthold, H.K. et al., Effect of policosanol on lipid levels among patients with hypercholesterolemia or combined hyperlipidemia: A randomized controlled trial. *JAMA*, 2006, 295(19): 2262–2269.

116. Dulin, M.F. et al., Policosanol is ineffective in the treatment of hypercholesterolemia: A randomized controlled trial. *Am J Clin Nutr*, 2006, 84(6): 1543–1548.

117. Matsuura, H., Saponins in garlic as modifiers of the risk of cardiovascular disease. *J Nutr*, 2001, 131(3s): 1000S–1005S.

118. Yeh, Y.Y. and L. Liu, Cholesterol-lowering effect of garlic extracts and organosulfur compounds. Human and animal studies. *J Nutr*, 2001, 131(3s): 989S–993S.

119. Borek, C., Garlic reduces dementia and heart-disease risk. *J Nutr*, 2006, 136(3 Suppl): 810S–812S.

120. Reuter, H.D., H.P. Koch, and L.D. Lawson, Therapeutic effects and applications of garlic and its preparations. *Garlic: The science and therapeutic application of Allium sativum*, Heinrich P. Koch and Larry D. Lawson. Baltimore : Williams & Wilkins 1996, 135–212.

121. Warshafsky, S., R.S. Kamer, and S.L. Sivak, Effect of garlic on total serum cholesterol. A meta-analysis. *Ann Intern Med*, 1993, 119(7 Pt 1): 599–605.

122. Silagy, C. and A. Neil, Garlic as a lipid lowering agent—A meta-analysis. *J R Coll Physicians Lond*, 1994, 28(1): 39–45.

123. Stevinson, C., M.H. Pittler, and E. Ernst, Garlic for treating hypercholesterolemia. A meta-analysis of randomized clinical trials. *Ann Intern Med*, 2000, 133(6): 420–429.

124. Gardner, C.D. et al., Effect of raw garlic vs commercial garlic supplements on plasma lipid concentrations in adults with moderate hypercholesterolemia: A randomized clinical trial. *Arch Intern Med*, 2007, 167(4): 346–353.

125. Berthold, H.K., T. Sudhop, and K. von Bergmann, Effect of a garlic oil preparation on serum lipoproteins and cholesterol metabolism: A randomized controlled trial. *JAMA*, 1998, 279(23): 1900–1902.

126. Gardner, C.D., L.M. Chatterjee, and J.J. Carlson, The effect of a garlic preparation on plasma lipid levels in moderately hypercholesterolemic adults. *Atherosclerosis*, 2001, 154(1): 213–220.

127. Isaacsohn, J.L. et al., Garlic powder and plasma lipids and lipoproteins: A multicenter, randomized, placebo-controlled trial. *Arch Intern Med*, 1998, 158(11): 1189–1194.

128. Superko, H.R. and R.M. Krauss, Garlic powder, effect on plasma lipids, postprandial lipemia, low-density lipoprotein particle size, high-density lipoprotein subclass distribution and lipoprotein(a). *J Am Coll Cardiol*, 2000, 35(2): 321–326.

129. Ernst, E., Cardiovascular effects of garlic (*Allium sativum*): A review. *Pharmatherapeutica*, 1987, 5(2): 83–89.

130. Satyavati, G.V., Gum guggul (*Commiphora mukul*)—The success story of an ancient insight leading to a modern discovery. *Indian J Med Res*, 1988, 87: 327–335.

131. Nityanand, S. and N.K. Kapoor, Cholesterol lowering activity of the various fractions of the guggal. *Indian J Exp Biol*, 1973, 11(5): 395–396.

132. Urizar, N.L. et al., A natural product that lowers cholesterol as an antagonist ligand for FXR. *Science*, 2002, 296(5573): 1703–1706.

133. Gopal, K. et al., Clinical trial of ethyl acetate extract of gum gugulu (gugulipid) in primary hyperlipidemia. *J Assoc Physicians India*, 1986, 34(4): 249–251.

134. Agarwal, R.C. et al., Clinical trial of gugulipid—A new hypolipidemic agent of plant origin in primary hyperlipidemia. *Indian J Med Res*, 1986, 84: 626–634.

135. Szapary, P.O. et al., Guggulipid for the treatment of hypercholesterolemia: A randomized controlled trial. *JAMA*, 2003, 290(6): 765–772.

136. US-FDA: United States Food and Drug Administration, FDA final rule for food labelling: Health claims: Soy protein and coronary heart disease (Federal Register 64: 57699–57733), 1999.

137. US-FDA: United States Food and Drug Administration, FDA final rule for food labelling: Health claims: Nuts and heart disease (Federal Register Docket No. 02P-0505), 2003.

138. Gardner, C.D. et al., The effect of a plant-based diet on plasma lipids in hypercholes-terolemic adults: A randomized trial. *Ann Intern Med*, 2005, 142(9): 725–733.
139. Lukaczer, D. et al., Effect of a low glycemic index diet with soy protein and phytoster-ols on CVD risk factors in postmenopausal women. *Nutrition*, 2006, 22(2): 104–113.
140. Jenkins, D.J. et al., Effects of a dietary portfolio of cholesterol-lowering foods vs lovas-tatin on serum lipids and C-reactive protein. *JAMA*, 2003, 290(4): 502–510.
141. Jenkins, D.J. et al., Assessment of the longer-term effects of a dietary portfolio of cho-lesterol-lowering foods in hypercholesterolemia. *Am J Clin Nutr*, 2006, 83(3): 582–591.
142. Kessler, R.C. et al., Long-term trends in the use of complementary and alternative medical therapies in the United States. *Ann Intern Med*, 2001, 135(4): 262–268.
143. Iwamoto, M. et al., Serum lipid profiles in Japanese women and men during consump-tion of walnuts. *Eur J Clin Nutr*, 2002, 56(7): 629–637.
144. Lovejoy, J.C. et al., Effect of diets enriched in almonds on insulin action and serum lipids in adults with normal glucose tolerance or type 2 diabetes. *Am J Clin Nutr*, 2002, 76(5): 1000–1006.
145. Sabate, J. et al., Serum lipid response to the graduated enrichment of a Step I diet with almonds: A randomized feeding trial. *Am J Clin Nutr*, 2003, 77(6): 1379–1384.
146. Tapsell, L.C. et al., Including walnuts in a low-fat/modified-fat diet improves HDL cholesterol-to-total cholesterol ratios in patients with type 2 diabetes. *Diabetes Care*, 2004, 27(12): 2777–2783.
147. Griel, A.E. et al., A macadamia nut-rich diet reduces total and LDL-cholesterol in mildly hypercholesterolemic men and women. *J Nutr*, 2008, 138(4): 761–767.

# 6

# Nutraceuticals and Atherosclerosis
## Human Trials

Lina Badimon, Teresa Padro, and Gemma Vilahur

## Contents

## Key features of coronary atherosclerotic plaque development

The following section briefly overviews the pathogenesis of the atherosclerotic process, which is required to give the proper background to place nutraceutical effects in context.

Atherosclerosis, which encompasses coronary, cerebrovascular, and peripheral arterial disease, is responsible for the majority of cases of cardiovascular disease (CVD) in both developing and developed countries. Atherosclerosis is an inflammatory disease that involves the arterial wall and is characterized by the progressive accumulation of lipids and macrophages/lymphocytes within the intima of large arteries.[1–3] The first step is the internalization of lipids (low-density lipoproteins [LDLs]) in the intima with endothelial activation/dysfunction, which leads to enhance the permeability of the endothelial layer and increase the expression of cytokines/chemokines and adhesion molecules.

These events increase LDL particles accumulation in the extracellular matrix where they aggregate/fuse, are retained by proteoglycans, and become targets for oxidative and enzymatic modifications. In turn, retained pro-atherogenic LDLs enhance selective leukocyte recruitment and attachment to the endothelial layer inducing their transmigration across the endothelium into the intima. While smooth muscle cell numbers decline with the severity of plaque progression, monocytes differentiate into macrophages, leading to foam cell formation.[1,4] Foam cells release growth factors, cytokines, metalloproteinases, and reactive oxygen species all of which perpetuate and amplify the vascular remodeling process that eventually results in a gradual narrowing of the lumen resulting in ischemic events distal to the arterial stenosis. However, these initial fatty streak lesions may also evolve into vulnerable plaques susceptible to rupture or erosion. The mechanisms responsible for determination of the development of vulnerable unstable plaques rather than stable ones is still unknown although there is evidence for the involvement of a number of key factors, including lipid infiltration and aggregation, oxidative stress and formation of oxidized LDL, diabetes, high or fluctuating blood sugar levels and formation of advanced glycation end products, amplification of the inflammatory process, and angiogenesis (intimal neovascularization).[5,6] Plaque disruption and the subsequent exposure of thrombogenic substrates initiate both platelet adhesion and aggregation on the exposed vascular surface and the activation of the clotting cascade leading to the clinical manifestation of the so-called atherothrombotic disease, acute myocardial infarction (MI), and sudden death.[1,7]

## Nutraceuticals and their healthy effect on atherosclerosis

The term nutraceuticals was coined from "nutrition" and "pharmaceutical," initially, to encourage clinical research and trials aimed at examining the true health effects of these substances. However, the common use of this term for marketing purposes has prompted to realize that there still lacks a firm regulatory definition.[8] Concerning atherosclerosis prevention, nutraceuticals include food and/or food-derived substances that exert health benefits and prevent and/or treat the onset of atherosclerosis and the occurrence of cardiovascular events in addition to their basic nutritional value. Indeed, atherosclerotic diseases occur in association with risk factors (e.g., inflammation, endothelial dysfunction, and oxidative stress)[7] that are amenable to prevention or treatment by nutraceutical interventions.

Epidemiological evidence has supported beneficial effects in populations consuming the Mediterranean-type diet supporting the "power" of diets low in saturated fat and rich in fruits and vegetables as well as a moderate wine consumption against the development and progression of CVD. Indeed, many nutrients and phytochemicals in fruits, vegetables, and wine, including fiber, vitamins, minerals, antioxidants, could be independently or jointly responsible for the apparent reduction in CVD risk. The advances in the knowledge of both the disease processes and healthy dietary components have provided new avenues to develop pharmaceutical and/or dietary strategies to halt the development of vascular disease. In this context, the expansion and growing popularity of nutraceuticals aimed at promoting heart health can be viewed as an example.

## Healthy fats: Monounsaturated and polyunsaturated fatty acids

Current recommendations on dietary fat emphasize quality rather than quantity.[9] Metabolic studies have long established that the dietary fatty acid composition, but not the total amount of fat, predicts serum cholesterol levels. Fatty acids can be divided into four general categories: saturated (SFA), monounsaturated (MUFA), polyunsaturated (PUFA), and *trans*-fatty acids (TFA). SFA and TFA are related with elevated cardiovascular risk whereas MUFA (ω-9) and PUFA (ω-3 and ω-6) are associated with a decreased risk of coronary heart disease (CHD).[10]

### Monounsaturated fatty acids

Ecological studies have suggested an inverse association between MUFA intake and total mortality, as well as with CHD death.[11] To this respect, the Seven Countries Study yielded the first convincing epidemiological evidence that mortality from CHD was particularly low in Mediterranean countries where olive oil, which is rich in MUFA, is the main dietary source of fat.[12] However, olive oil besides having a high level of MUFA (>75% oleic acid, cis18:1, n-9) contains other minor components with biological properties that make it more than a MUFA.[13] In fact, although data concerning olive oil intake and primary end points for CVD are scarce, olive oil has shown to reduce major risk factors for CVD (lipoprotein profile, blood pressure, glucose metabolism, and antithrombotic profile) as well as to positively modulate endothelial function, inflammation, and oxidative stress.[13] We have recently reported, in asymptomatic high cardiovascular risk subjects, that intake of traditional Mediterranean diet supplemented with virgin olive oil actively modulates the expression of key genes involved in vascular inflammation, foam cell formation, and thrombosis toward an anti-atherothrombotic profile.[14] Nevertheless, to be mentioned that, whereas in vitro and in vivo studies have supported that the use of olive oil reduces the development of atherosclerosis,[15] the use of hydroxytyrosol, the major antioxidant phenol contained in olive oil, has sometimes shown to promote atherosclerotic lesion development.[16] These data support the concept that phenolic component–enriched products, out of the original matrix, might not only be useless but also be harmful, and suggest that the formulation of functional foods should approximate—as much as possible—the natural environment in which active molecules are found.

The protective effect of MUFA against CHD was also supported by a regression analysis of data from the Nurses' Health Study of 80,082 women followed up for >14 years.[17] In contrast, however, two prospective studies found a positive association between intake of MUFA and CHD risk.[18,19] Such discrepancy might be explained because they did not adjust their results for important confounding variables as other types of fat simultaneously present on the diet (e.g., the beef, a major source of MUFA and SFA in the American diet). In fact, other epidemiological studies that have controlled for a number of potentially confounding variables also have reported protective effects of MUFA against CHD.[20] Moreover, results of a meta-analysis of 60 controlled trials published between 1970 and 1998 supported the beneficial effects of a high MUFA diet on serum lipid levels and LDL oxidation.[21-24] Besides, we have also reported in healthy subjects that MUFA-enriched diets prevent smooth muscle cells DNA synthesis likely reducing plaque-related smooth muscle cell proliferation.[25] Additionally, several intervention studies in human have supported that intake of MUFA-rich diet has additional non-lipid-related advantages, including favorable effects in preventing arrhythmias, lowering heart rate, blood pressure, platelets activity, coagulation, and fibrinolysis.[26-28]

## Polyunsaturated fatty acids

α-Linolenic acid (ALA) and linoleic acid (LA) belong to the ω-3 and ω-6 series of PUFA, respectively. LA and ALA are essential fatty acids that can be converted into long-chain PUFAs, such as arachidonic acid and eicosapentaenoic acid (EPA)/docosahexaenoic acid (DHA), respectively.[29] Food sources of ω-3 and ω-6 are found in seafood and fatty fish, most vegetable oils, cereals, and walnuts. Since the early studies in Greenland Eskimos, dietary intake of ω-3 fatty acids, introduced with fish, fish-derived products, or supplements, has been consistently associated with cardiovascular protection.[30] Indeed, ω-3 fatty acids have shown great promise in primary and secondary prevention of CVD[31] with estimated benefits that exceed the potential vascular risk associated with methylmercury, dioxins, polychlorinated biphenyls, and other environmental contaminants that frequently pollute marine food products.[32,33]

The two major ω-3 fatty acids that have been associated with cardiovascular benefit are EPA and DHA. Generally, very little ALA is converted to EPA, and even less to DHA, and therefore direct intake of the latter two is optimal. Data on the effects of ALA on CVD outcomes are limited. In cohort studies, low ALA intake has been associated with risk of fatal CHD[34,35] and sudden cardiac death.[36] In a recent study in a Costa Rica population (1819 cases with a first nonfatal acute MI [AMI] and 1819 population-based controls), ALA intake and plasma levels predicted a better prognosis, independent of fish and EPA/EDA intake, in the post-AMI population.[37]

Up to now, more than 25 published trials have evaluated the risk of CHD as a function of ω-3 PUFA plasma levels. Taken together, these studies showed that intake of fish oil is associated with CHD risk reduction.[38] A recent meta-analysis with pooled data from 19 observational studies[31] supported the notion that fish consumption is an important component of lifestyle modification for the prevention of total and fatal CHD. In addition, the authors of a meta-analysis of 11 prospective cohort studies (encompassing 222,364 persons with an average of 11.8 years of follow-up) concluded that each 20 g/day increase in fish intake is associated with a 7% lower risk of CHD mortality.[30] In support to these studies, the most compelling evidence for cardiovascular benefits of ω-3 PUFA comes from three large randomized trials (DART,[39] GISSI-Prevention trial,[40] JELIS[41]) that included 2033 men with recent MI, 11,323 post-MI patients, and 98,645 patients in primary (14,981) or secondary (83,664) prevention, respectively. In the DART trial, patients were randomly assigned to different dietary advice groups, whereas in the GISSI-Prevention trial patients were randomly assigned to receive an EPA plus DHA supplement or placebo on an open-label basis. These studies have documented that ω-3 PUFA lower cardiovascular risk in primary and more especially in secondary prevention of CHD. Thus, the investigators of the DART study concluded that a modest intake of fatty fish (2 or 3 portions per week) may reduce mortality in men who have recovered from MI. CHD mortality was lowered by 62% in the subgroup receiving a fish oil capsule containing 450 mg of EPA plus DHA daily.

A recent systematic review of the effects of marine ω-3 fatty acids on cardiovascular events from randomized controlled trials and clinical trials has indicated that marine ω-3, when administered as food or in supplements for at least 6 months, reduces cardiovascular events by 10%, cardiac death by 9%, and coronary events by 18%, while showing a trend for a lower total mortality (5%, $p=0.13$), especially in persons with high cardiovascular risk. These results, along with the existing evidence on the myriad of physiological effects of ω-3 on human health (coagulation, heart rate, heart rhythm, blood lipids, etc.), reinforce the American Heart Association recommendations for the intake of ω-3 in the prevention of CVD, especially in persons with high cardiovascular risk and secondary prevention.

The exact mechanisms by which ω-3 performs such atheroprotective functions are still in debate, but the main mechanisms proposed include plaque stabilization,[38,42] anti inflammatory,[43,44] blood pressure,[45] heart failure,[46] antiarrhythmic properties, or improving the lipid profile (mainly lowering plasma triglyceride levels).[28,47–49] The latter is accomplished by decreasing the production of hepatic triglycerides and increasing the clearance of plasma triglycerides, a mechanism probed for EPA and DHA not for ALA.[38]

Other studies, however, have not shown favorable results. Burr et al.,[50] based on a trial of 3114 men with angina, suggested that patients treated with fish oil capsules had a higher risk of sudden cardiac death than untreated control subjects.[50] A Norwegian study by Nilsen et al.[51] did not show a benefit of ω-3 PUFA supplementation in post-MI patients and in the recently presented OMEGA trial, consisting of 3851 patients with AMI from 104 centers in Germany, EPA/DHA did not show any further benefit on primary or secondary prevention when given together with other specific treatments (e.g., aspirin, clopidogrel, statins, β-blockers, and angiotensin-converting enzyme inhibitors).[52]

# Cholesterol-lowering natural agents: Sterols/stanols and red yeast rice

Phytosterols are plant-derived sterols or stanols (saturated sterols) with cholesterol-like chemical structure. Plant sterols/stanols reduce cholesterol absorption in the intestinal gut thereby reducing plasma LDL concentrations. Plant sterols/stanols are abundant in vegetable oils and olive oil, but also in fruits and nuts. However, recent advancements in food technology have seen the emergence of food products such as margarine, milk, yoghurt, and cereal products being enriched with plant sterols/stanols and promoted as a food which can help lower serum cholesterol. A meta-analysis of 41 trials showed that intake of 2 g/day of stanols or sterols reduced LDL concentrations by about 10%–11%[53] and that there were little additional effects at doses higher than 2.5 g/day. Furthermore, high-density lipoproteins (HDL) and/or very low-density lipoproteins (VLDL) were generally not affected by stanols/sterols intake. Yet, effects of sterols/stanols on LDLs have been found to be additive to diets and/or cholesterol-lowering drugs. As such, eating foods low in saturated fat and cholesterol and high in stanols/sterols has shown to further reduce LDL by 20% and adding sterols/stanols to statin medication seems more effective than doubling the statin dose.[54] On the other hand, similar efficacy has been observed between plant sterols and plant stanols when they are esterified, which is the form added to foods.[55] However, the food form may substantially affect LDL reduction. Serum LDL-cholesterol was significantly lowered when plant sterols were added to milk (15.9%) and yoghurt (8.6%), but significantly less when added to bread (6.5%) and cereal (5.4%).[56] Nevertheless, routine prescription of plant sterols/stanols has shown to be an effective strategy in the management of hypercholesterolemic patients in the clinical setting. In fact, the National Cholesterol Education Program Adult Treatment Panel III (NCEP ATP III) recommends, since 2001, phytosterol-enriched functional foods as part of an optimal dietetic prevention strategy in primary and secondary prevention of CVDs.[57] Moreover, the American Heart Association[58] and the *European Current Dietary Guidelines*[59] support plant sterols as a therapeutic option for individuals with elevated cholesterol levels. Up to now, however, it is unclear whether phytosterols have a positive effect on atherosclerosis progression and ultimate CVD.[53,60] Recent released guidelines have been more critical with food supplementation with phytosterols and have drawn attention to significant safety issues based on large epidemiological studies.[61] As such, results of the PROCAM study showed that patients afflicted with MI or sudden cardiac death had increased plant sterol concentrations.[62] Upper normal levels of plant sterols were also associated with a threefold increase of risk for coronary events among men in the highest tertile of coronary risk according to the PROCAM algorithm. Similar data are available for the plant sterol campesterol from the MONICA/KORA study. In this prospective study, campesterol correlated directly with the incidence of AMI.[63]

Another natural compound capable of reducing cholesterol levels is red yeast rice (*Monascus purpureus*). This fermented rice contains numerous monacolins which are naturally occurring HMG-CoA reductase inhibitors. Indeed, numerous studies have suggested a beneficial lipid-lowering effect from commercial preparations of this traditional supplement. For instance, a meta-analysis involving 9625 patients in 93 randomized trials, three different commercial preparations of red yeast rice produced a mean reduction in total cholesterol, LDL-cholesterol, triglyceride and a mean rise in HDL-cholesterol.[64] More recently, a double-blind, multicenter trial in China has demonstrated, in 4870 patients with a previous MI and high total cholesterol levels after 4.5 years follow-up, that Xuezhikang administration (a commercial red yeast rice preparation) at a dose of 0.6 g twice daily was associated with a reduction in the incidence of major coronary events, including nonfatal MI and death from CHD compared to placebo (5.7% vs. 10.4%).[65] Subsequent subgroup analyses of this trial have further supported these observations by confirming a reduction in cardiovascular outcomes amongst diabetics and in the elderly.[66] In addition, Xuezhikang has also shown, in small clinical trials, to effectively improve endothelial function in patients with CAD.[67] Finally, an extract of red yeast rice has recently demonstrated to be tolerated as well as pravastatin and achieved a comparable reduction of LDL-cholesterol in a population previously intolerant to statins.[68]

## Cereal grains and dietary fiber

High intake of dietary fiber (e.g., nondigestible polysaccharides [glucomannan], naturally occurring resistant starch and oligosaccharides, and lignins in plants) is associated with a reduced cardiovascular risk. Observational studies have consistently shown that subjects consuming relatively large amounts of dietary fiber have significantly lower rates of CHD,[69] stroke,[70] and peripheral vascular disease.[71] In a pooled analysis of 10 prospective cohorts,[72] each 10 g/day increment of energy-adjusted total dietary fiber was associated with a 14% decrease in risk of coronary events and a 27% decrease in risk of coronary death. However, in the Health Professionals Follow-up Study, only cereal fiber, not fruit or vegetable fiber, was inversely associated with risk of total stroke.[69] Also, Mozaffarian et al.,[73] based on the results of a prospective cohort study conducted over 8.6 years in a population of 3588 individuals (men/women) aged 65 years or older at baseline, reported an inverse association of cereal fiber consumption later in life with risk of total stroke and ischemic stroke and a trend toward lower risk of ischemic heart disease. Differing from the earlier results, the advice to increase cereal fiber intake did not affect recurrent MI or mortality (coronary or all-cause mortality) in the DART study.[39] Similarly, a pooled analysis of different studies suggests that cereal fiber by itself has low or no significant influence in CHD risk whereas a stronger inverse association was reported between whole-grain intake and risk for CHD.[74] However, based on the results of the prospective Iowa Women's Health

Study,[75] Jacobs et al. demonstrated that a similar amount of total cereal fiber had different associations with total mortality, depending on whether the fiber came from foods that contained primarily whole grain or refined grain. These results highlight the "whole-grain hypothesis,"[74] which argues that health benefits stem from more than just the fiber. The whole grain is nutritionally more important because it delivers a whole package of phytoprotective substances that might work synergistically to reduce cardiovascular risk.

Observational studies have consistently shown that major cardiovascular risk factors such as dyslipemia, hypertension, diabetes, and obesity are also less common in individuals with highest levels of fiber consumption, as recently reviewed by different authors.[76,77] To this respect, a meta-analysis evaluating 14 randomized clinical trials including a total of 531 patients concluded that glucomannan significantly reduces total cholesterol levels, LDL-cholesterol, and triglycerides when compared to the placebo. However, it exerted no effect on HDL-cholesterol and blood pressure.[78] On the other hand, the PREDIMED feeding trial substudy[79] has shown that the increase of dietary fiber intake associates with significant reductions in body weight, waist circumference, blood pressure, fasting glucose and a increase in HDL-cholesterol in a high-risk cohort of subjects with either type 2 diabetes or at least three CHD risk factors (current smoking habit, hypertension, HDL-cholesterol <40 mg/dL, body mass index >25 kg/m$^2$).

## Garlic

Several clinical reports, including meta-analyses, have revealed cholesterol-lowering effects of garlic supplementation in humans.[80,81] The mechanism of action is the inhibition of the 3-hydroxy-3-methyl-coenzyme A reductase activity with an additive effect on the statins.[82] Such reports have additionally impacted public awareness on the potential cardiovascular benefits of garlic. Garlic and garlic extracts have been postulated to impart cardiovascular benefits through multiple mechanisms. For example, recent studies of garlic extracts have shown it as a modulator of multiple cardiovascular risk factors, in addition to LDL lowering, including lowering blood pressure, reducing platelet aggregation and adhesion, preventing LDL oxidation, smoking-caused oxidative damage, and also directly suppressing atherosclerosis.[83-87] Moreover, in a randomized clinical trial in intermediate risk patients, garlic extract with B vitamins, folate, and L-arginine has shown to retard atherosclerosis progression.[88]

## Polyphenols

Several studies, although not all, have found an inverse association between polyphenol consumption and CVD mortality (Table 6.1). In fact, such beneficial effects may partly help to explain the protective CVD effects achieved by foods and beverages containing polyphenols.[89]

**Table 6.1** Clinical Studies
Linking Polyphenol Consumption
and CVD Outcome

| Lack Protection | Beneficial Effects |
| --- | --- |
| Hirvonen et al.[125] | Sesso et al.[136] |
| Hertog et al.[126] | Lin et al.[137] |
| Arts et al.[127] | Hertog et al.[138] |
| Geleijnse et al.[128] | |
| Knekt et al.[129] | |
| Ness and Powles[130] | |
| Hertog et al.[131] | |
| Hertog et al.[132] | |
| Rimm et al.[133] | |
| Keli et al.[134] | |
| Yochum et al.[135] | |

The beneficial effects derived from polyphenols appear to be mediated via a plethora of biochemical pathways and signaling mechanisms acting either independently or synergistically. Indeed, polyphenols have shown in in vivo studies to exert anti-atherosclerotic effects in the early stages of atherosclerosis development (e.g., decrease LDL oxidation), improve endothelial function, and increase nitric oxide release (potent vasodilator), modulate inflammation and lipid metabolism (i.e., hypolipidemic effect), improve antioxidant status, and protect against atherothrombotic episodes including myocardial ischemia and platelet aggregation. In addition, human clinical trials, albeit few and small, have also supported a benefit of polyphenols consumption on cardiovascular risk factors. For instance, patients suffering from coronary artery disease (CAD) have shown an improvement in endothelial function and on the coronary microcirculation.[90,91] Similarly, red wine consumption has shown to prevent the acute impairment of endothelial function that occurs following cigarette smoking or intake of high-fat meal[92,93] and to modulate monocyte migration in healthy subjects.[94] Indeed, it must be considered that some of the protective effects of wine appeared to be linked to ethanol per se. However, several studies have indicated, when comparing different beverage types, that wine seems to exert more protective effects than other forms of alcohols supporting a protective role for polyphenolic compounds. As to other polyphenol-related nonalcoholic drinks, it has been recently reported in patients with CAD and carotid artery stenosis that pomegranate juice consumption for 18 and 36 months, respectively, slowed atherosclerosis progression, assessed by carotid–intima thickness, likely through the potent antioxidant characteristics of pomegranate polyphenols.[95,96]

Tea, although contains few conventional nutrients, presents a high amount of polyphenols including catechins, flavonols, theaflavin, thearubigin.[97] Studies have shown that these substances delivered as an extract from decaffeinated tea (455 mg/day) may reduce pro-atherosclerotic factors in patients undergoing hemodialysis.[98] Cross-sectional, randomized controlled, and prospective population studies have shown that tea intake and/or increased dietary tea flavonoids reduced the risk of CVD, with consistent data demonstrating enhancement of nitric oxide production and concomitant improvement of endothelial function as well as reduction in total cholesterol levels and LDL-cholesterol.[99] In one particular population-based prospective cohort study (Ohsaki study),[100] 40,530 persons from Northern Japan were enrolled and data demonstrated an inverse relationship between CVD mortality and tea consumption.[101,102] Ras et al.[102] showed by meta-analysis of nine studies that tea consumption was directly associated with increased (40%) flow-mediated dilation of the brachial artery (a measurement of endothelial function), and Tinahones et al.[103] using the same methodology also showed concomitant reduction in oxLDL levels in patients taking green tea extract supplements demonstrating potential protection against CVD. In vivo studies have shown that green tea extracts appear to have insulin-like activities, with epigallocatechin gallate inhibiting intestinal glucose uptake via the sodium-dependent glucose transporter (SGLT1).[104] In vitro, tea extracts were shown to inhibit endothelial cell plasminogen activator inhibitor 1 (PAI-1; inhibits fibrinolysis and increases the risk of a deficient thrombolysis) through a pathway involving PI3K/Akt, again suggesting that they may contribute to cardiovascular protection.[105] Other studies have demonstrated that green tea extracts reduced plasma triglycerides, insulin resistance, and oxidative stress in insulin-resistant Wistar rats which suggests a potential protective role against diabetes-associated premature development of CVD in patients.[106]

## Antioxidant–vitamin supplementation

As stated earlier, according to the oxidative modification hypothesis in which reactive oxygen species and free radicals play a major role in the pathophysiology of atherosclerosis, supplementation with antioxidants was expected to protect against atherosclerosis.[22] In fact, dietary antioxidants such as vitamin E (α-tocopherol), β-carotene (provitamin A), vitamin A, and vitamin C (ascorbic acid) have been proven to protect against the development and progression of atherosclerosis in experimental models.[107–109] Observational prospective human cohort studies have also shown that a high dietary intake of fruits and vegetables is associated with a reduction in the incidence of CHD,[110] stroke,[23] and cardiovascular mortality in general.[111] Moreover, epidemiological studies have reported that high dietary intake of foods rich in vitamin E,[112] vitamin C,[113] and β-carotene[114] has been inversely associated with the incidence of CAD.

Different results are, however, obtained with vitamin and antioxidant supplementation. Evidence from the US Preventive Services Task Force has reported that although observational studies have shown an inverse correlation between dietary intake of vitamin E and the incidence of CAD, no such protective effects are reported for vitamin C and β-carotene supplementation.[115] In fact, a potential explanation for such vitamin E–specific protective effects may derive from its fat-soluble nature as well as its integration within LDL particles. As a consequence, there has been a major emphasis in conducting vitamin E supplementation randomized controlled trials attempting to confirm their protective role in both primary and secondary prevention. Three large primary prevention trials[116–118] and four large-scale secondary prevention trials[40,119–121] with vitamin E supplementation including 43,169 and 16,993 patients, respectively, have been performed until now. However, the majority of these prospective randomized controlled clinical trials with vitamin supplements have been disappointing.[122] Indeed, only one trial has shown a reduction in MI and cardiac events whereas all the others have shown no effect or detrimental effects. Likewise, controversy exists regarding the potential benefit associated with antioxidant supplementation in general. A total of 22 trials ($n = 134,590$ subjects)—of which 6 are primary, 12 secondary, and 3 both primary and secondary prevention—assessing the effects of antioxidant supplementation on the risk of CVD, coronary restenosis, and the progression of atherosclerotic lesions have been reported.[123] As seen in Table 6.2, the majority of the results (14 trials) have been disappointing failing to demonstrate any significant benefit with vitamin C, vitamin E, and β-carotene supplementation. Furthermore, within five trials such antioxidant supplementation was

**Table 6.2** Randomized Antioxidant Supplementation Clinical Trials

| Lack Protection | | Beneficial Effects | | Detrimental Effect (Increased Risk for All-Cause Mortality and/or Fatal Coronary Heart Disease) | |
|---|---|---|---|---|---|
| Primary Prevention | Secondary Prevention | Primary Prevention | Secondary Prevention | Primary Prevention | Secondary Prevention |
| HOPE[117] | HOPE[117] | ASAP[29] | SPACE[146] | ATBC[116] | HPS[144] |
| ATBC[116] | ATBC[116] | | IEISS[147] | CARET[124] | ATBC[116] |
| PPP[118] | GISSI[40] | | CHAOS | | HATS[142] |
| PHS[139] | MICRO-HOPE[48] | | CLAS[148] | | WAVE[143] |
| VEAPS[49] | HATS[142] | | PART[149] | | |
| CARET[124] | WAVE[143] | | ASAP[29] | | |
| SCPS[140] | HPS[144] | | Fang et al.[150] | | |
| SU.VI.MAX[141] | WACS[145] | | MVP[151] | | |

*Note:* Clinical trials based on antioxidant supplementation that have reported lack of protection, benefit or detrimental effects on the risk of suffering cardiovascular disease, coronary restenosis, and progression of atherosclerotic lesions.

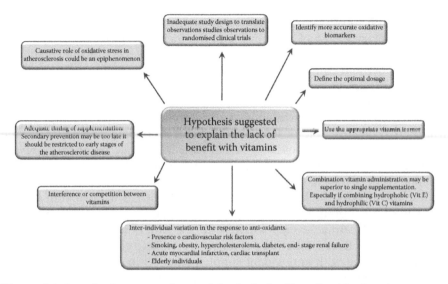

**Figure 6.1** *Hypothesis suggested to explain the lack of benefit with vitamins.*

associated with increased all-cause mortality, and two have shown higher risk of fatal CHD (ATBC and CARET).[116,124] The controversial results reported in clinical trials investigating the role of antioxidants in CVD may be attributed to several factors (Figure 6.1). Yet, taken all together, at present, the scientific data do not justify the use of antioxidant vitamin supplements for CVD risk reduction.

## Acknowledgments

This work was supported by PNS2010-16549 (to LB) from the Spanish Ministry of Science, CIBERobn CB06/03 and FIS PI101115 (to TP). We thank Fundación Jesús Serra-Fundación Investigación Cardiovascular (FIC), Barcelona, for their continuous support. G.V. is recipient of a grant from the Spanish Ministry of Science and Innovation (RyC-2009-5495; MICINN).

## References

1. Badimon, L., G. Vilahur, and T. Padro. 2009. Lipoproteins, platelets and atherothrombosis. *Rev Esp Cardiol* 62:1161–1178.
2. Badimon, L., R.F. Storey, and G. Vilahur. 2011. Update on lipids, inflammation and atherothrombosis. *Thromb Haemost* 105(Suppl 1):S34–S42.
3. Badimon, L., J. Martinez-Gonzalez, V. Llorente-Cortes et al. 2006. Cell biology and lipoproteins in atherosclerosis. *Curr Mol Med* 6:439–456.
4. Padro, T., R. Lugano, M. Garcia-Arguinzonis et al. 2012. LDL-induced impairment of human vascular smooth muscle cells repair function is reversed by HMG-CoA reductase inhibition. *PLoS One* 7:e38935.

5. Slevin, M., J. Krupinski, and L. Badimon. 2009. Controlling the angiogenic switch in developing atherosclerotic plaques: Possible targets for therapeutic intervention. *J Angiogenes Res* 1:4.

6. Slevin, M., A.B. Elasbali, M. Miguel Turu et al. 2006. Identification of differential protein expression associated with development of unstable human carotid plaques. *Am J Pathol* 168:1004–1021.

7. Badimon, L. and G. Vilahur. 2012. LDL-cholesterol versus HDL-cholesterol in the atherosclerotic plaque: Inflammatory resolution versus thrombotic chaos. *Ann N Y Acad Sci* 1254:18–32.

8. Zeisel, S.H. 1999. Regulation of "nutraceuticals." *Science* 285:1853–1855.

9. Hu, F.B., J.E. Manson, and W.C. Willett. 2001. Types of dietary fat and risk of coronary heart disease: A critical review. *J Am Coll Nutr* 20:5–19.

10. Erkkila, A.T., U.S. Schwab, V.D. de Mello et al. 2008. Effects of fatty and lean fish intake on blood pressure in subjects with coronary heart disease using multiple medications. *Eur J Nutr* 47:319–328.

11. Jacobs, D.R., Jr., P.G. McGovern, and H. Blackburn. 1992. The US decline in stroke mortality: What does ecological analysis tell us? *Am J Public Health* 82:1596–1599.

12. Keys, A., A. Menotti, M.J. Karvonen et al. 1986. The diet and 15-year death rate in the seven countries study. *Am J Epidemiol* 124:903–915.

13. Perez-Jimenez, F., G. Alvarez de Cienfuegos, L. Badimon et al. 2005. International conference on the healthy effect of virgin olive oil. *Eur J Clin Invest* 35:421–424.

14. Llorente-Cortes, V., R. Estruch, M.P. Mena et al. 2010. Effect of Mediterranean diet on the expression of pro-atherogenic genes in a population at high cardiovascular risk. *Atherosclerosis* 208:442–450.

15. Mangiapane, E.H., M.A. McAteer, G.M. Benson et al. 1999. Modulation of the regression of atherosclerosis in the hamster by dietary lipids: Comparison of coconut oil and olive oil. *Br J Nutr* 82:401–409.

16. Acin, S., M.A. Navarro, J.M. Arbones-Mainar et al. 2006. Hydroxytyrosol administration enhances atherosclerotic lesion development in apo E deficient mice. *J Biochem* 140:383–391.

17. Hu, F.B., M.J. Stampfer, J.E. Manson et al. 1997. Dietary fat intake and the risk of coronary heart disease in women. *N Engl J Med* 337:1491–1499.

18. Esrey, K.L., L. Joseph, and S.A. Grover. 1996. Relationship between dietary intake and coronary heart disease mortality: Lipid research clinics prevalence follow-up study. *J Clin Epidemiol* 49:211–216.

19. Posner, B.M., J.L. Cobb, A.J. Belanger et al. 1991. Dietary lipid predictors of coronary heart disease in men. The Framingham Study. *Arch Intern Med* 151:1181–1187.

20. Artaud-Wild, S.M., S.L. Connor, G. Sexton et al. 1993. Differences in coronary mortality can be explained by differences in cholesterol and saturated fat intakes in 40 countries but not in France and Finland. A paradox. *Circulation* 88:2771–2779.

21. Mensink, R.P., P.L. Zock, A.D. Kester et al. 2003. Effects of dietary fatty acids and carbohydrates on the ratio of serum total to HDL cholesterol and on serum lipids and apolipoproteins: A meta-analysis of 60 controlled trials. *Am J Clin Nutr* 77:1146–1155.

22. Diaz, M.N., B. Frei, J.A. Vita et al. 1997. Antioxidants and atherosclerotic heart disease. *N Engl J Med* 337:408–416.

23. He, F.J., C.A. Nowson, and G.A. MacGregor. 2006. Fruit and vegetable consumption and stroke: Meta-analysis of cohort studies. *Lancet* 367:320–326.

24. Moreno, J.A., J. Lopez-Miranda, P. Perez-Martinez et al. 2008. A monounsaturated fatty acid-rich diet reduces macrophage uptake of plasma oxidised low-density lipoprotein in healthy young men. *Br J Nutr* 100:569–575.

25. Mata, P., O. Varela, R. Alonso et al. 1997. Monounsaturated and polyunsaturated n-6 fatty acid-enriched diets modify LDL oxidation and decrease human coronary smooth muscle cell DNA synthesis. *Arterioscler Thromb Vasc Biol* 17:2088–2095.

26. Covas, M.I. 2007. Olive oil and the cardiovascular system. *Pharmacol Res* 55:175–186.

27. Lavie, C.J., R.V. Milani, M.R. Mehra et al. 2009. Omega-3 polyunsaturated fatty acids and cardiovascular diseases. *J Am Coll Cardiol* 54:585–594.
28. Breslow, J.L. 2006. n-3 fatty acids and cardiovascular disease. *Am J Clin Nutr* 83:1477S–1482S.
29. Salonen, J.T., K. Nyyssonen, R. Salonen et al. 2000. Antioxidant Supplementation in Atherosclerosis Prevention (ASAP) study: A randomized trial of the effect of vitamins E and C on 3-year progression of carotid atherosclerosis. *J Intern Med* 248:377–386.
30. He, K., Y. Song, M.L. Daviglus et al. 2004. Accumulated evidence on fish consumption and coronary heart disease mortality: A meta-analysis of cohort studies. *Circulation* 109:2705–2711.
31. Whelton, S.P., J. He, P.K. Whelton et al. 2004. Meta-analysis of observational studies on fish intake and coronary heart disease. *Am J Cardiol* 93:1119–1123.
32. Wang, C., W.S. Harris, M. Chung et al. 2006. n-3 Fatty acids from fish or fish-oil supplements, but not alpha-linolenic acid, benefit cardiovascular disease outcomes in primary- and secondary-prevention studies: A systematic review. *Am J Clin Nutr* 84:5–17.
33. Hooper, L., R.L. Thompson, R.A. Harrison et al. 2006. Risks and benefits of omega 3 fats for mortality, cardiovascular disease, and cancer: Systematic review. *BMJ* 332:752–760.
34. Pietinen, P., A. Ascherio, P. Korhonen et al. 1997. Intake of fatty acids and risk of coronary heart disease in a cohort of Finnish men. The Alpha-Tocopherol, Beta-Carotene Cancer Prevention Study. *Am J Epidemiol* 145:876–887.
35. Hu, F.B., M.J. Stampfer, J.E. Manson et al. 1999. Dietary intake of alpha-linolenic acid and risk of fatal ischemic heart disease among women. *Am J Clin Nutr* 69:890–897.
36. Albert, C.M., K. Oh, W. Whang et al. 2005. Dietary alpha-linolenic acid intake and risk of sudden cardiac death and coronary heart disease. *Circulation* 112:3232–3238.
37. Campos, H., A. Baylin, and W.C. Willett. 2008. Alpha-linolenic acid and risk of nonfatal acute myocardial infarction. *Circulation* 118:339–345.
38. Harris, W.S., M. Miller, A.P. Tighe et al. 2008. Omega-3 fatty acids and coronary heart disease risk: Clinical and mechanistic perspectives. *Atherosclerosis* 197:12–24.
39. Burr, M.L., A.M. Fehily, J.F. Gilbert et al. 1989. Effects of changes in fat, fish, and fibre intakes on death and myocardial reinfarction: Diet and reinfarction trial (DART). *Lancet* 2:757–761.
40. Gruppo Italiano per lo Studio della Sopravvivenza nell'Infarto miocardico. 1999. Dietary supplementation with n-3 polyunsaturated fatty acids and vitamin E after myocardial infarction: Results of the GISSI-Prevention trial. *Lancet* 354:447–455.
41. Yokoyama, M., H. Origasa, M. Matsuzaki et al. 2007. Effects of eicosapentaenoic acid on major coronary events in hypercholesterolaemic patients (JELIS): A randomised open-label, blinded endpoint analysis. *Lancet* 369:1090–1098.
42. Thies, F., J.M. Garry, P. Yaqoob et al. 2003. Association of n-3 polyunsaturated fatty acids with stability of atherosclerotic plaques: A randomised controlled trial. *Lancet* 361:477–485.
43. Massaro, M., E. Scoditti, M.A. Carluccio et al. 2010. Nutraceuticals and prevention of atherosclerosis: Focus on omega-3 polyunsaturated fatty acids and Mediterranean diet polyphenols. *Cardiovasc Ther* 28:e13–e19.
44. Serhan, C.N., N. Chiang, and T.E. Van Dyke. 2008. Resolving inflammation: Dual anti-inflammatory and pro-resolution lipid mediators. *Nat Rev Immunol* 8:349–361.
45. Morris, M.C., F. Sacks, and B. Rosner. 1993. Does fish oil lower blood pressure? A meta-analysis of controlled trials. *Circulation* 88:523–533.
46. Yamagishi, K., J.A. Nettleton, and A.R. Folsom. 2008. Plasma fatty acid composition and incident heart failure in middle-aged adults: The Atherosclerosis Risk in Communities (ARIC) Study. *Am Heart J* 156:965–974.
47. Mozaffarian, D., B.M. Psaty, E.B. Rimm et al. 2004. Fish intake and risk of incident atrial fibrillation. *Circulation* 110:368–373.

48. Lonn, E., S. Yusuf, B. Hoogwerf et al. 2002. Effects of vitamin E on cardiovascular and microvascular outcomes in high-risk patients with diabetes: Results of the HOPE study and MICRO-HOPE substudy. *Diabetes Care* 25:1919–1927.

49. Hodis, H.N., W.J. Mack, L. LaBree et al. 2002. Alpha-tocopherol supplementation in healthy individuals reduces low-density lipoprotein oxidation but not atherosclerosis: The Vitamin E Atherosclerosis Prevention Study (VEAPS). *Circulation* 106:1453–1459.

50. Burr, M.L., P.A. Ashfield-Watt, F.D. Dunstan et al. 2003. Lack of benefit of dietary advice to men with angina: Results of a controlled trial. *Eur J Clin Nutr* 57:193–200.

51. Nilsen, D.W., G. Albrektsen, K. Landmark et al. 2001. Effects of a high-dose concentrate of n-3 fatty acids or corn oil introduced early after an acute myocardial infarction on serum triacylglycerol and HDL cholesterol. *Am J Clin Nutr* 74:50–56.

52. Rauch, B., R. Schiele, S. Schneider et al. 2006. Highly purified omega-3 fatty acids for secondary prevention of sudden cardiac death after myocardial infarction—Aims and methods of the OMEGA-study. *Cardiovasc Drugs Ther* 20:365–75.

53. Katan, M.B., S.M. Grundy, P. Jones et al. 2003. Efficacy and safety of plant stanols and sterols in the management of blood cholesterol levels. *Mayo Clin Proc* 78:965–978.

54. Normen, L., D. Holmes, and J. Frohlich. 2005. Plant sterols and their role in combined use with statins for lipid lowering. *Curr Opin Investig Drugs* 6:307–316.

55. Hallikainen, M.A., E.S. Sarkkinen, H. Gylling et al. 2000. Comparison of the effects of plant sterol ester and plant stanol ester-enriched margarines in lowering serum cholesterol concentrations in hypercholesterolaemic subjects on a low-fat diet. *Eur J Clin Nutr* 54:715–725.

56. Clifton, P.M., M. Noakes, D. Sullivan et al. 2004. Cholesterol-lowering effects of plant sterol esters differ in milk, yoghurt, bread and cereal. *Eur J Clin Nutr* 58:503–509.

57. Expert Panel on Detection, Evaluation, and Treatment of High Blood Cholesterol in Adults. 2001. Executive Summary of The Third Report of The National Cholesterol Education Program (NCEP). *JAMA* 285:2486–2497.

58. Lichtenstein, A.H., L.J. Appel, M. Brands et al. 2006. Summary of American Heart Association Diet and Lifestyle Recommendations revision 2006. *Arterioscler Thromb Vasc Biol* 26:2186–2191.

59. European Commission. 2002. General view of the Scientific Committee on Food on the long-term effects of the intake of elevated levels of phytosterols from multiple dietary sources, with particular attention to the effects of beta-carotene.

60. Weingartner, O., M. Bohm, and U. Laufs. 2008. Plant sterols as dietary supplements for the prevention of cardiovascular diseases. *Dtsch Med Wochenschr* 133:1201–1204.

61. Weingartner, O., M. Bohm, and U. Laufs. 2009. Controversial role of plant sterol esters in the management of hypercholesterolaemia. *Eur Heart J* 30:404–409.

62. Assmann, G., P. Cullen, J. Erbey et al. 2006. Plasma sitosterol elevations are associated with an increased incidence of coronary events in men: Results of a nested case–control analysis of the Prospective Cardiovascular Munster (PROCAM) study. *Nutr Metab Cardiovasc Dis* 16:13–21.

63. Thiery, J., U. Ceglarek, G. Fiedler et al. 2006. Abstract 4099: Elevated campesterol serum levels—A significant predictor of incident myocardial infarction: Results of the population-based MONICA/KORA follow-up study 1994–2005. *Circulation* 114:II_884.

64. Liu, J., J. Zhang, Y. Shi et al. 2006. Chinese red yeast rice (*Monascus purpureus*) for primary hyperlipidemia: A meta-analysis of randomized controlled trials. *Chin Med* 1:4.

65. Lu, Z., W. Kou, B. Du et al. 2008. Effect of Xuezhikang, an extract from red yeast Chinese rice, on coronary events in a Chinese population with previous myocardial infarction. *Am J Cardiol* 101:1689–1693.

66. Li, J.J., Z.L. Lu, W.R. Kou et al. 2009. Beneficial impact of Xuezhikang on cardiovascular events and mortality in elderly hypertensive patients with previous myocardial infarction from the China Coronary Secondary Prevention Study (CCSPS). *J Clin Pharmacol* 49:947–956.

67. Zhao, S.P., L. Liu, Y.C. Cheng et al. 2004. Xuezhikang, an extract of cholestin, protects endothelial function through antiinflammatory and lipid-lowering mechanisms in patients with coronary heart disease. *Circulation* 110:915–920.
68. Halbert, S.C., B. French, R.Y. Gordon et al. Tolerability of red yeast rice (2,400 mg twice daily) versus pravastatin (20 mg twice daily) in patients with previous statin intolerance. *Am J Cardiol* 105:198–204.
69. Liu, S., J.E. Buring, H.D. Sesso et al. 2002. A prospective study of dietary fiber intake and risk of cardiovascular disease among women. *J Am Coll Cardiol* 39:49–56.
70. Oh, K., F.B. Hu, E. Cho et al. 2005. Carbohydrate intake, glycemic index, glycemic load, and dietary fiber in relation to risk of stroke in women. *Am J Epidemiol* 161.161–169.
71. Merchant, A.T., F.B. Hu, D. Spiegelman et al. 2003. Dietary fiber reduces peripheral arterial disease risk in men. *J Nutr* 133:3658–3663.
72. Pereira, M.A., E. O'Reilly, K. Augustsson et al. 2004. Dietary fiber and risk of coronary heart disease: A pooled analysis of cohort studies. *Arch Intern Med* 164:370–376.
73. Mozaffarian, D., S.K. Kumanyika, R.N. Lemaitre et al. 2003. Cereal, fruit, and vegetable fiber intake and the risk of cardiovascular disease in elderly individuals. *JAMA* 289:1659–1666.
74. Anderson, J.W., T.J. Hanna, X. Peng et al. 2000. Whole grain foods and heart disease risk. *J Am Coll Nutr* 19:291S–299S.
75. Jacobs, D.R., M.A. Pereira, K.A. Meyer et al. 2000. Fiber from whole grains, but not refined grains, is inversely associated with all-cause mortality in older women: The Iowa Women's Health Study. *J Am Coll Nutr* 19:326S–330S.
76. Anderson, J.W., P. Baird, R.H. Davis, Jr. et al. 2009. Health benefits of dietary fiber. *Nutr Rev* 67:188–205.
77. Seal, C.J. 2006. Whole grains and CVD risk. *Proc Nutr Soc* 65:24–34.
78. Sood, N., W.L. Baker, and C.I. Coleman. 2008. Effect of glucomannan on plasma lipid and glucose concentrations, body weight, and blood pressure: Systematic review and meta-analysis. *Am J Clin Nutr* 88:1167–1175.
79. Estruch, R., M.A. Martinez-Gonzalez, D. Corella et al. 2009. Effects of dietary fibre intake on risk factors for cardiovascular disease in subjects at high risk. *J Epidemiol Community Health* 63:582–588.
80. Neil, H.A., C.A. Silagy, T. Lancaster et al. 1996. Garlic powder in the treatment of moderate hyperlipidaemia: A controlled trial and meta-analysis. *J Roy Coll Physicians Lond* 30:329–334.
81. Warshafsky, S., R.S. Kamer, and S.L. Sivak. 1993. Effect of garlic on total serum cholesterol. A meta-analysis. *Ann Intern Med* 119:599–605.
82. Ackermann, R.T., C.D. Mulrow, G. Ramirez et al. 2001. Garlic shows promise for improving some cardiovascular risk factors. *Arch Intern Med* 161:813–824.
83. Dillon, S.A., R.S. Burmi, G.M. Lowe et al. 2003. Antioxidant properties of aged garlic extract: An in vitro study incorporating human low density lipoprotein. *Life Sciences* 72:1583–1594.
84. Borek, C. 2001. Antioxidant health effects of aged garlic extract. *J Nutr* 131:1010S–1015S.
85. Rahman, K. and D. Billington. 2000. Dietary supplementation with aged garlic extract inhibits ADP-induced platelet aggregation in humans. *J Nutr* 130:2662–2665.
86. Apitz-Castro, R., J.J. Badimon, and L. Badimon. 1994. A garlic derivative, ajoene, inhibits platelet deposition on severely damaged vessel wall in an in vivo porcine experimental model. *Thromb Res* 75:243–249.
87. Apitz-Castro, R., J.J. Badimon, and L. Badimon. 1992. Effect of ajoene, the major antiplatelet compound from garlic, on platelet thrombus formation. *Thromb Res* 68:145–155.
88. Budoff, M.J., N. Ahmadi, K.M. Gul et al. 2009. Aged garlic extract supplemented with B vitamins, folic acid and L-arginine retards the progression of subclinical atherosclerosis: A randomized clinical trial. *Prev Med* 49:101–107.
89. St. Leger, A.S., A.L. Cochrane, and F. Moore. 1979. Factors associated with cardiac mortality in developed countries with particular reference to the consumption of wine. *Lancet* 1:1017–1020.

90. Hozumi, T., K. Sugioka, K. Shimada et al. 2006. Beneficial effect of short term intake of red wine polyphenols on coronary microcirculation in patients with coronary artery disease. *Heart* 92:681–682.

91. Perez-Jimenez, J. and F. Saura-Calixto. 2008. Grape products and cardiovascular disease risk factors. *Nutr Res Rev* 21:158–173.

92. Papamichael, C., E. Karatzis, K. Karatzi et al. 2004. Red wine's antioxidants counteract acute endothelial dysfunction caused by cigarette smoking in healthy nonsmokers. *Am Heart J* 147:E5.

93. Cuevas, A.M., V. Guasch, O. Castillo et al. 2000. A high-fat diet induces and red wine counteracts endothelial dysfunction in human volunteers. *Lipids* 35:143–148.

94. Imhof, A., R. Blagieva, N. Marx et al. 2008. Drinking modulates monocyte migration in healthy subjects: A randomised intervention study of water, ethanol, red wine and beer with or without alcohol. *Diab Vasc Dis Res* 5:48–53.

95. Davidson, M.H., K.C. Maki, M.R. Dicklin et al. 2009. Effects of consumption of pomegranate juice on carotid intima-media thickness in men and women at moderate risk for coronary heart disease. *Am J Cardiol* 104:936–942.

96. Aviram, M., M. Rosenblat, D. Gaitini et al. 2004. Pomegranate juice consumption for 3 years by patients with carotid artery stenosis reduces common carotid intima-media thickness, blood pressure and LDL oxidation. *Clin Nutr* 23:423–433.

97. Chacko, S.M., P.T. Thambi, R. Kuttan et al. 2010. Beneficial effects of green tea: A literature review. *Chin Med* 5:13.

98. Hsu, S.P., M.S. Wu, C.C. Yang et al. 2007. Chronic green tea extract supplementation reduces hemodialysis-enhanced production of hydrogen peroxide and hypochlorous acid, atherosclerotic factors, and proinflammatory cytokines. *Am J Clin Nutr* 86:1539–1547.

99. Hodgson, J.M. and K.D. Croft. 2010. Tea flavonoids and cardiovascular health. *Mol Aspects Med* 31:495–502.

100. Kuriyama, S., T. Shimazu, K. Ohmori et al. 2006. Green tea consumption and mortality due to cardiovascular disease, cancer, and all causes in Japan: The Ohsaki study. *JAMA* 296:1255–1265.

101. Kuriyama, S. 2008. The relation between green tea consumption and cardiovascular disease as evidenced by epidemiological studies. *J Nutr* 138:1548S–1553S.

102. Ras, R.T., P.L. Zock, and R. Draijer. 2011. Tea consumption enhances endothelial-dependent vasodilation: A meta-analysis. *PLoS One* 6:e16974.

103. Tinahones, F.J., M.A. Rubio, L. Garrido-Sanchez et al. 2008. Green tea reduces LDL oxidability and improves vascular function. *J Am Coll Nutr* 27:209–213.

104. Kobayashi, Y., M. Suzuki, H. Satsu et al. 2000. Green tea polyphenols inhibit the sodium-dependent glucose transporter of intestinal epithelial cells by a competitive mechanism. *J Agric Food Chem* 48:5618–5623.

105. Liu, J., C. Ying, Y. Meng et al. 2009. Green tea polyphenols inhibit plasminogen activator inhibitor-1 expression and secretion in endothelial cells. *Blood Coag Fibrinolysis: Int J Haemostasis Thrombosis* 20:552–557.

106. Hininger-Favier, I., R. Benaraba, S. Coves et al. 2009. Green tea extract decreases oxidative stress and improves insulin sensitivity in an animal model of insulin resistance, the fructose-fed rat. *J Am Coll Nutr* 28:355–361.

107. Carr, A.C., B.Z. Zhu, and B. Frei. 2000. Potential antiatherogenic mechanisms of ascorbate (vitamin C) and alpha-tocopherol (vitamin E). *Circ Res* 87:349–354.

108. Steinberg, D., S. Parthasarathy, T.E. Carew et al. 1989. Beyond cholesterol. Modifications of low-density lipoprotein that increase its atherogenicity. *N Engl J Med* 320:915–924.

109. Berliner, J.A. and J.W. Heinecke. 1996. The role of oxidized lipoproteins in atherogenesis. *Free Rad Biol Med* 20:707–727.

110. Liu, S., I.M. Lee, U. Ajani et al. 2001. Intake of vegetables rich in carotenoids and risk of coronary heart disease in men: The Physicians' Health Study. *Int J Epidemiol* 30:130–135.

111. Gaziano, J.M., J.E. Manson, L.G. Branch et al. 1995. A prospective study of consumption of carotenoids in fruits and vegetables and decreased cardiovascular mortality in the elderly. *Ann Epidemiol* 5:255–260.
112. Rimm, E.B., M.J. Stampfer, A. Ascherio et al. 1993. Vitamin E consumption and the risk of coronary heart disease in men. *N Engl J Med* 328:1450–1456.
113. Osganian, S.K., M.J. Stampfer, E. Rimm et al. 2003. Vitamin C and risk of coronary heart disease in women. *J Am Coll Cardiol* 42:246–252.
114. Osganian, S.K., M.J. Stampfer, E. Rimm et al. 2003. Dietary carotenoids and risk of coronary artery disease in women. *Am J Clin Nutr* 77:1390–1399.
115. Morris, C.D. and S. Carson. 2003. Routine vitamin supplementation to prevent cardiovascular disease: A summary of the evidence for the U.S. Preventive Services Task Force. *Ann Intern Med* 139:56–70.
116. The Alpha-Tocopherol, Beta Carotene Cancer Prevention Study Group. 1994. The effect of vitamin E and beta carotene on the incidence of lung cancer and other cancers in male smokers. *N Engl J Med* 330:1029–1035.
117. Yusuf, S., G. Dagenais, J. Pogue et al. 2000. Vitamin E supplementation and cardiovascular events in high-risk patients. The Heart Outcomes Prevention Evaluation Study Investigators. *N Engl J Med* 342:154–160.
118. de Gaetano, G. 2001. Low-dose aspirin and vitamin E in people at cardiovascular risk: A randomised trial in general practice. Collaborative Group of the Primary Prevention Project. *Lancet* 357:89–95.
119. Rapola, J.M., J. Virtamo, S. Ripatti et al. 1998. Effects of alpha tocopherol and beta carotene supplements on symptoms, progression, and prognosis of angina pectoris. *Heart* 79:454–458.
120. Rapola, J.M., J. Virtamo, S. Ripatti et al. 1997. Randomised trial of alpha-tocopherol and beta-carotene supplements on incidence of major coronary events in men with previous myocardial infarction. *Lancet* 349:1715–1720.
121. Stephens, N.G., A. Parsons, P.M. Schofield et al. 1996. Randomised controlled trial of vitamin E in patients with coronary disease: Cambridge Heart Antioxidant Study (CHAOS). *Lancet* 347:781–786.
122. Brigelius-Flohe, R., D. Kluth, and A. Banning. 2005. Is there a future for antioxidants in atherogenesis? *Mol Nutr Food Res* 49:1083–1089.
123. Katsiki, N. and C. Manes. 2009. Is there a role for supplemented antioxidants in the prevention of atherosclerosis? *Clin Nutr* 28:3–9.
124. Omenn, G.S., G.E. Goodman, M.D. Thornquist et al. 1996. Effects of a combination of beta carotene and vitamin A on lung cancer and cardiovascular disease. *N Engl J Med* 334:1150–1155.
125. Hirvonen, T., P. Pietinen, M. Virtanen et al. 2001. Intake of flavonols and flavones and risk of coronary heart disease in male smokers. *Epidemiology* 12:62–67.
126. Hertog, M.G., D. Kromhout, C. Aravanis et al. 1995. Flavonoid intake and long-term risk of coronary heart disease and cancer in the seven countries study. *Arch Intern Med* 155:381–386.
127. Arts, I.C., P.C. Hollman, E.J. Feskens et al. 2001. Catechin intake and associated dietary and lifestyle factors in a representative sample of Dutch men and women. *Eur J Clin Nutr* 55:76–81.
128. Geleijnse, J.M., L.J. Launer, D.A. Van der Kuip et al. 2002. Inverse association of tea and flavonoid intakes with incident myocardial infarction: The Rotterdam Study. *Am J Clin Nutr* 75:880–886.
129. Knekt, P., R. Jarvinen, A. Reunanen et al. 1996. Flavonoid intake and coronary mortality in Finland: A cohort study. *BMJ* 312:478–481.
130. Ness, A.R. and J.W. Powles. 1997. Fruit and vegetables, and cardiovascular disease: A review. *Int J Epidemiol* 26:1–13.
131. Hertog, M.G., E.J. Feskens, P.C. Hollman et al. 1993. Dietary antioxidant flavonoids and risk of coronary heart disease: The Zutphen Elderly Study. *Lancet* 342:1007–1011.

132. Hertog, M.G., E.J. Feskens, and D. Kromhout. 1997. Antioxidant flavonols and coronary heart disease risk. *Lancet* 349:699.

133. Rimm, E.B., M.B. Katan, A. Ascherio et al. 1996. Relation between intake of flavonoids and risk for coronary heart disease in male health professionals. *Ann Intern Med* 125:384–389.

134. Keli, S.O., M.G. Hertog, E.J. Feskens et al. 1996. Dietary flavonoids, antioxidant vitamins, and incidence of stroke: The Zutphen study. *Arch Intern Med* 156:637–642.

135. Yochum, L., L.H. Kushi, K. Meyer et al. 1999. Dietary flavonoid intake and risk of cardiovascular disease in postmenopausal women. *Am J Epidemiol* 149:943–949.

136. Sesso, H.D., J.M. Gaziano, S. Liu et al. 2003. Flavonoid intake and the risk of cardiovascular disease in women. *Am J Clin Nutr* 77:1400–1408.

137. Lin, J., K.M. Rexrode, F. Hu et al. 2007. Dietary intakes of flavonols and flavones and coronary heart disease in US women. *Am J Epidemiol* 165:1305–1313.

138. Hertog, M.G., P.M. Sweetnam, A.M. Fehily et al. 1997. Antioxidant flavonols and ischemic heart disease in a Welsh population of men: The Caerphilly Study. *Am J Clin Nutr* 65:1489–1494.

139. Hennekens, C.H., J.E. Buring, J.E. Manson et al. 1996. Lack of effect of long-term supplementation with beta carotene on the incidence of malignant neoplasms and cardiovascular disease. *N Engl J Med* 334:1145–1149.

140. Greenberg, E.R., J.A. Baron, M.R. Karagas et al. 1996. Mortality associated with low plasma concentration of beta carotene and the effect of oral supplementation. *JAMA* 275:699–703.

141. Zureik, M., P. Galan, S. Bertrais et al. 2004. Effects of long-term daily low-dose supplementation with antioxidant vitamins and minerals on structure and function of large arteries. *Arterioscler Thromb Vasc Biol* 24:1485–1491.

142. Eidelman, R.S., D. Hollar, P.R. Hebert et al. 2004. Randomized trials of vitamin E in the treatment and prevention of cardiovascular disease. *Arch Intern Med* 164:1552–1556.

143. Bleys, J., E.R. Miller, 3rd, R. Pastor-Barriuso et al. 2006. Vitamin–mineral supplementation and the progression of atherosclerosis: A meta-analysis of randomized controlled trials. *Am J Clin Nutr* 84:880–887; quiz 954–955.

144. Heart Protection Study Collaborative Group. 2002. MRC/BHF Heart Protection Study of antioxidant vitamin supplementation in 20,536 high-risk individuals: A randomised placebo-controlled trial. *Lancet* 360:23–33.

145. Cook, N.R., C.M. Albert, J.M. Gaziano et al. 2007. A randomized factorial trial of vitamins C and E and beta carotene in the secondary prevention of cardiovascular events in women: Results from the Women's Antioxidant Cardiovascular Study. *Arch Intern Med* 167:1610–1618.

146. Boaz, M., S. Smetana, T. Weinstein et al. 2000. Secondary prevention with antioxidants of cardiovascular disease in endstage renal disease (SPACE): Randomised placebo-controlled trial. *Lancet* 356:1213–1218.

147. Singh, R.B., M.A. Niaz, S.S. Rastogi et al. 1996. Usefulness of antioxidant vitamins in suspected acute myocardial infarction (the Indian experiment of infarct survival-3). *Am J Cardiol* 77:232–236.

148. Tomoda, H., M. Yoshitake, K. Morimoto et al. 1996. Possible prevention of postangioplasty restenosis by ascorbic acid. *Am J Cardiol* 78:1284–1286.

149. Yokoi, H., H. Daida, Y. Kuwabara et al. 1997. Effectiveness of an antioxidant in preventing restenosis after percutaneous transluminal coronary angioplasty: The Probucol Angioplasty Restenosis Trial. *J Am Coll Cardiol* 30:855–862.

150. Fang, J.C., S. Kinlay, J. Beltrame et al. 2002. Effect of vitamins C and E on progression of transplant-associated arteriosclerosis: A randomised trial. *Lancet* 359:1108–1113.

151. Tardif, J.C., G. Cote, J. Lesperance et al. 1997. Probucol and multivitamins in the prevention of restenosis after coronary angioplasty. Multivitamins and Probucol Study Group. *N Engl J Med* 337:365–372.

# 7

# Nutraceuticals in Cardiovascular Diseases
## Mechanisms and Application

Srinivas M. Tipparaju, Jared Tur, and Siva K. Panguluri

## Contents

## Introduction

In the present nutraceutical market, a vast range of products are used for many diseases either for primary treatment or for supplemental treatment. Among these nutraceutical products, a major portion is currently being used for cardiovascular diseases (CVDs). Some of the popular examples for such use include garlic, fish oil, curcumin, vitamins (niacin, ascorbic acid, tocopherol), and supplements such as ubiquinone (CoQ10). Therefore, it is imperative to understand the exact mechanism and the benefits of these components of nutraceuticals that are being used by patients and individuals. However, the challenges that are faced for such approach include the chemical characterization of these products and whether the use of whole or a component

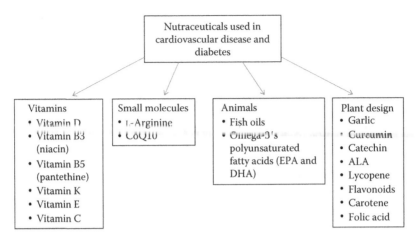

**Figure 7.1** *Nutraceutical types in heart and diabetes. The flowchart depicts the different class of nutraceuticals that we used in cardiovascular diseases and diabetes. The major classes include vitamins, small molecules, animal products, and plant-derived products. Specific examples in each class are given for clarity.*

of the natural compound is benefiting the user is still being evaluated by researchers. Therefore, in this chapter, we will look at the present understanding developed in these areas. Use of most popular nutraceuticals along with its use and mechanism and research-based understanding will be elucidated (Figure 7.1).

## Cardiovascular diseases

Diseases related to heart and the vasculatures are commonly defined as CVD; however, vascular disease can also lead to brain disorders termed as cerebrovascular diseases (stroke). The use of nutraceutical has been very popular in these areas pertaining to prevention, subsiding the disease, and also cures at early stages of disease. Therefore, the role of nutraceutical in CVD is important and significant in the present-day use for many diseases. The burden of CVD is significant in most western societies and industrialized countries. Recent data from AHA suggest that diseases related to cardiovascular and cerebrovascular cause highest mortality in the US population and are, therefore, the number one cause of death in the United States. Recent data also suggest that in the developing world the risk of heart disease is on the rise and more people are affected with this disease. CVD can be mainly divided into four groups: atherosclerosis, hypertension, arrhythmias, and myocardial ischemia. Atherosclerosis is caused by increased accumulation of cholesterol in the arteries and blood vessels. Abnormal cholesterol metabolism caused by hyperlipidemia leads to increased cholesterol in the blood-forming plaque buildup. If untreated, the plaques become calcified and form foam cells ultimately resulting in severe blockade of the blood and rupture. Therefore, high cholesterol along with other forms of hyperlipidemia need to be treated and kept under normal limits for avoiding plaque buildup and disease progression.

Hypertension is disease of the blood vessels, which is a measure of high blood pressure (>120 systolic pressure and >90 diastolic pressure in mm Hg). Increase in the blood pressure is defined as hypertension and a decrease in the pressure is called as hypotension (<65 mm Hg diastolic pressure and <95 mm Hg systolic pressure). Chronic increase in the blood pressure leads to enlargement of the heart called as hypertrophy. The enlarged heart is compromised in many ways such that the energy and oxygen demands of the heart for its functioning are increased and, therefore, over a period of time is susceptible to compensatory remodeling and later to decompensation and failure. Hypertrophy is classically defined into two types referred as concentric and eccentric hypertrophy. Concentric hypertrophy causes the heart to have a decreased cavity or lumen size such that the volume of blood filled is decreased, whereas in eccentric hypertrophy the muscle size is reduced along with increase in the length of the cardiomyocytes due to stretch causing an overall increase in size of the heart, which is enlarged like an over-inflated balloon. Hypertrophy in the heart is caused by increased hyperplasia (Figure 7.2).

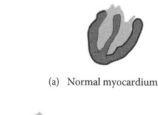

(a) Normal myocardium

(b) Concentric hypertrophic myocardium

(c) Eccentric hypertrophic myocardium

**Figure 7.2** *Pictorial depiction of classical hypertrophy and types. (a) The normal heart is having proportional sizes of right and left ventricles which is divided into two separate chambers by the ventricular septum. The muscle mass of the left ventricle is higher compared with the right ventricle. Hypertrophy leads to enlargement of heart (concentric or eccentric) when the muscle mass of the heart is increased and referred as concentric hypertrophy (b) in which the lumen or chamber size is decreased and therefore the amount of blood pumped is also decreased. (c) In eccentric hypertrophy, the walls become thin resulting in increase in cavity size but the muscle size is diminished and therefore the force with which the heart can pump is diminished.*

Arrhythmias are defined as abnormal beating of the heart. The normal functioning of the heart results in 70 beats per minute and, therefore, each beat is slightly <1 s. The heart is an electrical organ and its primary role is to pump blood to the body. To achieve this goal, the heart works as an electrical synchronous unit. The electrical synchrony and the rhythm are basically generated due to the concerted action of ion channel that are located in each cardiomyocyte. The major voltage ion channels that are defined in the heart are calcium (ICa), sodium (INa), and potassium (IK). The electrical system in the heart is complex; however, under normal circumstances, it functions in synchrony and results in the heartbeat. However, during abnormal or disease states there is disruption in the conduction and propagation of the electrical signaling leading to arrhythmias. Research in this area has led to the understanding that nutraceuticals or active components of nutraceuticals can modulate the ion channels directly via signal transduction pathways. This in turn leads to changes in the kinetics and function of the ion channel that translates into change in heart rhythm and function. Therefore, in-depth understanding of nutraceuticals is an important and upcoming field in the area of cardiovascular sciences. The voltage-gated potassium channels are a major family of ion channel proteins that regulate the shape and duration of action potential in function I the heart. Among the Kv1–12 family, the Kv1 and Kv4 family proteins bind to Kvβ subunits (Shaker channel proteins); the Kvβ proteins have catalytic properties and bind pyridine nucleotides (NADPH, NADP$^+$, NADH, NAD$^+$) with high affinity [1]. In our previous study, we identified the substrate specificity and catalytic efficacy of the Kvβ2 subunit and reported substrates such as 12-oxo-ETE (12-oxo-5Z,8Z,10E,14Z-eicosatetraenoic acid) and environmental chemicals such as acrolein as substrates of Kvβ2 [2]. We also found that methylglyoxal, which is an advanced glycosylation end product found in coffee and hot beverage drinks [3], is a substrate for Kvβ subunit suggesting that the cytosolic subunits of voltage-gated potassium channels (Kvβ1–3) [4] may have a physiological role that remains unknown. To evaluate the physiological and biological substrates that modulate the Kv channel properties in a previous study, we utilized electrospray mass ionization (ESI-MS) technique for product identification generated by reacting Kvβ2 with oxidized PAPC (1-palmitoyl-2-arachidonoyl-*sn*-glycero-3-phosphocholine). Our studies clearly identified several important carbonyl-containing compounds such as 1-palmitoyl-2-(5-oxovaleroyl)-*sn*-glycero-3-phosphorylcholine (POVPC) converted to PHVPC, and 1-palmitoyl-2-(5,6)-epoxyisoprostane E2-*sn*-glycero-3-phosphocholine (PEIPC) to 2H-PEIPC [5]. Therefore, it is likely that Kvβ subunits can bind to other chemicals that are present in nutraceutical systems. Additional research and development is needed to fully understand the role of Kv channels as to "how nutraceuticals may be affecting the cellular functioning in the heart." The components of fish oil are thought to protect against CVDs; however, the precise mechanisms are unknown; therefore, previous studies tested and demonstrated that EPA and DHA compete with arachidonic acid (AA) for conversion by CYP enzymes resulting in active metabolites. These studies showed that CYP enzymes efficiently converted

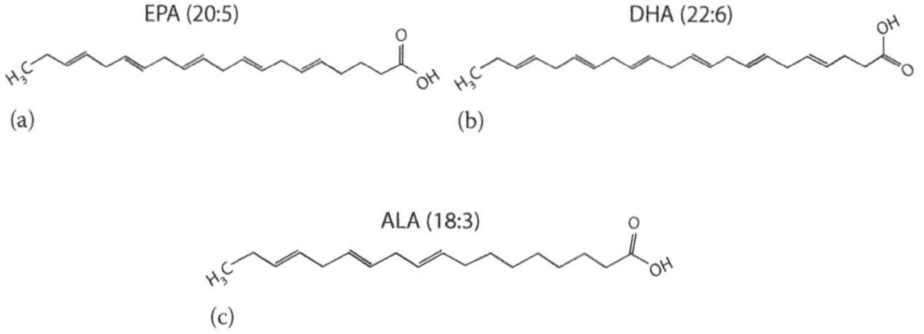

**EPA (20:5)**

(a)

**DHA (22:6)**

(b)

**ALA (18:3)**

(c)

**Figure 7.3** *Chemical structures of the commonly used fish oil compounds. (a) The eicosapentaenoic acid (EPA) also noted as (20:5) which represents with number of carbon and the unsaturation linkages (double bonds). (b) The docosahexaenoic acid or DHA is also noted with 22 carbon length chain with six unsaturated carbon–carbon bonds (double bonds); α-linoleic acid (ALA) shown in (c) is 18 carbon long-chain compound with three unsaturated bonds.*

EPA and DHA (components of dietary ω-3 fatty acids) that could provide benefit for CVDs [6] (Figure 7.3).

## Myocardial ischemia

The heart is a highly vascularized organ in which there are macro- and microvascular network of vessels that supply oxygenated blood to the heart itself. However, during abnormal vascular architecture, development of plaque or vasoconstriction of the vessels disrupts the supply of blood to the heart. Under those circumstances, the heart becomes ischemic, resulting in loss of nutrients and oxygen to the affected area. Acute and chronic ischemic conditions lead to scar development resulting in necrosis. The necrosis portion of the heart can be nonfunctional or act as substrate for arrhythmogenic events; therefore, understanding the mechanisms of drugs and nutraceuticals that can minimize the risk of arrhythmogenic region development and progression is highly desirable.

Polyphenols exert antioxidant effects and preserve the nitric oxide activity and also function as a anti-inflammatory by modulating matric metalloproteases. Polyphenols prevent mitochondrial damage, and are also used in pharmacological preconditioning and involved in epigenetic modifications such as histone acyl transferases. Recent studies show that polyphenols prevent against hypertrophy, vascular remodeling, and fibrosis after myocardial infraction [7]. Resveratrol a polyphenolic compound that is known for its antioxidant properties, is principally derived from grape juice and is present in red wine. Recent studies reveal that resveratrol targets specific proteins such as N-ribosyl dihydronicotinamide:quinone oxidoreductase and offers cardioprotection [8]. Consistent studies from several laboratories show that the protective agent resveratrol shows protective effects against oxidized or modified low-density lipoprotein (LDL) and offers lowered platelet aggregation decreasing the risk of clot formation and plugging of atherosclerotic lesions in the blood vessels. Also, it is noticed that the proliferation of smooth muscle cell is decreased by the use of resveratrol. Ultimately, there is enormous scientific evidence suggesting the direct role of resveratrol in preventing against the damage caused by increased atherogenesis in the vascular and endothelium.

The cardioprotective effects of resveratrol are via non-genomic and genomic mechanisms. The detailed description is noted in the review by Wu and Hsieh [8]; however, it is important to note that there are resveratrol target proteins, for example, NQO2 protein, that are involved in signaling cascade and ultimately acting in protecting pathways. At the genomic mechanisms level, there is promoter recognition and transcription complex formation leading to transcriptional control activation or deactivation of responsive gene expression.

The human heart is a highly metabolic organ. There are high demands for energy, which in normal physiology are met by fatty oxidation and metabolism; however, during metabolic alterations or endocrine changes, the heart adapts into a more carbohydrate-utilizing organ along with fatty acids. Disease states such as diabetes are a classic example of such situations where the heart utilizes carbohydrates. The metabolic source ultimately results in generation of ATP (adenosine triphosphate), which is the source of energy for cardiac function. Altered metabolism and increased use of pharmaceutical drugs lead to depletion of micro- and macronutrients in humans, and, therefore, replenishment of such nutrient sources is a continuous need. Among these commonly are CoQ10, flavonoids, ω-3 fatty acids, L-carnitine, vitamins, L-arginine.

## CoQ10

CoQ10 (ubiquinone) is an important member of the mitochondrial electron transport chain (ETC). Some patients with prolonged use of statins use or therapy have been noted with a depletion of CoQ10 in the blood [9] the study concludes that routine CoQ10 supplementation for all patients taking statin to prevent myotoxicity is not recommended. But some patients in the subpopulation may be at a long-term risk of statin therapy and may be supplemented with CoQ10 therapy as supplement.

## Garlic

The use of garlic has been widely accepted for its antihyperlipidemic activity in which the bad cholesterol (LDL) is decreased and helps improve the good cholesterol (high-density lipoprotein [HDL]). The main chemical constituents that are responsible for this property are allinin and allicin. Epidemiological studies suggest that the consumption of garlic in diet has an inverse correlation with progression of CVD. It is viewed that garlic inhibits the synthesis of lipids in the body, decreases the platelet aggregation and blood pressure, and helps cells as antioxidants [10]. Recent studies using a norepinephrine-induced hypertrophy of the heart in rat model treated with garlic prevented cardiomyopathy and cell death. The plausible mechanism may be in part mediated by nitric oxide (NO) and hydrogen sulfide ($H_2S$) [11]. Pharmacological blockade of NO and $H_2S$ significantly inhibited the beneficial effects of garlic extracts. Therefore, the authors concluded that garlic is beneficial in hypertrophy.

In traditional Hindu and Indian medicine called Ayurveda, the description for the use of garlic is recommended for the treatment of heart ailments and arthritis (joint pains).

## Vitamin D

The physiological role of vitamin D is for significant interest for the scientific and clinical community. A small prospective study with the use of vitamin D supplementation for reducing the risk of CVD was conducted in 114 subjects. The study measured blood flow parameters, and associated with body mass index blood pressure, glucose. The study concludes that vitamin D supplementation does not improve endothelial function, arterial stiffness, and inflammation. Recent epidemiological data clearly suggest the link between cardiovascular and diabetic risks in patients with low levels of vitamin D. However, systematic case-controlled and randomized studies are necessary to prove the beneficial roles of vitamin D and decrease the burden of toxic effects that may potentially exacerbate the therapeutic use clinically. Toxicity of vitamin D can lead to hypercalcemia, hypophosphatemia, and may lead to secondary complication [12].

## L-Arginine

The use of L-arginine is primarily for its role as substrate for nitric oxide synthesis. Studies elucidating the use of L-arginine reveal that it provides increased bioavailability of authentic NO that causes vasorelaxation and increased blood flow in the vasculature of the heart. Biochemically, L-arginine is converted to L-citrulline in the presence of nitric oxide synthase (NOS), $Ca^{2+}$, tetrahydrobiopterin; NO is released as a by-product; and the downstream events are triggered for its binding to soluble guanylyl cyclase and activation of cGMP causing vasorelaxation (Figure 7.4). Using streptozotocine diabetic mice and treating with L-arginine caused restoration of NO levels and prevented tissue accumulation of sorbitol along with increased accumulation of glutathiolation of aldose reductase. The study by West et al. clearly demonstrated that L-arginine treatment in diabetic mice decreased superoxide production and reduced the levels of serum triglycerides and cell adhesion molecule (ICAM); the report attests to the possibility that L-arginine supports the increased bioavailability and corrects the major biochemical abnormalities in diabetic model of mice [13]. Studies also demonstrated that dietary supplementation of L-arginine is used to up-regulate NO and inhibited the development of CVD. Novel drug delivery systems using allograft with L-arginine polymers

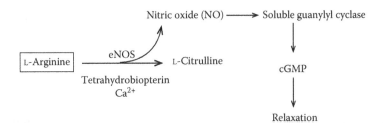

**Figure 7.4** *Biosynthesis of nitric oxide from L-arginine. The schematic depiction illustrates the biosynthesis of nitric oxide (NO) from L-arginine which is converted to L-citrulline in the presence of enzyme nitric oxide synthase (NOS), calcium, and tetrahydrobiopterin (cofactor). NO is released in the endothelium, diffuses into the vascular cell, binds to soluble guanylyl cyclase (sGS), activates cyclic guanosine monophosphate (cGMP), and causes relaxation of the smooth muscle located in the blood vessel.*

provides benefit over gene delivery of inducible NOS (iNOS) or endothelial NOS (eNOS). Study by Kown et al. shows that the L-arginine polymers may be more efficacious than L-arginine amino acid therapy [14].

## Diabetes and its pathological effects

Diabetes is one of the growing problems in developed countries such as the United States and developing countries like India and China. According to World Health Organisation (WHO), 6.4% of the world's adult population live with diabetes and 1 in every 10 is diabetic. According to the International Diabetes Foundation, India being the first in diabetic population which is followed by China and the United States.

According to the National Diabetes Information Clearinghouse (NIDC), diabetes affects 25.8 million people of all ages in the United States, which accounts for 8.3% of the total US population. Diabetes was also listed as one of the underlying cause of heart diseases with 68% of deaths from diabetes-related heart diseases. American Diabetes Association (ADA) also listed that there is two- to fourfold higher risk of heart disease deaths with diabetes and also two- to fourfold higher risks for heart stroke in addition to other diseases such as kidney diseases, nervous system disorders, and blindness. The total cost of diagnosis for diabetes in the United States is more than $174 billion.

Obesity is an important health problem that is closely associated with diabetes. The incidence of this condition has tripled during the past two decades and the trend continues unchecked. Nearly two-thirds of adults in the United States are overweight (35%) or obese (30%), with women, minorities, and the lower-income accounting for a disproportionate share of this population. Overweight and obesity are leading causes of hypertension, stroke, heart disease, chronic musculoskeletal problems, type 2 diabetes, and certain types of cancers.

Most people know the adverse health consequences of an unhealthy diet and excess body weight, yet the ranks of the overweight and obese increase each year. According to the NIH strategic plan for obesity research, one reason for rising obesity rates may lie in the abundant choices of relatively inexpensive calorically dense foods that are convenient and taste good. Increasing physical activity and eating healthy diet are strongly recommended by most public health bodies for reducing body fat [15]. Many people find it difficult in complying their lifestyle changes to reduce body fat, especially long term. One study showed that more than half of the body weight lost by exercise or diet is regained after 1 year [16]. Therefore, in addition to these recommendations from public health bodies, more acceptable and sustainable strategies to reduce obesity must be found.

## Prevention and side effects of current treatment strategies

Prevention and control of diabetes including diet, insulin, and oral medication to lower blood glucose levels are the major treatment and management currently under standard practice. Occurrence of diabetic ketoacidosis (DKA) and weight gain are a major drawback of insulin treatments in addition to the cost involved in the medication. On average between 10% and 20% of the patients stop diabetes medications due to side effects. Most common side effects of these drugs include hypoglycemia (usually minor, but can be fatal if not treated in time), weight gain, gastrointestinal side effects, edema, and increase in LDL in addition to heart diseases, anemia, and some allergic reactions.

Use of natural compounds from plant and animal sources is one of the growing areas in the field of natural products and alternate medicine. These are commonly called as nutraceuticals as they are natural products with pharmaceutical potential. Although there are many important nutraceuticals available for diabetes and diabetes-related heart diseases including but not limited to ω-3 fatty acids, vitamin D, folic acid, and other probiotics, in this chapter we give specific emphasis on ω-3 fatty acids.

## Long-chain ω-3 fatty acids

The American Heart Association (AHA) recommends consumption of ω-3 fatty acids to prevent cardiovascular complications based on evidences from epidemiological, observational, and clinical trial studies [17,18]. Many investigators have been exploring the benefits of long-chain ω-3 polyunsaturated fatty acids (n-3 PUFA), especially eicosapentaenoic acid (EPA) and docosahexaenoic acid (DHA) for reducing cardiovascular complications [15]. The studies have also suggested that consumption of n-3 PUFA is also associated with various health benefits including lowering plasma triacylglycerol (TAG), improving insulin sensitivity, reduced blood pressure, inflammation, thrombosis, and arrhythmia in addition to lowering the risk of CVD and diabetes [19]. Many reports suggest that the use of n-3 PUFA or fish oil by pregnant women have been associated with an increase in birth weight, reduced risk of premature birth, and other complications during pregnancy [20]. Moreover, these n-3 PUFA are known to be the limiting factors for the growth and development of newborns and young children [21]. Many investigators have reported that incorporation of n-3 PUFA into high-fat obesogenic diet–fed rodents reduces body fat accumulation [22–26]. In another study, greater incorporation of n-3 PUFA and reduced incorporation of n-6 PUFA, which has the potential to influence gene expression that would alter the metabolic activity leading to reduction of body fat accumulation, was observed in ob/ob (diabetic) mice fed with cod liver oil [27]. A study by Hill et al. [28] showed that subjects fed with fish oil supplementation along with regular exercise reduced body fat with other beneficial effect on plasma TAG, HDL, cholesterol, and endothelium-dependent vasodilation. Many of such studies in animals and humans showing the effects of n-3 PUFA on body weight and/or body composition suggests that the effects are independent of energy intake [24,25,28,29]. But the recent sub-analysis of a larger weight loss studies ($n = 278$) showed that weight loss achieved through caloric restriction when fish or fish oil was included in the diet, was associated with greater satiety immediately after and 120 min after a test meal [30].

## Pregnant and lactating women

Long-chain PUFA such as DHA and AA are known to be important for the growth and development of fetus and infant. These two PUFA deposition is high during the last trimester of pregnancy and first month of birth [31], whereas insufficient amounts of PUFA can lead to adverse effects [32]. Essential PUFA such as n-3 and n-6 that are required for the fetus are also supplied during pregnancy by preferential placental transfer [33,34]. The studies also suggested that PUFA, especially n-3 series such as DHA and EPA, may have beneficial effects on pregnancy outcomes, including duration of gestation and infant birth weight [35,36]. In addition, studies also suggested that supplementation of n-3 PUFA to lactating women increases breast milk n-3 PUFA content [37], and supplementation of n-3 during pregnancy or lactation or both improves infant's cognition and visual development [38].

Few studies also reported that women who reported belching and unpleasant taste attributed to the oil capsule was significantly higher in the fish oil group than the olive oil or no oil group [39]. In this study, the authors also reported that although there are no significant differences between fish oil group and other groups in side effects such as prolongation of labor or the need of surgery, there is a significant increase in blood loss at delivery in fish oil group. In contrast, other studies by Smuts et al. [40] showed that women with one or more adverse effects during pregnancy are higher in normal egg group than in the DHA-supplemented group. Gynecological adverse effects in general, and labor-related adverse effects in particular are less common in the group supplemented with DHA. Despite several issues and concerns regarding the safety of increasing n-3 and n-6 PUFA during pregnancy or lactation, the available literature suggests that PUFA have many important beneficial effects for diabetic, obese, and pregnant women and more studies are, therefore, warranted for the complete understanding of their mode of action(s).

## Hyperlipidemia

Hyperlipidemia is described as an increase in the amount of fat such as cholesterol and triglycerides within the blood. High fat within the blood can lead to heart disease, pancreatitis, and other health issues. Some patients demonstrate higher accumulations of cholesterol and triglycerides within their blood due to diet and sedentary lifestyle and/or genetics. According to the CDC, 14.1% (2007–2010) of adults age 20 years and older in the United States had high serum total cholesterol greater than or equal to 240 mg/dL. Total serum cholesterol in the blood is made up of LDL, very low-density lipoprotein (VLDL), HDL, chylomicrons, in addition triglycerides and lipoprotein (a). A typical breakdown of serum cholesterol is carried by 65% LDL, 20% HDL, 14% VLDL, and 1% chylomicrons [41]. According to the AHA, a serum total cholesterol level of <200 mg/dL is desirable; LDL levels of <100 mg/dL and HDL levels of 60 mg/dL or greater while triglyceride levels should be <150 mg/dL.

LDLs are often remarked as the "bad" cholesterol because as more accumulation occurs within the arteries a buildup can begin to form, plague (this process is known as atherosclerosis), which narrows the arteries making them less flexible and makes one more susceptible to blockage by clot formation leading to possible stroke or heart attack. VLDLs are responsible for the transportation of endogenous triglycerides [41].

HDLs on the other hand are often remarked as the "good" cholesterol, believed to carry cholesterol away from the arteries and back to the liver, where it can be passed from the body. HDLs may even help remove cholesterol that has already begun forming plague on the arteries.

Chylomicrons are produced in the small intestine and function to transport ingested dietary triglycerides, cholesterol, and other lipids [41].

ω-3 Fatty acids have gained much interest in recent years particularly after the observation made among Greenland's Eskimo population demonstrating low incidences of CVD believed to be due to a diet of seafood containing n-3 PUFA [42]. The major dietary components of ω-3 fatty acids include EPA, DHA, both components found in fish, and α-linolenic acid (ALA), found mainly in nuts and seeds, as well as vegetable oil (Figure 7.3). Numerous clinical studies have demonstrated that ALA may not be as effective at lowering CVD when compared with EPA and DHA together [43].

Eicosanoids, the bioactive molecule derived from ω-6 and ω-3 fatty acids, are broken down by δ-6 desaturase in ω-3 fatty acids to form EPA and DHA. EPA and DHA are predominantly anti inflammatory agents and inhibit platelet aggregation (clotting). A randomized open-label, blind end-point study utilized 18,645 patients with 6.5 mmol/L total cholesterol treated daily with 1800 mg EPA and demonstrated reduced LDL cholesterol, unstable angina, and nonfatal coronary events [44]. n-3 PUFA were also noted with a reduction in blood pressure, particularly in patients with hypertension and clinical atherosclerotic disease or hypercholesterolemia [45].

Currently, the US Food and Drug Administration (FDA) has approved a ω-3 PUFA ethyl ester formulation, Lovaza™ (GlaxoSmithKline) at a dosage of 4 g/day for the treatment of very high triglyceride levels greater than 500 mg/dL [46]. Three to four grams per day of n-3 fatty acids from fish oil (EPA and DHA) have been demonstrated to reduce elevated triglyceride levels by up to 30%–40%; however, as VLDL concentrations decrease, LDL cholesterol tends to rise [47].

## Conclusions

The use of nutraceuticals is very popular among patients with CVDs. Recent awareness about the benefits, supplemental use, and availability of the nutraceuticals in commercial markets and advertising by nutraceutical industry has led to increased utilization and consumption of these products. The lack of understanding of the mechanisms of the active ingredients or the whole product is important for developing and using these compounds and its components for therapeutic use or benefits. From a pharmacist and healthcare standpoint research regarding the efficacy, pharmacokinetics and its drug interactions is an effective way to promote or may be able to avoid its use under certain conditions. Also, drug interactions can be minimized with the use of nutraceuticals if the interactions are studied in-depth using research-based animal models for elucidating mechanisms and their benefits in diseases.

## Acknowledgment

The authors acknowledge the funding support (HL102171, USF COP seed grant, and startup funding to SMT).

## References

1. Tipparaju SM, Liu SQ, Barski OA, Bhatnagar A. NADPH binding to beta-subunit regulates inactivation of voltage-gated K(+) channels. *Biochem Biophys Res Commun* 2007, 359(2):269–276.
2. Tipparaju SM, Barski OA, Srivastava S, Bhatnagar A. Catalytic mechanism and substrate specificity of the beta-subunit of the voltage-gated potassium channel. *Biochemistry* 2008, 47(34):8840–8854.
3. Wang J, Chang T. Methylglyoxal content in drinking coffee as a cytotoxic factor. *J Food Sci* 2010, 75(6):H167–H171.
4. Barski OA, Tipparaju SM, Bhatnagar A. The aldo–keto reductase superfamily and its role in drug metabolism and detoxification. *Drug Metab Rev* 2008, 40(4):553–624.

5. Xie Z, Barski OA, Cai J, Bhatnagar A, Tipparaju SM. Catalytic reduction of carbonyl groups in oxidized PAPC by Kvbeta2 (AKR6). *Chem Biol Interact* 2011, 191(1–3):255–260.
6. Arnold C, Markovic M, Blossey K, Wallukat G, Fischer R, Dechend R, Konkel A, von Schacky C, Luft FC, Muller DN et al. Arachidonic acid-metabolizing cytochrome P450 enzymes are targets of {omega}-3 fatty acids. *J Biol Chem* 2010, 285(43):32720–32733.
7. Jiang F, Chang CW, Dusting GJ. Cytoprotection by natural and synthetic polyphenols in the heart: Novel mechanisms and perspectives. *Curr Pharm Des* 2010, 16(37):4103–4112.
8. Wu JM, Hsieh TC. Resveratrol: A cardioprotective substance. *Ann N Y Acad Sci* 2011, 1215:16–21.
9. Levy HB, Kohlhaas HK. Considerations for supplementing with coenzyme Q10 during statin therapy. *Ann Pharmacother* 2006, 40(2):290–294.
10. Rahman K, Lowe GM. Garlic and cardiovascular disease: A critical review. *J Nutr* 2006, 136(3 Suppl):736S–740S.
11. Lieben Louis XM, Murphy RM, Thandapilly SJM, Yu LD, Netticadan TD. Garlic extracts prevent oxidative stress, hypertrophy and apoptosis in cardiomyocytes: A role for nitric oxide and hydrogen sulfide. *BMC Complement Altern Med* 2012, 12(1):140.
12. Brandenburg VM, Vervloet MG, Marx N. The role of vitamin D in cardiovascular disease: From present evidence to future perspectives. *Atherosclerosis* 2012, 225(2):253–263.
13. West MB, Ramana KV, Kaiserova K, Srivastava SK, Bhatnagar A. L-Arginine prevents metabolic effects of high glucose in diabetic mice. *FEBS Lett* 2008, 582(17):2609–2614.
14. Kown MH, van der Steenhoven T, Uemura S, Jahncke CL, Hoyt GE, Rothbard JB, Robbins RC. L-Arginine polymer mediated inhibition of graft coronary artery disease after cardiac transplantation. *Transplantation* 2001, 71(11):1542–1548.
15. Buckley JD, Howe PR. Long-chain omega-3 polyunsaturated fatty acids may be beneficial for reducing obesity—A review. *Nutrients* 2010, 2(12):1212–1230.
16. Curioni CC, Lourenco PM. Long-term weight loss after diet and exercise: A systematic review. *Int J Obesity* 2005, 29(10):1168–1174.
17. Kris-Etherton PM, Harris WS, Appel LJ. Fish consumption, fish oil, omega-3 fatty acids, and cardiovascular disease. *Circulation* 2002, 106(21):2747–2757.
18. Lichtenstein AH, Appel LJ, Brands M, Carnethon M, Daniels S, Franch HA, Franklin B, Kris-Etherton P, Harris WS, Howard B et al. Diet and lifestyle recommendations revision 2006: A scientific statement from the American Heart Association Nutrition Committee. *Circulation* 2006, 114(1):82–96.
19. Martinez-Victoria E, Yago MD. Omega 3 polyunsaturated fatty acids and body weight. *Br J Nutr* 2012, 107(Suppl 2):S107–S116.
20. Drouillet P, Kaminski M, De Lauzon-Guillain B, Forhan A, Ducimetiere P, Schweitzer M, Magnin G, Goua V, Thiebaugeorges O, Charles MA. Association between maternal seafood consumption before pregnancy and fetal growth: Evidence for an association in overweight women. The EDEN mother–child cohort. *Paediatr Perinat Epidemiol* 2009, 23(1):76–86.
21. Innis SM. Human milk: Maternal dietary lipids and infant development. *Proc Nutr Soc* 2007, 66(3):397–404.
22. Baillie RA, Takada R, Nakamura M, Clarke SD. Coordinate induction of peroxisomal acyl-CoA oxidase and UCP-3 by dietary fish oil: A mechanism for decreased body fat deposition. *Prostaglandins Leukot Essent Fatty Acids* 1999, 60(5–6):351–356.
23. Belzung F, Raclot T, Groscolas R. Fish oil n-3 fatty acids selectively limit the hypertrophy of abdominal fat depots in growing rats fed high-fat diets. *Am J Physiol—Regul Integr Comp Physiol* 1993, 264(6):R1111–R1118.
24. Cunnane SC, McAdoo KR, Horrobin DF. n-3 Essential fatty acids decrease weight gain in genetically obese mice. *Br J Nutr* 1986, 56(1):87–95.
25. Hainault I, Carolotti M, Hajduch E, Guichard C, Lavau M. Fish oil in a high lard diet prevents obesity, hyperlipemia, and adipocyte insulin resistance in rats. *Ann N Y Acad Sci* 1993, 683:98–101.

26. Ruzickova J, Rossmeisl M, Prazak T, Flachs P, Sponarova J, Veck M, Tvrzicka E, Bryhn M, Kopecky J. Omega-3 PUFA of marine origin limit diet-induced obesity in mice by reducing cellularity of adipose tissue. *Lipids* 2004, 39(12):1177–1185.

27. Clarke SD. Polyunsaturated fatty acid regulation of gene transcription: A mechanism to improve energy balance and insulin resistance. *Br J Nutr* 2000, 83(Suppl 1):S59–S66.

28. Hill AM, Buckley JD, Murphy KJ, Howe PR. Combining fish-oil supplements with regular aerobic exercise improves body composition and cardiovascular disease risk factors. *Am J Clin Nutr* 2007, 85(5):1267–1274.

29. Huang XF, Xin X, McLennan P, Storlien L. Role of fat amount and type in ameliorating diet-induced obesity: Insights at the level of hypothalamic arcuate nucleus leptin receptor, neuropeptide Y and pro-opiomelanocortin mRNA expression. *Diabetes Obes Metab* 2004, 6(1):35–44.

30. Thorsdottir I, Tomasson H, Gunnarsdottir I, Gisladottir E, Kiely M, Parra MD, Bandarra NM, Schaafsma G, Martinez JA. Randomized trial of weight-loss-diets for young adults varying in fish and fish oil content. *Int J Obesity* 2007, 31(10):1560–1566.

31. Innis SM. Essential fatty acids in growth and development. *Prog Lipid Res* 1991, 30(1):39–103.

32. Koletzko B. Lipid supply and metabolism in infancy. *Curr Opin Clin Nutr Metab Care* 1998, 1(2):171–177.

33. Larque E, Demmelmair H, Berger B, Hasbargen U, Koletzko B. In vivo investigation of the placental transfer of (13)C-labeled fatty acids in humans. *J Lipid Res* 2003, 44(1):49–55.

34. Al MD, van Houwelingen AC, Kester AD, Hasaart TH, de Jong AE, Hornstra G. Maternal essential fatty acid patterns during normal pregnancy and their relationship to the neonatal essential fatty acid status. *Br J Nutr* 1995, 74(1):55–68.

35. Olsen SF, Joensen HD. High liveborn birth weights in the Faroes: A comparison between birth weights in the Faroes and in Denmark. *J Epidemiol Community Health* 1985, 39(1):27–32.

36. Elias SL, Innis SM. Infant plasma trans, n-6, and n-3 fatty acids and conjugated linoleic acids are related to maternal plasma fatty acids, length of gestation, and birth weight and length. *Am J Clin Nutr* 2001, 73(4):807–814.

37. Jensen CL, Maude M, Anderson RE, Heird WC. Effect of docosahexaenoic acid supplementation of lactating women on the fatty acid composition of breast milk lipids and maternal and infant plasma phospholipids. *Am J Clin Nutr* 2000, 71(1 Suppl):292S–299S.

38. Decsi T, Koletzko B. n-3 Fatty acids and pregnancy outcomes. *Curr Opin Clin Nutr Metab Care* 2005, 8(2):161–166.

39. Olsen SF, Sorensen JD, Secher NJ, Hedegaard M, Henriksen TB, Hansen HS, Grant A. Randomised controlled trial of effect of fish-oil supplementation on pregnancy duration. *Lancet* 1992, 339(8800):1003–1007.

40. Smuts CM, Huang M, Mundy D, Plasse T, Major S, Carlson SE. A randomized trial of docosahexaenoic acid supplementation during the third trimester of pregnancy. *Obstet Gynecol* 2003, 101(3):469–479.

41. Sardesai VM. *Introduction to Clinical Nutrition*. New York: Marcel Dekker, 1998.

42. DeFilippis AP, Sperling LS. Understanding omega-3's. *Am Heart J* 2006, 151(3):564–570.

43. Breslow JL. n-3 Fatty acids and cardiovascular disease. *Am J Clin Nutr* 2006, 83(6):S1477–S1482.

44. Yokoyama M, Origasa H, Matsuzaki M, Matsuzawa Y, Saito Y, Ishikawa Y, Oikawa S, Sasaki J, Hishida H, Itakura H et al. Effects of eicosapentaenoic acid on major coronary events in hypercholesterolaemic patients (JELIS): A randomised open-label, blinded endpoint analysis. *Lancet* 2007, 369(9567):1090–1098.

45. Morris MC, Sacks F, Rosner B. Does fishoil lower blood pressure? A meta-analysis of controlled trials. *Circulation* 1993, 88(2):523–533.

46. Lavie CJ, Milani RV, Mehra MR, Ventura HO. Omega-3 polyunsaturated fatty acids and cardiovascular diseases. *J Am Coll Cardiol* 2009, 54(7):585–594.

47. Harris WS. n-3 Fatty acids and serum lipoproteins: Human studies. *Am J Clin Nutr* 1997, 65(5):1645S–1654S.

# IV

# Metabolic Syndrome: Obesity, Diabetes, and Hypertension

# 8

# Potential of Nutraceuticals against Obesity Complications

Kylie Conroy, Isobel Davidson, and Mary Warnock

## Contents

## Obesity and its consequences

Defined as the excessive accumulation of body fat that presents a risk to health (WHO 2011), obesity is the result of the interaction between environmental factors, such as increased calorie intake and reduced levels of physical activity, and genetics with heritability ranging from 30% to 70% (Lyon and Hirschhorn 2005). Regarded as the epidemic of the twenty-first century, over a half of the population in many countries are overweight or obese (OECD 2009). In 2005, approximately one-third of the world's population was estimated to be overweight or obese (Kelly et al. 2008). Once considered a problem only in high-income countries, obesity is now also increasing in low- and middle-income countries, particularly in urban settings (WHO 2011).

Obesity has been regarded as a disease by the World Health Organisation (WHO) since its inception in 1948; however, it was not deemed a public health problem by WHO until 1997 (James 2008). The most widely used method of classifying overweight and obesity is by body mass index (BMI = weight/height$^2$), and those with a BMI of 25–29.99 kg/m$^2$ are classified as overweight and those with a BMI of $\geq$30 kg/m$^2$ as obese (WHO 2011). However, although BMI correlates with body fat it is not a direct measure and results may be skewed by a high muscle mass such as can be seen in some athletes.

Therefore, it is necessary to further distinguish obesity by fat distribution on the body, whether it is peripheral or abdominal, such as that determined by using a measure of an individual's waist circumference (WC) and/or waist/hip ratio (WHR).

Such measurements are useful as abdominal, or central, obesity in particular is associated with a cluster of risk factors for cardiovascular disease (CVD) and type 2 diabetes mellitus (T2DM), collectively termed the metabolic syndrome (Alberti et al. 2009). This is due to differences in the activity of fat depots in their secretion of factors such as adipokines and cytokines. Also associated with the presence and severity of obesity is nonalcoholic fatty liver disease (NAFLD) (de Alwis and Day 2008), which itself has a strong association with metabolic syndrome (Targher et al. 2010). These diseases have increased in parallel with each other that highlights the recent theory as to their etiology that is a systemic, subclinical inflammation, which is thought to arise from vadipose tissue dysfunction (Vachharajani and Granger 2009). This form of chronic inflammation has no known infectious agent nor is it the result of a well-defined autoimmune response.

This pro-inflammatory state may explain why disease risk is not only increased for those who are obese. Analysis of data from the Framingham Heart study has found that relative risk for hypertension and CVD is increased in the overweight (Wilson et al. 2002). Furthermore, assessment of data from the National Health and Nutrition Examination Survey clearly demonstrates a high prevalence of these diseases in those who are overweight (Must et al. 1999). In a further study assessing the same factors in women, in the Nurses Health Study, and men, in the Health Professional Follow-up Study, risk was increased even at a BMI of 22 kg/m$^2$, which is generally regarded as normal (Field et al. 2001).

Antiobesity drugs, targeted to reduce energy intake by reducing appetite (e.g., Sibutramine) or the absorption of energy-dense nutrients (e.g., Orlistat), have been intensively researched; however, few have made it to the point where they are prescribed to patients. Furthermore, a recent meta-analysis of studies assessing the long-term effectiveness of these antiobesity drugs found that their use is associated with only a modest reduction in weight and no clear effects on cardiovascular risk profiles (Rucker et al. 2007). The side effects of these drugs, combined with their uncertain long-term effects, have highlighted the importance of investigating alternatives.

In a systematic review of current obesity prevention strategies, Wieringa et al. (2008) identified a key area of intervention to be nutritional factors; however, the authors also argued for a limited role for functional foods due to their inability to affect underlying dietary habits (Wieringa et al. 2008). In support of this however, plasma levels of vitamins and antioxidants are lower in obese subjects than in the normal weight subjects (Aasheim et al. 2008; Panagiotakos et al. 2005) and an inverse relationship has been shown between serum total antioxidant capacity and WC (Chrysohoou et al. 2007).

This is possibly due to high intakes of energy-dense, nutrient-poor foods, which can replace nutrient-rich foods in the diet (Drewnowski 2004; Mendoza et al. 2007). Research has further indicated the modulatory effects of vitamins and antioxidants on immune system function (Wintergerst et al. 2007) and it may, therefore, be that these reduced levels have a role in the development of chronic inflammation and ultimately disease, in obesity (Bullo et al. 2007). Thus, functional foods and nutraceuticals may provide a powerful, cost-effective adjunct to traditional weight loss methods by ameliorating the development of comorbidities associated with overweight and obesity.

## Adipose tissue and its pathogenic development

Mammals have two different types of adipose tissue with contrasting physiological roles. White adipose tissue (WAT) is involved in energy storage and brown adipose tissue (BAT) is designed to generate heat via energy expenditure (Hahn and Novak 1975). Research into the function of WAT over the last decade, prompted by the increase in obesity, has initiated a paradigm shift and it is now regarded as an endocrine organ, secreting numerous products, collectively called "adipokines." Well over 50 adipokines have so far been identified with key physiological roles in energy metabolism, inflammatory response, and cardiovascular homeostasis, and, therefore, any disturbance in their balance initiates events which impact negatively on these areas.

WAT may be further subdivided into subcutaneous and visceral fat. Subcutaneous fat (SAT) is found in the hypodermis, forming a layer under the skin and visceral adipose tissue (VAT) envelopes the internal organs in the abdominal cavity (Wronska and Kmiec 2012). These two depots differ in regards to their cellular, molecular, and physiological composition and have different prognostic and clinical implications. VAT, compared to SAT, is highly vasculated and innervated with its blood flow draining directly into the liver via the portal vein. Moreover, it has a greater number of immune cells and of large adipocytes, which are more metabolically active, sensitive to lipolysis, and insulin-resistant, and a lower pre-adipocyte-differentiating capacity (Wronska and Kmiec 2012). VAT also has a greater mortality predictive capacity than SAT (Kuk et al. 2006; McNeely et al. 2012).

Although mature adipocytes comprise the major fraction of adipose tissue, also contained within this are pre-adipocytes, fibroblasts, pericytes, endothelial cells, and macrophages (Ouchi et al. 2011). Expansion of fat mass during excess energy intake is provided through an increase in the number and size of adipocytes, and an increase in adipocyte size is generally linked to greater metabolic and pro-inflammatory secretory activity (Skurk et al. 2007). The remodeling of adipose tissue during expansion results in key changes to the microenvironment, including an increase in the number and size of adipocytes. This leads to a reduction in vasculature as angiogenesis is unable to keep pace with tissue growth, as has been found in obese humans (Goossens et al. 2011; McQuaid et al. 2011). Fibrotic changes to WAT has also been found in obese humans (Henegar et al. 2008; Mutch et al. 2009), indicating chronic inflammatory activity. An increase in adipose tissue fibrosis has also been proposed to restrict adipocyte growth and expandability leading to ectopic fat accumulation, particularly in the liver and skeletal muscle, and an adverse metabolic profile (Khan et al. 2009; Virtue and Vidal-Puig 2010).

Adipose tissue expansion is also linked with an incursion of immune cells including macrophages, lymphocytes, and mast cells (Anderson et al. 2010; Ouchi et al. 2011). There are two types of macrophage: M1, which secrete pro-inflammatory cytokines and M2, or alternatively activated, which have an anti inflammatory cytokine profile; the latter are normally resident within adipose tissue (Mantovani et al. 2004). In obese states, the balance between M2 and M1 macrophages shifts with M1 or classically activated macrophages dominating which have an enhanced secretion of pro-inflammatory cytokines (Aron-Wisnewsky et al. 2009). However, it has recently been suggested that M2 macrophages may play a role in the fibrotic development of adipose tissue as they have been found localized to areas of adipose tissue fibrosis and in the presence of adipocytes secrete fibrotic factors (Spencer et al. 2010).

It is these and possibly other as yet unidentified factors, which are thought to promote the shift in adipose tissue endocrine output from anti inflammatory to one which is pro-inflammatory. Subsequent to this increase in inflammatory cytokine production is the development of major (hypercholesterolemia, hypertension, hyperglycemia) and emerging (atherogenic dyslipidemia, insulin resistance, pro-inflammatory state, prothrombotic state) risk factors for CVD and T2DM (Grundy 2004), two primary complications of obesity. The presence of three or more of these risk factors in an individual with increased BMI and WC has been defined as the metabolic syndrome and in these individuals risk for all-cause mortality is increased by 1.5-fold (Mottillo et al. 2010).

## Insulin resistance and T2DM

An increase in VAT mass has been correlated with the development of insulin resistance, more so than total or subcutaneous fat mass (Hoffstedt et al. 1997; Kelley et al. 2000). This may be due to basal lipolytic activity of VAT being higher than SAT. In addition, it is also less sensitive to insulin's anti-lipolytic effects (Ahmadian et al. 2010). This has been proposed to increase delivery of free fatty acids (FFA) to the liver and subsequently raise systemic FFA and glucose levels (Miyazaki et al. 2002). There is clear evidence showing strong associations between insulin resistance in obesity and high circulating FFAs (Savage et al. 2007), and further investigation has revealed the ability of FFA infusions to cause peripheral insulin resistance in humans and animals (Barazzoni et al. 2012; Liu et al. 2009).

Recent meta-analysis studies have found tea (Huxley et al. 2009; Jing et al. 2009) and coffee (van Dam and Hu 2005) consumption to reduce the risk of T2DM. In the Hoorn study (van Dam et al. 2004), a population cohort comprising 2280 Dutch men and women, a daily intake of five cups of coffee was associated with a lower fasting insulin level. In a cross-sectional study of nondiabetic adults, positive relationships were found between caffeinated coffee consumption and insulin sensitivity and decaffeinated coffee and β-cell function (Loopstra-Masters et al. 2011). These effects are thought to be due to the high antioxidant and polyphenolic activity and not the caffeine content of the coffee (Oba et al. 2010). Indeed caffeine has been shown to have a negative impact on insulin sensitivity, although tolerance to caffeine's effects could develop (Keijzers et al. 2002; Petrie et al. 2004).

The primary phenolic compound of interest in coffee is chlorogenic acid (CGA), the concentration of which in coffee can be as high as 2.5 mM. Coffee drinkers are thought to ingest 500–1000 mg of CGA daily (Clifford 2000; Olthof et al. 2001). Animal studies have found CGA consumption to reduce blood glucose concentrations (Bassoli et al. 2008), in addition to lowering body weight and insulin levels (Cho et al. 2010). In humans, despite epidemiological evidence, few intervention studies investigating the effects of coffee and CGA consumption on glucose homeostasis have been undertaken. In a double-blind crossover study, an oral glucose tolerance test (OGTT) was performed in overweight participants who had, 30 min previously, consumed a one-off dose of 12 g decaffeinated coffee (containing 264 mg CGA) or 1 g CGA (van Dijk et al. 2009). This study found decaffeinated coffee to have no effect on plasma glucose or insulin concentrations. However, CGA consumption significantly reduced glucose concentrations in 15 min following OGTT, with plasma insulin concentrations also being significantly reduced both before and after the OGTT (van Dijk et al. 2009). Consumption of 10 g of CGA-enriched instant coffee (containing 90–100 mg CGA) by 12 healthy participants prior to an OGTT found it significantly reduced glucose absorption compared to the control sucrose solution (Thom 2007). Further investigation in obese individuals who consumed 11 g CGA-enriched coffee daily for 12 weeks found that this led to a 5.4 kg reduction in weight (Thom 2007); however, it was not investigated as to whether this was related to any improvements in the OGTT.

Few longer-term intervention studies have been conducted; however, an early uncontrolled study showed that 14 days of decaffeinated coffee consumption decreased plasma glucose (Naismith et al. 1970). In a randomized crossover study, very high coffee consumption for 4 weeks had no effect on fasting glucose but did, however, increase fasting insulin concentrations (van Dam et al. 2004); however, the daily intake of coffee during the test period was not clearly described. These actions of CGA on glucose metabolism have been hypothesized to be mediated in two main ways: first, by the delay of intestinal glucose absorption through the inhibition of glucose-6-phosphate translocase 1 and second by stimulation of incretin hormone glucagon-like peptide 1 (GLP-1) secretion, which maintains pancreatic β-cell responsiveness to elevated plasma glucose levels (McCarty 2005). Certainly, decaffeinated coffee consumption was found to reduce glucose absorption compared to caffeinated coffee in addition to increasing plasma GLP-1 concentrations (Johnston et al. 2003).

In contrast with coffee, tea and its polyphenolic constituents has been the subject of great interest. The major teas consumed are black tea (wholly fermented), green tea (minimal fermentation), and oolong tea (partially fermented) and these are all products of the *Camellia sinensis* plant species (Yang and Landau 2000). Animal and in vitro research suggests that tea and its constituents positively effect glucose metabolism and diabetes development through several mechanisms (Kao et al. 2006), and these insulin-enhancing effects are thought to be primarily due to the polyphenolic catechin, epigallo-catechin gallate (EGCG) (Anderson and Polansky 2002). However, other compounds with phenolic and antioxidant activity have been described, namely the theaflavins and thearubigins.

Extracts of these teas have been produced and tested for their effects on glucose homeostasis in humans; these are preferred over the use of pure EGCG as they have been shown to be more stable (Kaszkin et al. 2004). In healthy individuals, consumption of 1.5 g green tea powder (84 mg EGCG) in 150 mL hot water prior to an OGTT significantly reduced blood glucose levels compared to hot water alone at 30 and 120 min (Tsuneki et al. 2004). However, participant characteristics were not outlined. In another study, healthy, normal weight men were supplemented with three capsules of green tea extract (136 mg EGCG) three times before an OGTT. Plasma glucose levels were unchanged; however, insulin levels were significantly lower than the placebo (Venables et al. 2008). Contrastingly, consumption of green tea (leaf brewed in hot water; 32.4 mg EGCG) with a meal of white bread and turkey was found to significantly increase postprandial plasma glucose at 120 min with no further effects on plasma insulin (Josic et al. 2010). Therefore, acute intake of green tea may improve insulin sensitivity in healthy adults and control postprandial hyperglycemia; however, only when taken prior to a meal.

Longer-term interventions have had mixed results. In a randomized controlled trial, healthy, overweight to obese males consumed two capsules of green tea extract daily for 2 weeks (>97% EGCG), which had no effect on glucose tolerance or insulin sensitivity or secretion (Brown et al. 2009). Moreover, a randomized crossover trial, comprising normal weight individuals with borderline T2DM or overt T2DM who consumed a packet of green tea powder/extract daily (456 mg EGCG), found that at 2 months, although blood glucose and insulin levels were lower than baseline these were not significantly different from controls (Fukino et al. 2005). However, in a follow-on study by the same research group, significant reductions in hemoglobin A1c (HbA1c) levels (provides average measure of plasma glucose concentration over time) were found (Fukino et al. 2007). In a randomized, placebo-controlled, double-blind trial where participants consumed one capsule of a green tea extract (857 mg EGCG) or placebo daily for 16 weeks also found significant reductions in HbA1c; however, this was only when compared within groups as no significance was found when compared to placebo. Moreover, no significant effects, either between or within groups, were noted for fasting glucose or insulin (Hsu et al. 2011).

Oolong tea has also been investigated for its possible effects on glucose metabolism. No effects on fasting plasma glucose or insulin were noted in a study comprising healthy men who consumed oolong tea daily for 5 days (Baer et al. 2011). However, in a randomized crossover study, participants with T2DM consumed 1500 mL oolong tea (386 mg EGCG) daily for 30 days, which resulted in a significant reduction in plasma glucose compared to consumption of water alone (Hosoda et al. 2003). No other parameters of diabetic control were assessed. The effects of tea on lowering plasma glucose may be mediated through inhibition of carbohydrate digestive enzymes, as has been found in rats (Kobayashi et al. 2000; Shimizu et al. 2000). Moreover, EGCG and other tea polyphenols have been found to enhance insulin action in adipocytes and myocytes (Wu et al. 2004a,b).

Other nutraceuticals, which have been investigated include cinnamon; in a systematic review of the literature, only five studies were found investigating its effects in T2DM patients (Kirkham et al. 2009). Of these, only two reported significant improvements in fasting blood glucose compared to the placebo (Khan et al. 2003; Mang et al. 2006); in another study involving insulin-resistant individuals, significant reductions in fasting blood glucose were found in those supplemented with cinnamon extract compared to placebo (Ziegenfuss et al. 2006). However, in a recent large study involving 109 T2DM patients who took one 9 g capsule of cinnamon daily for 90 days in addition to their usual care, significant reductions in HbA1c levels were found (Crawford 2009). In healthy humans, encapsulated cinnamon taken for daily 14 days (3 g) has been found to positively affect responses to an OGTT, with a significant improvement in blood glucose levels at day 1 following intake, and at day 14 a significant improvement in blood insulin levels and insulin sensitivity (Solomon and Blannin 2009).

There is clear evidence from animal experiments suggesting that curcumin and turmeric may have beneficial effects in diabetes and insulin resistance (Aggarwal 2010). However, only a few human intervention studies have been undertaken, with mixed results. In a randomized crossover trial, the effects of 6 g encapsulated turmeric on blood glucose and insulin in healthy subjects following an OGTT were investigated (Wickenberg et al. 2010). Insulin was the only parameter affected and was significantly increased compared to the placebo intervention. In a further study, the daily consumption of seven 2.8 g capsules of turmeric had no effect on fasting plasma glucose in healthy participants (Tang et al. 2008), although this is most likely due to the study population.

In summary, although human intervention studies into the effects of nutraceuticals show improvements in aspects of glucose metabolism and T2DM these results are not consistent and therefore further investigations are required in order to gain a better understanding of their effects in abrogating the development of insulin resistance and the progression to T2DM.

## Atherosclerosis and CVD

As noted earlier, elevated visceral adiposity is associated with an increase in inflammatory cytokine production leading to hypertension, hyperglycemia, and dyslipidemia, factors with known atherogenic properties. Moreover, endothelial dysfunction has been found in obesity and is itself an independent risk factor for CVD (Pasimeni et al. 2006). The effect of diet quality on cardiovascular health has long been known. Systematic meta-analysis of the research indicates that diets high in fruit and vegetables are cardioprotective (Mente et al. 2009; Tyrovolas and Panagiotakos 2010) and a recent investigation into urinary total polyphenol excretion (TPE) and hypertension has found that those with the highest level of urinary TPE have a low prevalence of hypertension compared to those with the lowest excretion. Systolic and diastolic blood pressure was further negatively associated with urinary TPE in this study (Medina-Remon et al. 2011). These effects have been attributed to their high fiber content but also to their antioxidant polyphenolic content (Arts and Hollman 2005); with this view, research has been undertaken to elucidate specific plant-derived products, which may modulate cardiovascular risk.

Red wine, and its primary polyphenol, resveratrol, has been of primary interest as epidemiological studies have found a lower risk of CVD mortality in wine drinkers (Theobald et al. 2000). Although some studies suggest that the benefit of red wine may be related to the moderate intake of alcohol, which increases high-density lipoprotein (HDL)-cholesterol, lowers hypertension, and is antithrombotic (Gronbak 2009), numerous in vitro and animal studies have illustrated the strong cardioprotective effects of resveratrol (Petrovski et al. 2011). In a randomized, triple-blind, placebo-controlled trial comprising three parallel arms (grape extract containing resveratrol 8 mg [GE-RES], a grape extract [GE] with a similar polyphenolic content but lacking resveratrol, and placebo [maltodextrin]), patients with risk factors for CVD and taking statins were asked to take one capsule daily for 6 months and then two capsules daily for further 6 months (Tome-Carneiro et al. 2012). It was found that consumption of GE-RES led to a significant reduction in markers of inflammation and an improvement in fibrinolytic status, compared to GE or placebo. In a randomized crossover trial, obese participants with untreated borderline hypertension were given acute doses of placebo or encapsulated resveratrol (30, 90, and 270 mg) at weekly intervals. One hour following consumption, arterial low-mediated dilatation of the brachial artery (FMD), a biomarker of endothelial function and cardiovascular health, was assessed. Plasma resveratrol significantly and dose dependently increased, which was also related to a significant dose-dependent increase in FMD (Wong et al. 2011).

However, in a double-blind, placebo-controlled, three-period crossover trial with participants who had high normal blood pressure or hypertension, supplementation with either 280 or 560 mg red wine polyphenols did not significantly lower peripheral or central blood pressure (Botden et al. 2012). Resveratrol is found primarily in the skin and seeds of grapes, which has led to the development of grape-seed extracts. Four week supplementation with a muscadine grape-seed extract (1300 mg daily) of adults with coronary heart disease or more than one cardiovascular risk factor was found to significantly increase brachial artery diameter (Mellen et al. 2010). However, it had no effect on FMD or on markers of inflammation or lipid peroxidation. Regardless, in a recent meta-analysis of randomized controlled trials conducted up until 2010. It was found that grape-seed extract was able to significantly lower systolic blood pressure; nonetheless, no overall significant effects on plasma lipids or markers of inflammation were noted (Feringa et al. 2011).

Other polyphenol-rich foods have also been investigated for their potential to reduce cardiovascular risk factors and abrogate the development of CVD. Recent meta-analysis studies of both acute and long-term randomized trials involving cocoa polyphenols have shown it to have promising effects in increasing FMD and improving insulin resistance (Hooper et al. 2012; Shrime et al. 2011). A small randomized, single-blind, crossover study of overweight and obese subjects ($n = 14$) found that polyphenol-rich chocolate reduced fasting blood glucose levels, and systolic and diastolic blood pressure, with 500 and 1000 mg polyphenol doses being equally effective (Almoosawi et al. 2010). In a similar placebo-controlled, double-blind study, daily supplementation for 8 weeks with 45 g of high-polyphenol chocolate (85% cocoa solids; 16.6 mg epicatechins) of 12 T2DM patients was found to have no effects on blood glucose; however, HDL-cholesterol was significantly increased compared to the control (Mellor et al. 2010). Similarly, in a larger randomized crossover feeding trial comprising patients with T2DM or hypertension, dyslipidemia, and obesity, daily consumption of 40 g of cocoa powder (495.5 mg total polyphenols; 46.1 mg epicatechin) with 500 mL skimmed milk for 4 weeks resulted in significant increases in HDL-cholesterol and significant decreases in oxidized low-density lipoprotein (oxLDL). These changes were also associated with higher urinary cocoa polyphenol metabolites (Khan et al. 2012). The beneficial effect of polyphenols has also been noted in healthy individuals. Consumption for 3 weeks of six capsules per day containing green tea extract (25 mg EGCG) by healthy men significantly reduced the ratio of total/HDL-cholesterol compared to baseline; however, no effects on blood lipids or blood pressure were noted (Frank et al. 2009).

A number of studies have investigated the effects of pomegranate juice consumption on markers of CVD. In a small 1 year study comprising 10 participants with carotid artery stenosis, consumption of 50 mL pomegranate juice per day resulted in significant reductions in LDL oxidation by 90% (Aviram et al. 2004). In diabetic patients, consumption of 50 mL/day of concentrated pomegranate juice for 4 weeks significantly improved the levels of markers (Rock et al. 2008). In this small study, consumption of 50 mL of pomegranate juice per day for 4 weeks was able to increase PON1–HDL association, which ameliorates serum lipid oxidation by hydrolyzing oxidized lipids (Fuhrman et al. 2010), which could limit the development of coronary artery disease.

To summarize, the etiology of CVD is complex, involving many factors. Although polyphenols found in red wine and grapes, in particular resveratrol, show promise in abrogating CVD development, this has been limited to animal and in vitro studies with results from human studies remaining inconsistent. Nonetheless, human intervention studies involving polyphenol-enriched cocoa have found consistent benefits in improving markers of CVD risk which warrants further longer-term investigations.

## Nonalcoholic fatty liver disease

This disease spectrum, occurring in non-drinkers, encompasses hepatic steatosis, at its mildest, through nonalcoholic steatohepatitis (NASH) to NAFLD-associated cirrhosis, end-stage liver disease, and possible progression to hepatocellular carcinoma (de Alwis and Day 2008). Insulin resistance is a known factor in the development of NAFLD and this in combination with increased FFA delivery to hepatocytes results in increased lipogenesis, decreased FFA oxidation and lipid transport in these cells, thereby initiating hepatic steatosis (McCullough 2006). Progression from this to NASH is thought to result from increased hepatic oxidative stress and lipid peroxidation (Day 2002; Koruk et al. 2004; Sumida et al. 2003).

Studies have, therefore, investigated the potential of antioxidant vitamins to alleviate NASH development. Nascent research using vitamin E has found mixed results (Adams and Angulo 2003; Harrison et al. 2003); moreover, health issues were related with high doses of vitamin E (Miller et al. 2005). A high coffee intake has been associated with the reduced progression of liver disease (Gutiérrez-Grobe et al. 2012; Johnson et al. 2011; Molloy et al. 2011); however, it is uncertain whether these benefits are due to caffeine or polyphenols present within coffee. In a rat model of diet-induced liver disease, supplementation of a high-fat diet with 5% caffeinated coffee extract abrogated the development of NAFLD without changing abdominal obesity or dyslipidemia (Panchal et al. 2012). Nonetheless, in a similar model daily decaffeinated coffee consumption (equating to six cups of espresso coffee for a 70 kg person) resulted in a reduction in hepatic fat accumulation, systemic and liver oxidative stress, and liver inflammation (Vitaglione et al. 2011). Due to the potential negative health effects of caffeine (Heckman et al. 2010) and its previously mentioned effects on insulin sensitivity, further research comparing the effects of coffee polyphenols and caffeine alone is warranted in order to ascertain which compound provides the beneficial effects.

In a small 3 month intervention study ($n = 15$), grape-seed extract, whose major polyphenol is resveratrol, was found to improve a parameter of liver function and reduce the grade of steatosis (Khoshbaten et al. 2010). However, the grape-seed extract was not standardized and nor was the polyphenol content described. In a rat model of liver disease, 10 mg of resveratrol daily for 4 weeks attenuated overall NAFLD severity (Bujanda et al. 2008). Investigation into the mechanism of resveratrol activity has found that it reduces plasma fatty acid availability by increasing their oxidation and possibly by reducing their flux from plasma, thereby decreasing liver fat accumulation (Gómez-Zorita et al. 2012).

Oleuropein, the main polyphenol of olive oil when given to rats fed a high-fat diet, was also found to have benefits in diet-induced NAFLD by reducing hepatic fat accumulation (Park et al. 2010). Olive leaf extract, high in oleuropein, was shown to have similar effects in the spontaneously hypertensive rat fed a high-fat diet along with a reduction in liver oxidative stress (Omagari et al. 2010). Gene expression analysis of murine hepatic tissue following treatment with oleuropein during a high-fat diet found that genes involved in hepatic fatty acid uptake and transport and those regulating oxidative stress responses, detoxification of lipid peroxidation products, and pro-inflammatory cytokine production were reduced (Kim et al. 2010). This indicates the multifactorial effects of oleuropein in attenuating NAFLD development. Other compounds that have shown benefits in animal models of NAFLD include black rice extract (Jang et al. 2012) and lotus root extract (Tsuruta et al. 2011).

In summary, although little human research into the potential benefits of polyphenols in abrogating the development or progression of NAFLD has been undertaken, a number of animal studies provide clear evidence of their effectiveness. Olive-derived polyphenols and coffee in particular show great potential; however, more remains to be understood. These results warrant the development of well-planned human intervention studies in order to fully assess their potential role in the treatment of NAFLD.

## Conclusions

Obesity is implicated in the development of a wide variety of diseases; however, those outlined within this chapter are the major comorbidities it is associated with. The health effects seen in humans of various forms of fruits and vegetables along with leaf and flower infusions have led to a vast amount of research expense, in both time and money, investigating their benefits in extract form. The use of nutraceuticals in modulating the mechanisms contributing to the comorbidities of obesity is the focus of numerous studies. The results that have emerged are mixed; however, beneficial effects have been found in certain areas. Many have found nutraceuticals able to positively modulate these mechanisms and have a great potential to reduce overweight and obesity-related comorbidities. Nonetheless much research, especially that utilizing well-structured randomized controlled trials, is still necessary before these become mainstream interventions in the treatment of obesity-related diseases.

# References

Aasheim, E. T., Hofso, D., Hjelmesaeth, J., Birkeland, K. I., and Bohmer, T. 2008. Vitamin status in morbidly obese patients: A cross-sectional study. *American Journal of Clinical Nutrition*, 87(2), 362–369.

Adams, L. A. and Angulo, P. 2003. Vitamins E and C for the treatment of NASH: Duplication of results but lack of demonstration of efficacy. *American Journal of Gastroenterology*, 98(11), 2348–2350.

Aggarwal, B. B. 2010. Targeting inflammation-induced obesity and metabolic diseases by curcumin and other nutraceuticals. *Annual Review of Nutrition*, 30(1), 173–199.

Ahmadian, M., Wang, Y., and Sul, H. S. 2010. Lipolysis in adipocytes. *The International Journal of Biochemistry & Cell Biology*, 42(5), 555–559.

Alberti, K. G. M. M., Eckel, R. H., Grundy, S. M., Zimmet, P. Z., Cleeman, J. I., Donato, K. A., Fruchart, J. C., James, W. P., Loria, C. M., and Smith, S. C., Jr. 2009. Harmonizing the metabolic syndrome: A joint interim statement of the International Diabetes Federation Task Force on Epidemiology and Prevention; National Heart, Lung, and Blood Institute; American Heart Association; World Heart Federation; International Atherosclerosis Society; and International Association for the Study of Obesity. *Circulation*, 120(16), 1640–1645.

Almoosawi, S., Al-Dujaili, E. A. S., and Fyfe, L. 2010. Polyphenol-rich dark chocolate: Effect on fasting capillary glucose, total cholesterol, blood pressure and glucocorticoids in healthy overweight and obese subjects. *Proceedings of the Nutrition Society*, 69, E29.

Anderson, E. K., Gutierrez, D. A., and Hasty, A. H. 2010. Adipose tissue recruitment of leukocytes. *Current Opinion in Lipidology*, 21(3), 172–177.

Anderson, R. A. and Polansky, M. M. 2002. Tea enhances insulin activity. *Journal of Agricultural and Food Chemistry*, 50(24), 7182–7186.

Aron-Wisnewsky, J., Tordjman, J., Poitou, C., Darakhshan, F., Hugol, D., Basdevant, A., Aissat, A., Guerre-Millo, M., and Clément, K. 2009. Human adipose tissue macrophages: M1 and M2 cell surface markers in subcutaneous and omental depots and after weight loss. *Journal of Clinical Endocrinology & Metabolism*, 94(11), 4619–4623.

Arts, I. C. and Hollman, P. C. 2005. Polyphenols and disease risk in epidemiologic studies. *The American Journal of Clinical Nutrition*, 81(1), 317S–325S.

Aviram, M., Rosenblat, M., Gaitini, D., Nitecki, S., Hoffman, A., Dornfeld, L., Volkova, N. et al. 2004. Pomegranate juice consumption for 3 years by patients with carotid artery stenosis reduces common carotid intima–media thickness, blood pressure and LDL oxidation. *Clinical Nutrition*, 23(3), 423–433.

Baer, D., Novotny, J., Harris, G., Stote, K., Clevidence, B., and Rumpler, W. 2011. Oolong tea does not improve glucose metabolism in non-diabetic adults. *European Journal of Clinical Nutrition*, 65, 87–93.

Barazzoni, R., Zanetti, M., Cappellari, G., Semolic, A., Boschelle, M., Codarin, E., Pirulli, A., Cattin, L., and Guarnieri, G. 2012. Fatty acids acutely enhance insulin-induced oxidative stress and cause insulin resistance by increasing mitochondrial reactive oxygen species (ROS) generation and nuclear factor-KB inhibitor (IKB)-nuclear factor-KB (NFKB) activation in rat muscle, in the absence of mitochondrial dysfunction. *Diabetologia*, 55(3), 773–782.

Bassoli, B. K., Cassolla, P., Borba-Murad, G. R., Constantin, J., Salgueiro-Pagadigorria, C. L., Bazotte, R. B., da Silva, R. S. r. d. S. F., and de Souza, H. M. 2008. Chlorogenic acid reduces the plasma glucose peak in the oral glucose tolerance test: Effects on hepatic glucose release and glycaemia. *Cell Biochemistry and Function*, 26(3), 320–328.

Botden, I. P. G., Draijer, R., Westerhof, B. E., Rutten, J. H. W., Langendonk, J. G., Sijbrands, E. J. G., Danser, A. H. J., Zock, P. L., and van den Meiracker, A. H. 2012. Red wine polyphenols do not lower peripheral or central blood pressure in high normal blood pressure and hypertension. *American Journal of Hypertension*, 25(6), 718–723.

Brown, A. L., Lane, J., Coverly, J., Stocks, J., Jackson, S., Stephen, A., Bluck, L., Coward, A., and Hendrickx, H. 2009. Effects of dietary supplementation with the green tea polyphenol epigallocatechin-3-gallate on insulin resistance and associated metabolic risk factors: Randomized controlled trial. *British Journal of Nutrition*, 101(6), 886–894.

Bujanda, L., Hijona, E., Larzabal, M., Beraza, M., Aldazabal, P., García-Urkia, N., Sarasqueta, C. et al. 2008. Resveratrol inhibits nonalcoholic fatty liver disease in rats. *BMC Gastroenterology*, 8, 40.

Bullo, M., Casa-Agustench, P., Amigo-Correig, P., Aranceta, J., and Salas-Salvado, J. 2007. Inflammation, obesity and comorbidities: The role of diet. *Public Health Nutrition*, 10(10A), 1164–1172.

Cho, A. S., Jeon, S. M., Kim, M. J., Yeo, J., Seo, K. I., Choi, M. S., and Lee, M. K. 2010. Chlorogenic acid exhibits anti-obesity property and improves lipid metabolism in high-fat diet-induced obese mice. *Food and Chemical Toxicology*, 48(3), 937–943.

Chrysohoou, C., Panagiotakos, D. B., Pitsavos, C., Skoumas, I., Papademetriou, L., Economou, M., and Stefanadis, C. 2007. The implication of obesity on total antioxidant capacity in apparently healthy men and women: The ATTICA study. *Nutrition, Metabolism and Cardiovascular Diseases*, 17(8), 590–597.

Clifford, M. N. 2000. Chlorogenic acids and other cinnamates—Nature, occurrence, dietary burden, absorption and metabolism. *Journal of the Science of Food and Agriculture*, 80(7), 1033–1043.

Crawford, P. 2009. Effectiveness of cinnamon for lowering hemoglobin A1C in patients with type 2 diabetes: A randomized, controlled trial. *The Journal of the American Board of Family Medicine*, 22(5), 507–512.

Day, C. P. 2002. Pathogenesis of steatohepatitis. *Best Practice & Research Clinical Gastroenterology*, 16(5), 663–678.

de Alwis, N. M. W. and Day, C. P. 2008. Non-alcoholic fatty liver disease: The mist gradually clears. *Journal of Hepatology*, 48(Suppl 1), S104–S112.

Drewnowski, A. 2004. Obesity and the food environment: Dietary energy density and diet costs. *American Journal of Preventive Medicine*, 27(3 Suppl 1), 154–162.

Feringa, H. H. H., Laskey, D. A., Dickson, J. E., and Coleman, C. I. 2011. The effect of grape seed extract on cardiovascular risk markers: A meta-analysis of randomized controlled trials. *Journal of the American Dietetic Association*, 111(8), 1173–1181.

Field, A. E., Coakley, E. H., Must, A., Spadano, J. L., Laird, N., Dietz, W. H., Rimm, E., and Colditz, G. A. 2001. Impact of overweight on the risk of developing common chronic diseases during a 10-year period. *Archives of Internal Medicine*, 161(13), 1581–1586.

Frank, J., George, T. W., Lodge, J. K., Rodriguez-Mateos, A. M., Spencer, J. P. E., Minihane, A. M., and Rimbach, G. 2009. Daily consumption of an aqueous green tea extract supplement does not impair liver function or alter cardiovascular disease risk biomarkers in healthy men. *The Journal of Nutrition*, 139(1), 58–62.

Fukino, Y., Ikeda, A., Maruyama, K., Aoki, N., Okubo, T., and Iso, H. 2007. Randomized controlled trial for an effect of green tea-extract powder supplementation on glucose abnormalities. *European Journal of Clinical Nutrition*, 62(8), 953–960.

Fukino, Y., Shimbo, M., Aoki, N., Okubo, T., and Iso, H. 2005. Randomized controlled trial for an effect of green tea consumption on insulin resistance and inflammation markers. *Journal of Nutritional Science and Vitaminology*, 51(5), 335–342.

Fuhrman, B., Volkova, N., and Aviram, M. 2010. Pomegranate juice polyphenols increase recombinant paraoxonase-1 binding to high-density lipoprotein: Studies in vitro and in diabetic patients. *Nutrition*, 26(4), 59–66.

Gómez-Zorita, S., Fernández-Quintela, A., Macarulla, M. T., Aguirre, L., Hijona, E., Bujanda, L., Milagro, F., Martínez, J. A., and Portillo M. P. 2012. Resveratrol attenuates steatosis in obese Zucker rats by decreasing fatty acid availability and reducing oxidative stress. *British Journal of Nutrition*, 107(2), 202–210.

Goossens, G. H., Bizzarri, A., Venteclef, N., Essers, Y., Cleutjens, J. P., Konings, E., Jocken, J. W. E. et al. 2011. Increased adipose tissue oxygen tension in obese compared with lean men is accompanied by insulin resistance, impaired adipose tissue capillarization, and inflammation/clinical perspective. *Circulation*, 124(1), 67–76.

Gronbak, M. 2009. The positive and negative health effects of alcohol and the public health implications. *Journal of Internal Medicine*, 265(4), 407–420.

Grundy, S. M. 2004. Obesity, metabolic syndrome, and cardiovascular disease. *Journal of Clinical Endocrinology & Metabolism*, 89(6), 2595–2600.

Gutiérrez-Grobe, Y., Chávez-Tapia, N., Sánchez-Valle, V., Gavilanes-Espinar, J., Ponciano-Rodríguez, G., Uribe, M., and Méndez-Sánchez, N. 2012. High coffee intake is associated with lower grade nonalcoholic fatty liver disease: The role of peripheral antioxidant activity. *Annals of Hepatology*, 11(3), 350–355.

Hahn, P. and Novak, M. 1975. Development of brown and white adipose tissue. *Journal of Lipid Research*, 16(2), 79–91.

Harrison, S. A., Torgerson, S., Hayashi, P., Ward, J., and Schenker, S. 2003. Vitamin E and vitamin C treatment improves fibrosis in patients with nonalcoholic steatohepatitis. *American Journal of Gastroenterology*, 98(11), 2485–2490.

Heckman, M., Weil, J., and Gonzalez de Mejia, E. 2010. Caffeine (1,3,7-trimethylxanthine) in foods: A comprehensive review on consumption, functionality, safety, and regulatory matters. *Journal of Food Science*, 75(3), R77–R87.

Henegar, C., Tordjman, J., Achard, V., Lacasa, D., Cremer, I., Guerre-Millo, M., Poitou, C. et al. 2008. Adipose tissue transcriptomic signature highlights the pathological relevance of extracellular matrix in human obesity. *Genome Biology*, 9(1), R14.

Hoffstedt, J., Arner, P., Hellers, G., and Lönnqvist, F. 1997. Variation in adrenergic regulation of lipolysis between omental and subcutaneous adipocytes from obese and non-obese men. *Journal of Lipid Research*, 38(4), 795–804.

Hooper, L., Kay, C., Abdelhamid, A., Kroon, P. A., Cohn, J. S., Rimm, E. B., and Cassidy, A. 2012. Effects of chocolate, cocoa, and flavan-3-ols on cardiovascular health: A systematic review and meta-analysis of randomized trials. *The American Journal of Clinical Nutrition*, 95(3), 740–751.

Hosoda, K., Wang, M. F., Liao, M. L., Chuang, C. K., Iha, M., Clevidence, B., and Yamamoto, S. 2003. Antihyperglycemic effect of oolong tea in type 2 diabetes. *Diabetes Care*, 26(6), 1714–1718.

Hsu, C., Liao, Y., Lin, S., Tsai, T., Huang, C., and Chou, P. 2011. Does supplementation with green tea extract improve insulin resistance in obese type 2 diabetics? A randomised, double-blind, and placebo-controlled clinical study. *Alternative Medical Review*, 16(2), 157–163.

Huxley, R., Lee, C. M. Y., Barzi, F., Timmermeister, L., Czernichow, S., Perkovic, V., Grobbee, D. E., Batty, D., and Woodward, M. 2009. Coffee, decaffeinated coffee, and tea consumption in relation to incident type 2 diabetes mellitus: A systematic review with meta-analysis. *Archives of Internal Medicine*, 169(22), 2053–2063.

Jang, H., Park, M., Kim, H., Lee, Y., Hwang, K., Park, J., Park, D., and Kwon, O. 2012. Black rice (*Oryza sativa* L.) extract attenuates hepatic steatosis in C57BL/6 J mice fed a high-fat diet via fatty acid oxidation. *Nutrition and Metabolism*, 9, 27.

James, W. P. T. 2008. WHO recognition of the global obesity epidemic. *International Journal of Obesity*, 32(S7), S120–S126.

Jing, Y., Han, G., Hu, Y., Bi, Y., Li, L., and Zhu, D. 2009. Tea consumption and risk of type 2 diabetes: A meta-analysis of cohort studies. *Journal of General Internal Medicine*, 24(5), 557–562.

Johnson, S., Koh, W., Wang, R., Govindarajan, S., Yu, M., and Yuan, J. 2011. Coffee consumption and reduced risk of hepatocellular carcinoma: Findings from the Singapore Chinese Health Study. *Cancer Causes Control*, 22(3), 503–510.

Johnston, K. L., Clifford, M. N., and Morgan, L. M. 2003. Coffee acutely modifies gastrointestinal hormone secretion and glucose tolerance in humans: Glycemic effects of chlorogenic acid and caffeine. *The American Journal of Clinical Nutrition*, 78(4), 728–733.

Josic, J., Olsson, A., Wickeberg, J., Lindstedt, S., and Hlebowicz, J. 2010. Does green tea affect postprandial glucose, insulin and satiety in healthy subjects: A randomised controlled trial. *Nutrition Journal*, 9, 63.

Kao, Y. H., Chang, H. H., Lee, M. J., and Chen, C. L. 2006. Tea, obesity, and diabetes. *Molecular Nutrition & Food Research*, 50(2), 188–210.

Kaszkin, M., Beck, K. F., Eberhardt, W., and Pfeilschifter, J. 2004. Unravelling green tea's mechanisms of action: More than meets the eye. *Molecular Pharmacology*, 65(1), 15–17.

Keijzers, G. B., De Galan, B. E., Tack, C. J., and Smits, P. 2002. Caffeine can decrease insulin sensitivity in humans. *Diabetes Care*, 25(2), 364–369.

Kelley, D. E., Thaete, F. L., Troost, F., Huwe, T., and Goodpaster, B. H. 2000. Subdivisions of subcutaneous abdominal adipose tissue and insulin resistance. *American Journal of Physiology—Endocrinology and Metabolism*, 278(5), E941–E948.

Kelly, T., Yang, W., Chen, C. S., Reynolds, K., and He, J. 2008. Global burden of obesity in 2005 and projections to 2030. *International Journal of Obesity*, 32(9), 1431–1437.

Khan, A., Safdar, M., Ali Khan, M. M., Khattak, K. N., and Anderson, R. A. 2003. Cinnamon improves glucose and lipids of people with type 2 diabetes. *Diabetes Care*, 26(12), 3215–3218.

Khan, N., Monagas, M., Andres-Lacueva, C., Casas, R., Urpi-Sardá, M., Lamuela-Raventos, R. M., and Estruch, R. 2012. Regular consumption of cocoa powder with milk increases HDL cholesterol and reduces oxidized LDL levels in subjects at high-risk of cardiovascular disease. *Nutrition, Metabolism and Cardiovascular Diseases*, 22(12), 1046–1053.

Khan, T., Muise, E. S., Iyengar, P., Wang, Z. V., Chandalia, M., Abate, N., Zhang, B. B., Bonaldo, P., Chua, S., and Scherer, P. E. 2009. Metabolic dysregulation and adipose tissue fibrosis: Role of collagen VI. *Molecular and Cellular Biology*, 29(6), 1575–1591.

Khoshbaten, M., Aliasgarzadeh, A., Masnadi, K., Farhang, S., Tarzamani, M., Babaei, H., Kiani, J., Zaare, M., and Najafipoor, F. 2010. Grape seed extract to improve liver function in patients with nonalcoholic fatty liver change. *Saudi Journal of Gastroenterology*, 16(3), 194–197.

Kim, Y., Choi, Y., and Park, T. 2010. Hepatoprotective effect of oleuropein in mice: Mechanisms uncovered by gene expression profiling. *Biotechnology Journal*, 5(9), 950–960.

Kirkham, S., Akilen, R., Sharma, S., and Tsiami, A. 2009. The potential of cinnamon to reduce blood glucose levels in patients with type 2 diabetes and insulin resistance. *Diabetes, Obesity and Metabolism*, 11(12), 1100–1113.

Kobayashi, Y., Suzuki, M., Satsu, H., Arai, S., Hara, Y., Suzuki, K., Miyamoto, Y., and Shimizu, M. 2000. Green tea polyphenols inhibit the sodium-dependent glucose transporter of intestinal epithelial cells by a competitive mechanism. *Journal of Agricultural and Food Chemistry*, 48(11), 5618–5623.

Koruk, M., Taysi, S., Savas, M. C., Yilmaz, O., Akcay, F., and Karakok, M. 2004. Oxidative stress and enzymatic antioxidant status in patients with nonalcoholic steatohepatitis. *Annals of Clinical & Laboratory Science*, 34(1), 57–62.

Kuk, J. L., Katzmarzyk, P. T., Nichaman, M. Z., Church, T. S., Blair, S. N., and Ross, R. 2006. Visceral fat is an independent predictor of all-cause mortality in men. Obesity, 14(2), 336–341.

Liu, Z., Liu, J., Jahn, L. A., Fowler, D. E., and Barrett, E. J. 2009. Infusing lipid raises plasma free fatty acids and induces insulin resistance in muscle microvasculature. *Journal of Clinical Endocrinology & Metabolism*, 94(9), 3543–3549.

Loopstra-Masters, R., Liese, A., Haffner, S., Wagenknecht, L., and Hanley, A. 2011. Associations between the intake of caffeinated and decaffeinated coffee and measures of insulin sensitivity and beta cell function. *Diabetologia*, 54(2), 320–328.

Lyon, H. N. and Hirschhorn, J. N. 2005. Genetics of common forms of obesity: A brief overview. *American Journal of Clinical Nutrition*, 82(1), 215S–217S.

Mang, B., Wolters, M., Schmitt, B., Kelb, K., Lichtinghagen, R., Stichtenoth, D. O., and Hahn, A. 2006. Effects of a cinnamon extract on plasma glucose, HbA1c, and serum lipids in diabetes mellitus type 2. *European Journal of Clinical Investigation*, 36(5), 340–344.

Mantovani, A., Sica, A., Sozzani, S., Allavena, P., Vecchi, A., and Locati, M. 2004. The chemokine system in diverse forms of macrophage activation and polarization. *Trends in Immunology*, 25(12), 677–686.

McCarty, M. F. 2005. A chlorogenic acid-induced increase in GLP-1 production may mediate the impact of heavy coffee consumption on diabetes risk. *Medical Hypotheses*, 64(4), 848–853.

McCullough, A. J. 2006. Pathophysiology of nonalcoholic steatohepatitis. *Journal of Clinical Gastroenterology*, 40(Suppl 1), S17–S29.

McNeely, M. J., Shofer, J. B., Leonetti, D. L., Fujimoto, W. Y., and Boyko, E. J. 2012. Associations among visceral fat, all-cause mortality, and obesity-related mortality in Japanese Americans. *Diabetes Care*, 35(2), 296–298.

McQuaid, S. E., Hodson, L., Neville, M. J., Dennis, A. L., Cheeseman, J., Humphreys, S. M., Ruge, T. et al. 2011. Downregulation of adipose tissue fatty acid trafficking in obesity. *Diabetes*, 60(1), 47–55.

Medina-Remon, A., Zamora-Ros, R., Rotches-Ribalta, M., Andres-Lacueva, C., Martinez-Gonzalez, M. A., Covas, M. I., Corella, D. et al. 2011. Total polyphenol excretion and blood pressure in subjects at high cardiovascular risk. *Nutrition, Metabolism and Cardiovascular Diseases*, 21(5), 323–331.

Mellen, P. B., Daniel, K. R., Brosnihan, K. B., Hansen, K. J., and Herrington, D. M. 2010. Effect of muscadine grape seed supplementation on vascular function in subjects with or at risk for cardiovascular disease: A randomized crossover trial. *Journal of the American College of Nutrition*, 29(5), 469–475.

Mellor, D. D., Sathyapalan, T., Kilpatrick, E. S., Beckett, S., and Atkin, S. L. 2010. High-cocoa polyphenol-rich chocolate improves HDL cholesterol in type-2 diabetes patients. *Diabetic Medicine*, 27(11), 1318–1321.

Mendoza, J. A., Drewnowski, A., and Christakis, D. A. 2007. Dietary energy density is associated with obesity and the metabolic syndrome in U.S. adults. *Diabetes Care*, 30(4), 974–979.

Mente, A., de Koning, L., Shannon, H. S., and Anand, S. S. 2009. A systematic review of the evidence supporting a causal link between dietary factors and coronary heart disease. *Archives of Internal Medicine*, 169(7), 659–669.

Miller, E. R., Pastor-Barriuso, R., Dalal, D., Riemersma, R. A., Appel, L. J., and Guallar, E. 2005. Meta-analysis: High-dosage vitamin E supplementation may increase all-cause mortality. *Annals of Internal Medicine*, 142(1), 37–46.

Miyazaki, Y., Glass, L., Triplitt, C., Wajcberg, E., Mandarino, L. J., and DeFronzo, R. A. 2002. Abdominal fat distribution and peripheral and hepatic insulin resistance in type 2 diabetes mellitus. *American Journal of Physiology—Endocrinology and Metabolism*, 283(6), E1135–E1143.

Molloy, J. W., Calcagno, C. J., Williams, C. D., Jones, F. J., Torres, D. M., and Harrison, S. A. 2011. Association of coffee and caffeine consumption with fatty liver disease, nonalcoholic steatohepatitis, and degree of hepatic fibrosis. *Hepatology*, 55(2), 429–436.

Mottillo, S., Filion, K. B., Genest, J., Joseph, L., Pilote, L., Poirier, P., Rinfret, S., Schiffrin, E. L., and Eisenberg, M. J. 2010. The metabolic syndrome and cardiovascular risk: A systematic review and meta-analysis. *Journal of the American College of Cardiology*, 56(14), 1113–1132.

Must, A., Spadano, J., Coakley, E. H., Field, A. E., Colditz, G., and Dietz, W. H. 1999. The Disease burden associated with overweight and obesity. *JAMA: The Journal of the American Medical Association*, 282(16), 1523–1529.

Mutch, D. M., Tordjman, J., Pelloux, V., Hanczar, B., Henegar, C., Poitou, C., Veyrie, N., Zucker, J. D., and Clément, K. 2009. Needle and surgical biopsy techniques differentially affect adipose tissue gene expression profiles. *The American Journal of Clinical Nutrition*, 89(1), 51–57.

Naismith, D., Akinyanju, P., Szanto, S., and Yudkin, J. 1970. The effect in volunteers of coffee and decaffeinated coffee on blood glucose, insulin, plasma lipids and some factors involved in blood clotting. *Annals of Nutrition and Metabolism*, 12(3), 144–151.

Oba, S., Nagata, C., Nakamura, K., Fujii, K., Kawachi, T., Takatsuka, N., and Shimizu, H. 2010. Consumption of coffee, green tea, oolong tea, black tea, chocolate snacks and the caffeine content in relation to risk of diabetes in Japanese men and women. *British Journal of Nutrition*, 103, 453–459.

OECD. 2009. *Health at a Glance 2009*. OECD Publishing. doi: 10.1787/health_glance-2009-en

Olthof, M. R., Hollman, P. C. H., and Katan, M. B. 2001. Chlorogenic acid and caffeic acid are absorbed in humans. *The Journal of Nutrition*, 131(1), 66–71.

Omagari, K., Kato, S., Tsuneyama, K., Hatta, H., Sato, M., Hamasaki, M., Sadakane, Y. et al. 2010. Olive leaf extract prevents spontaneous occurrence of non-alcoholic steatohepatitis in SHR/NDmcr-cp rats. *Pathology*, 42(1), 66–72.

Ouchi, N., Parker, J. L., Lugus, J. J., and Walsh, K. 2011. Adipokines in inflammation and metabolic disease. *Nature Reviews Immunology*, 11(2), 85–97.

Panagiotakos, D. B., Pitsavos, C., Yannakoulia, M., Chrysohoou, C., and Stefanadis, C. 2005. The implication of obesity and central fat on markers of chronic inflammation: The ATTICA study. *Atherosclerosis*, 183(2), 308–315.

Panchal, S. K., Poudyal, H., Waanders, J., and Brown, L. 2012. Coffee extract attenuates changes in cardiovascular and hepatic structure and function without decreasing obesity in high-carbohydrate, high-fat diet-fed male rats. *Journal of Nutrition*, 142(4), 690–697.

Park, S., Choi, Y., Um, S. J., Yoon, S. K., and Park, T. 2010. Oleuropein attenuates hepatic steatosis induced by high-fat diet in mice. *Journal of Hepatology*, 54(5), 984–993.

Pasimeni, G., Ribaudo, M. C., Capoccia, D., Rossi, F., Bertone, C., Leonetti, F., and Santiemma, V. 2006. Non-invasive evaluation of endothelial dysfunction in uncomplicated obesity: Relationship with insulin resistance. *Microvascular Research*, 71(2), 115–120.

Petrie, H. J., Chown, S. E., Belfie, L. M., Duncan, A. M., McLaren, D. H., Conquer, J. A., and Graham, T. E. 2004. Caffeine ingestion increases the insulin response to an oral-glucose-tolerance test in obese men before and after weight loss. *The American Journal of Clinical Nutrition*, 80(1), 22–28.

Petrovski, G., Gurusamy, N., and Das, D. K. 2011. Resveratrol in cardiovascular health and disease. *Annals of the New York Academy of Sciences*, 1215(1), 22–33.

Rock, W., Rosenblat, M., Miller-Lotan, R., Levy, A.P., Elias, M., and Aviram, M. 2008. Consumption of wonderful variety pomegranate juice and extract by diabetic patients increases paraoxonase 1 association with high-density lipoprotein and stimulates its catalytic activities. *Journal of Agricultural and Food Chemistry*, 56(18), 8704–8713.

Rucker, D., Padwal, R., Li, S. K., Curioni, C., and Lau, D. C. W. 2007. Long term pharmacotherapy for obesity and overweight: Updated meta-analysis. *British Medical Journal*, 335(7631), 1194–1199.

Savage, D. B., Petersen, K. F., and Shulman, G. I. 2007. Disordered lipid metabolism and the pathogenesis of insulin resistance. *Physiological Reviews*, 87(2), 507–520.

Shimizu, M., Kobayashi, Y., Suzuki, M., Satsu, H., and Miyamoto, Y. 2000. Regulation of intestinal glucose transport by tea catechins. *BioFactors*, 13(1–4), 61–65.

Shrime, M. G., Bauer, S. R., McDonald, A. C., Chowdhury, N. H., Coltart, C. E. M., and Ding, E. L. 2011. Flavonoid-rich cocoa consumption affects multiple cardiovascular risk factors in a meta-analysis of short-term studies. *The Journal of Nutrition*, 141(11), 1982–1988.

Skurk, T., Alberti-Huber, C., Herder, C., and Hauner, H. 2007. Relationship between adipocyte size and adipokine expression and secretion. *Journal of Clinical Endocrinology Metabolism*, 92(3), 1023–1033.

Solomon, T. and Blannin, A. 2009. Changes in glucose tolerance and insulin sensitivity following 2-weeks of daily cinnamon ingestion in healthy humans. *European Journal of Applied Physiology*, 105(6), 969–976.

Spencer, M., Yao-Borengasser, A., Unal, R., Rasouli, N., Gurley, C. M., Zhu, B., Peterson, C. A., and Kern, P. A. 2010. Adipose tissue macrophages in insulin-resistant subjects are associated with collagen VI and fibrosis and demonstrate alternative activation. *American Journal of Physiology—Endocrinology and Metabolism*, 299(6), E1016–E1027.

Sumida, Y., Nakashima, T., Yoh, T., Furutani, M., Hirohama, A., Kakisaka, Y., Nakajima, Y. et al. 2003. Serum thioredoxin levels as a predictor of steatohepatitis in patients with nonalcoholic fatty liver disease. *Journal of Hepatology*, 38(1), 32–38.

Tang, M., Larson-Meyer, D. E., and Liebman, M. 2008. Effect of cinnamon and turmeric on urinary oxalate excretion, plasma lipids, and plasma glucose in healthy subjects. *The American Journal of Clinical Nutrition*, 87(5), 1262–1267.

Targher, G., Day, C. P., and Bonora, E. 2010. Risk of cardiovascular disease in patients with nonalcoholic fatty liver disease. *New England Journal of Medicine*, 363(14), 1341–1350.

Theobald, H., Bygren, L., Carstensen, J., and Engfeldt, P. 2000. A moderate intake of wine is associated with reduced total mortality and reduced mortality from cardiovascular disease. *Journal of Studies on Alcohol*, 61(5), 652–656.

Thom, E. 2007. The effect of chlorogenic acid enriched coffee on glucose absorption in healthy volunteers and its effect on body mass when used long-term in overweight and obese people. *The Journal of International Medical Research*, 35(6), 900–908.

Tome-Carneiro, J., Gonzalvez, M., Larrosa, M., Yanez-Gasco, M. J., Garcia-Almagro, F. J., Ruiz-Ros, J. A., Garcia-Conesa, M. T., Tomas-Barberan, F. A., and Espin, J. C. 2012. One-year consumption of a grape nutraceutical containing resveratrol improves the inflammatory and fibrinolytic status of patients in primary prevention of cardiovascular disease. *The American Journal of Cardiology*, 110(3), 356–363.

Tsuneki, H., Ishizuka, M., Terasawa, M., Wu, J., Sasaoka, T., and Kimura, I. 2004. Effect of green tea on blood glucose levels and serum proteomic patterns in diabetic (db/db) mice and on glucose metabolism in healthy humans. *BMC Pharmacology*, 4, 18.

Tsuruta, Y., Nagao, K., Kai, S., Tsuge, K., Yoshimura, T., Koganemaru, K., and Yanagital, T. 2011. Polyphenolic extract of lotus root (edible rhizome of *Nelumbo nucifera*) alleviates hepatic steatosis in obese diabetic db/db mice. *Lipids Health Disease*, 10, 202.

Tyrovolas, S. and Panagiotakos, D. B. 2010. The role of Mediterranean type of diet on the development of cancer and cardiovascular disease, in the elderly: A systematic review. *Maturitas*, 65(2), 122–130.

Vachharajani, V. and Granger, D. N. 2009. Adipose tissue: A motor for the inflammation associated with obesity. *IUBMB Life*, 61(4), 424–430.

van Dam, R., Dekker, J., Nijpels, G., Stehouwer, C., Bouter, L., and Heine, R. 2004. Coffee consumption and incidence of impaired fasting glucose, impaired glucose tolerance, and type 2 diabetes: The Hoorn Study. *Diabetologia*, 47(12), 2152–2159.

van Dam, R. M. and Hu, F. B. 2005. Coffee consumption and risk of type 2 diabetes: A systematic review. *JAMA: The Journal of the American Medical Association*, 294(1), 97–104.

van Dam, R. M., Pasman, W. J., and Verhoef, P. 2004. Effects of coffee consumption on fasting blood glucose and insulin concentrations. *Diabetes Care*, 27(12), 2990–2992.

van Dijk, S. J., Olthof, M. R., Meeuse, J. C., Seebus, E., Heine, R. J., and van Dam, R. M. 2009. Acute effects of decaffeinated coffee and the major coffee components chlorogenic acid and trigonelline on glucose tolerance. *Diabetes Care*, 32(6), 1023–1025.

Venables, M. C., Hulston, C. J., Cox, H. R., and Jeukendrup, A. E. 2008. Green tea extract ingestion, fat oxidation, and glucose tolerance in healthy humans. *The American Journal of Clinical Nutrition*, 87(3), 778–784.

Virtue, S. and Vidal-Puig, A. 2010. Adipose tissue expandability, lipotoxicity and the metabolic syndrome—An allostatic perspective. *Biochimica et Biophysica Acta (BBA)—Molecular and Cell Biology of Lipids*, 1801(3), 338–349.

Vitaglione, P., Morisco1, F., Mazzone, G., Amoruso, D., Ribecco, M., Romano, A., Fogliano1, V., Caporaso, N., and D'Argenio, G. 2011. Coffee reduces liver damage in a rat model of steatohepatitis: The underlying mechanisms and the role of polyphenols and melanoidins. *Hepatology*, 52(5), 1652–1661.

WHO. 2011. Obesity and overweight. Available at: http://www.who.int/mediacentre/factsheets/fs311/en/ (accessed on 6/2011).

Wickenberg, J., Ingemansson, L., and Hlebowicz, J. 2010. Effects of *Curcuma longa* (turmeric) on postprandial plasma glucose and insulin in healthy subjects. *Nutrition Journal*, 9(1), 1–5.

Wieringa, N. F., Van Der Windt, H. J., Zuiker, R. R. M., Dijkhuizen, L., Verkerk, M. A., Vonk, R. J., and Swart, J. A. A. 2008. Positioning functional foods in an ecological approach to the prevention of overweight and obesity. *Obesity Reviews*, 9(5), 464–473.

Wilson, P. W. F., D'Agostino, R. B., Sullivan, L., Parise, H., and Kannel, W. B. 2002. Overweight and obesity as determinants of cardiovascular risk: The Framingham experience. *Archives of Internal Medicine*, 162(16), 1867–1872.

Wintergerst, E. S., Maggini, S., and Hornig, D. H. 2007. Contribution of selected vitamins and trace elements to immune function. *Annals of Nutrition and Metabolism*, 51(4), 301–323.

Wong, R. H. X., Howe, P. R. C., Buckley, J. D., Coates, A. M., Kunz, I., and Berry, N. M. 2011. Acute resveratrol supplementation improves flow-mediated dilatation in overweight/obese individuals with mildly elevated blood pressure. *Nutrition, Metabolism and Cardiovascular Diseases*, 21(11), 851–856.

Wronska, A. and Kmiec, Z. 2012. Structural and biochemical characteristics of various white adipose tissue depots. *Acta Physiologica*, 205(2), 194–208.

Wu, L. Y., Juan, C. C., Ho, L. T., Hsu, Y. P., and Hwang, L. S. 2004a. Effect of green tea supplementation on insulin sensitivity in Sprague–Dawley rats. *Journal of Agricultural and Food Chemistry*, 52(3), 643–648.

Wu, L. Y., Juan, C. C., Hwang, L., Hsu, Y. P., Ho, P. H., and Ho, L. T. 2004b. Green tea supplementation ameliorates insulin resistance and increases glucose transporter IV content in a fructose-fed rat model. *European Journal of Nutrition*, 43(2), 116–124.

Yang, C. S. and Landau, J. M. 2000. Effects of tea consumption on nutrition and health. *The Journal of Nutrition*, 130(10), 2409–2412.

Ziegenfuss, T., Hofheins, J., Mendel, R., Landis, J., and Anderson, R. 2006. Effects of a water-soluble cinnamon extract on body composition and features of the metabolic syndrome in pre-diabetic men and women. *Journal of the International Society of Sports Nutrition*, 3, 45–53.

# 9

# Role of Nutraceutical Supplements in the Prevention and Treatment of Hypertension

Mark C. Houston

## Contents

# Introduction

Hypertension is a consequence of the interaction of genetics and environment. Macro- and micronutrients are crucial in the regulation of blood pressure (BP) and subsequent target organ damage (TOD). Nutrient–gene interactions, subsequent gene expression, oxidative stress, inflammation, and autoimmune vascular dysfunction have positive or negative influences on vascular biology in humans. Endothelial activation with endothelial dysfunction (ED) and vascular smooth muscle dysfunction (VSMD) initiates and perpetuates essential hypertension.

Macro- and micronutrient deficiencies are very common in the general population and may be even more common in patients with hypertension and cardiovascular disease (CVD) due to genetic, environmental causes, and prescription drug use. These deficiencies will have an enormous impact on present and future cardiovascular health and outcomes such as hypertension, myocardial infarction (MI), stroke, and renal disease. The diagnosis and treatment of these nutrient deficiencies will reduce BP and improve vascular health, ED, vascular biology, and cardiovascular events.

# Epidemiology

The epidemiology underscores the etiologic role of diet and associated nutrient intake in hypertension. The transition from the Paleolithic diet to our modern diet has produced an epidemic of nutritionally related diseases (Table 9.1). Hypertension, atherosclerosis, coronary heart disease (CHD), MI, congestive heart failure (CHF), cerebrovascular accidents (CVA), renal disease, type 2 diabetes mellitus (DM), metabolic syndrome (MS), and obesity are some of these diseases [1,2]. Table 9.1 contrasts intake of nutrients involved in BP regulation during the Paleolithic era and modern time. Evolution from a preagricultural, hunter–gatherer milieu to an agricultural, refrigeration society has imposed an unnatural and unhealthful nutritional selection process. In sum, diet has changed more than genetics can adapt.

**Table 9.1** Dietary Intake of Nutrients Involved in Vascular Biology: Comparing and Contrasting the Diet of Paleolithic and Contemporary Humans

| Nutrients and Dietary Characteristics | Paleolithic Intake | Modern Intake |
|---|---|---|
| Sodium | <50 mmol/day (1.2 g) | 175 mmol/day (4 g) |
| Potassium | >10,000 meq/day (256 g) | 150 meq/day (6 g) |
| Sodium/potassium ratio | <0.13/day | >0.67/day |
| Protein | 37% | 20% |
| Carbohydrate | 41% | 40%–50% |
| Fat | 22% | 30%–40% |
| Polyunsaturated/saturated fat ratio | 1.4 | 0.4 |
| Fiber | >100 g/day | 9 g/day |

The human genetic makeup is 99.9% that of our Paleolithic ancestors, yet our nutritional, vitamin, and mineral intakes are vastly different [3]. The macronutrient and micronutrient variations, radical oxygen species (ROS), inflammatory mediators, cell adhesion molecules (CAMs), signaling molecules, and autoimmune dysfunction contribute to the higher incidence of hypertension and other CVDs through complex nutrient–gene interactions and nutrient–caveolae interactions in the endothelium [4–7]. Reduction in nitric oxide (NO) bioavailability and endothelial activation initiate the vascular dysfunction and hypertension. Poor nutrition, coupled with obesity and a sedentary lifestyle, has resulted in an exponential increase in nutritionally related diseases. In particular, the high $Na^+/K^+$ ratio of modern diets has contributed to hypertension, stroke, CHD, CHF, and renal disease [3,8] as have the relatively low intake of omega-3 polyunsaturated fatty acid (PUFA) and increase in omega-6 PUFA saturated fat and trans-FAs [9].

# Pathophysiology

Vascular biology assumes a pivotal role in the initiation and perpetuation of hypertension and TOD. Oxidative stress, inflammation, and autoimmune dysfunction of the vascular system are the primary pathophysiologic and functional mechanisms that induce vascular disease. All three of these are closely interrelated and establish a deadly combination that leads to ED, VSMD, hypertension, vascular disease, atherosclerosis, and CVD.

## Oxidative stress

Oxidative stress, with an imbalance between ROS and the antioxidant defense mechanisms, contributes to the etiology of hypertension in animals [10] and humans [11,12]. ROS are generated by multiple cellular sources, including NADPH oxidase, mitochondria, xanthine oxidase, uncoupled endo-thelium-derived NO synthase (U-eNOS), cyclooxygenase, and lipo-oxygenase [11]. Superoxide anion is the predominant ROS produced by these tissues. Hypertensive patients have impaired endogenous and exogenous antioxi-dant defense mechanisms [13], an increased plasma oxidative stress, and an exaggerated oxidative stress response to various stimuli [13,14]. Hypertensive subjects also have lower plasma ferric reducing ability of plasma (FRAP), lower vitamin C levels, and increased plasma 8-isoprostanes, which correlate with both systolic BP (SBP) and diastolic BP (DBP). Various single-nucleotide polymorphisms (SNPs) in genes that codify for antioxidant enzymes are directly related to hypertension [15]. These include NADPH oxidase, xan-thine oxidase, superoxide dismutase (SOD) 3, catalase, GPx 1 (glutathione peroxidase), and thioredoxin. Antioxidant deficiency and excess free radi-cal production have been implicated in human hypertension in numerous epidemiologic, observational, and interventional studies [13,14,16] (see Table 9.2). ROS directly damage endothelial cells; degrade NO; influence eicosanoid metabolism; oxidize LDL, lipids, proteins, carbohydrates, DNA, and organic molecules; increase catecholamines; damage the genetic machin-ery; and influence gene expression and transcription factors (1,11,12,13,14). The interrelations of neurohormonal systems, oxidative stress, and CVD are shown in Figure 9.1. The increased oxidative stress, inflammation, and auto-immune vascular dysfunction in human hypertension result from a combina-tion of increased generation of ROS, an exacerbated response to ROS, and a decreased antioxidant reserve [13–18].

## Inflammation

The link between inflammation and hypertension has been suggested in both cross-sectional and longitudinal studies [19]. Increases in high-sensitivity C-reactive protein (HS-CRP) as well as other inflammatory cytokines occur in hypertension and hypertensive-related TOD, such as increased carotid intima–media thickness (IMT) [20]. HS-CRP predicts future cardiovascular events [19,20]. Elevated HS-CRP is both a risk marker and risk factor for hypertension and CVD [21,22]. Increases in HS-CRP of over 3 μg/mL may increase BP in just a few days that is directly proportional to the increase in HS-CRP [21,22]. NO and eNOS are inhibited by HS-CRP [21,22]. The AT2R, which normally coun-terbalances AT1R, is downregulated by HS-CRP [21,22]. Angiotensin II (A-II) upregulates many of the cytokines, especially interleukin 6 (IL-6), CAMs, and chemokines by activating nuclear factor kappa-beta (NF-kb) leading to vaso-constriction. These events, along with the increases in oxidative stress and endothelin-1, elevate BP [19].

## Autoimmune dysfunction

Innate and adaptive immune responses are linked to hypertension and hypertension-induced CVD through at least three mechanisms: Icytokine

**Table 9.2** Oxidative Stress Induces ED, Vascular Disease, and Hypertension

The Cytotoxic Reactive Oxygen Species and the Natural Defense Mechanisms

| Reactive Oxygen Species | | Antioxidant Defense Mechanisms | |
|---|---|---|---|
| *Free radicals* | | *Enzymatic scavengers* | |
| $O_2^{\bullet-}$ | Superoxide anion radical | SOD | Superoxide dismutase |
| $OH^{\bullet}$ | Hydroxyl radical | | $2O_2^{\bullet-} + 2H^+ \rightarrow H_2O_2 + O_2$ |
| $ROO^{\bullet}$ | Lipid peroxide (peroxyl) | CAT | Catalase (peroxisomal-bound) |
| $RO^{\bullet}$ | Alkoxyl | | $2H_2O_2 \rightarrow O_2 + H_2O$ |
| $RS^{\bullet}$ | Thiyl | GTP | Glutathione peroxidase |
| $NO^{\bullet}$ | Nitric oxide | | $2GSH + H_2O_2 \rightarrow GSSG + 2H_2O$ |
| $NO_2^{\bullet}$ | Nitrogen dioxide | | $2GSH + ROOH \rightarrow GSSG + ROH + 2H_2O$ |
| $ONOO^-$ | Peroxynitrite | | |
| $CCl_3$ | Trichloromethyl | *Nonenzymatic scavengers* | |
| | | Vitamin A | |
| *Non-radicals* | | Vitamin C (ascorbic acid) | |
| $H_2O_2$ | Hydrogen peroxide | Vitamin E ($\alpha$-tocopherol) | |
| $HOCl$ | Hypochlorous acid | $\beta$-carotene | |
| $ONOO^-$ | Peroxynitrite | Cysteine | |
| $^1O_2$ | Singlet oxygen | Coenzyme Q | |
| | | Uric acid | |
| | | Flavonoids | |
| | | Sulfhydryl group | |
| | | Thioether compounds | |

*Notes:* The superscripted bold dot indicates an unpaired electron and the negative charge indicates a gained electron. GSH, reduced glutathione; GSSG, oxidized glutathione; R, lipid chain. Singlet oxygen is an unstable molecule due to the two electrons present in its outer orbit spinning in opposite directions.

Host-protective factors include enzymatic and nonenzymatic defenses influenced by d et and nutrients.

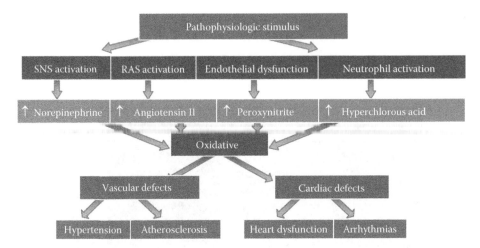

**Figure 9.1** *Neurohormonal and oxidative stress system interaction on cardiac and vascular muscle.* Note: *SNS, sympathetic nervous system; RAS, renin angiotensin [aldosterone] system.*

production, central nervous system stimulation, and renal damage. This includes salt-sensitive hypertension with increased renal inflammation as a result of T cell imbalance, dysregulation of CD+4 and CD+8 lymphocytes, and chronic leukocytosis with increased neutrophils and reduced lymphocytes [23,24]. Macrophages and various T cell subtypes regulate BP, invade the arterial wall, activate Toll-like receptors (TLRs), and induce autoimmune vascular damage [25,26]. A-II activates immune cells (T cells, macrophages, and dendritic cells) and promotes cell infiltration into target organs [26]. CD4+T lymphocytes express AT1R and PPAR gamma receptors, when activated, and release TNF alpha, interferon, and interleukins within the vascular wall [26]. Interleukin 17 (IL-17) produced by T cells may play a pivotal role in the genesis of hypertension caused by A-II [26].

## Treatment and prevention

Many of the natural compounds in food, certain nutraceutical supplements, vitamins, antioxidants, or minerals function in a similar fashion to a specific class of antihypertensive drugs. Although the potency of these natural compounds may be less than the antihypertensive drug, when used in combination with other nutrients and nutraceutical supplements, the antihypertensive effect is additive or synergistic. Table 9.3 summarizes these natural compounds into the major antihypertensive drug classes such as diuretics, beta blockers, central alpha agonists, calcium channel blocker (CCB), angiotensin-converting enzyme inhibitor (ACEI), and angiotensin receptor blockers (ARBs).

**Table 9.3** Natural Antihypertensive Compounds Categorized by Antihypertensive Class

| Antihypertensive Therapeutic Class (Alphabetical Listing) | Foods and Ingredients Listed by Therapeutic Class | Nutrients and Other Supplements Listed by Therapeutic Class |
|---|---|---|
| ACEIs | Egg yolk<br>Fish (specific)<br>Bonito<br>Dried salted fish<br>Fish sauce<br>Sardines<br>Tuna<br>Garlic<br>Gelatin<br>Hawthorne berry<br>Milk products (specific)<br>Casein<br>Sour milk<br>Whey (hydrolyzed)<br>Sake<br>Sea vegetables (kelp)<br>Wheat germ (hydrolyzed)<br>Zein (corn protein) | Omega-3 FAs<br>Pycnogenol<br>Zinc |
| ARBs | Celery<br>Fiber<br>Garlic | Coenzyme Q 10<br>GLA<br>Resveratrol<br>Potassium<br>Vitamin C<br>Vitamin B6 (pyridoxine) |
| Beta blockers | Hawthorne berry | |
| CCBs | Celery<br>Garlic<br>Hawthorn berry | ALA<br>Calcium<br>Magnesium<br>N-acetylcysteine<br>Omega-3 FAs<br>EPA<br>DHA<br>Vitamin B6<br>Vitamin C<br>Vitamin E |
| Central alpha agonists (Reduce SNS activity) | Celery<br>Fiber<br>Garlic<br>Protein | Coenzyme Q 10<br>GLA<br>Potassium<br>Restriction of sodium<br>Taurine<br>Vitamin C<br>Vitamin B6<br>Zinc |

*(continued)*

**Table 9.3 (continued)** Natural Antihypertensive Compounds Categorized by Antihypertensive Class

| Antihypertensive Therapeutic Class (Alphabetical Listing) | Foods and Ingredients Listed by Therapeutic Class | Nutrients and Other Supplements Listed by Therapeutic Class |
|---|---|---|
| Direct vasodilators | Celery<br>Cooking oils with<br>MUFAs<br>Fiber<br>Garlic<br>Soy | Alpha linolenic acid<br>Arginine<br>Calcium<br>Flavonoids<br>Magnesium<br>Omega-3 FAs<br>Potassium<br>Taurine<br>Vitamin C<br>Vitamin E |
| Diuretics | Celery<br>Hawthorn berry<br>Protein | Calcium<br>Coenzyme Q 10<br>Fiber<br>GLA<br>L-carnitine<br>Magnesium<br>Potassium<br>Taurine<br>Vitamin B6<br>Vitamin C |

## Sodium (Na$^\pm$) reduction

The average sodium intake in the United States is 5,000 mg/day with some areas of the country consuming 15,000–20,000 mg/day [27]. However, the minimal requirement for sodium is probably about 500 mg/day [27]. Epidemiologic, observational, and controlled clinical trials demonstrate that an increased sodium intake is associated with higher BP [27]. A reduction in sodium intake in hypertensive patients, especially the salt-sensitive patients, will significantly lower BP by 4–6/2–3 mmHg that is proportional to the severity of sodium restriction and may prevent or delay hypertension in high-risk patients [28,29].

Salt sensitivity (≥10% increase in mean arterial pressure (MAP) with salt loading) occurs in about 51% of hypertensive patients and is a key factor in determining the cardiovascular, cerebrovascular, renal, and BP response to dietary salt intake [30]. Cardiovascular events are more common in the salt-sensitive patients than in salt-resistant ones, independent of BP [31].

Sodium intake per day in hypertensive patients should be between 1500 and 2000 mg. Sodium restriction improves BP reduction in those patients that are on pharmacologic treatment [32,33]. Reducing dietary sodium intake may reduce damage to the brain, heart, kidney, and vasculature through mechanisms dependent on the small BP reduction as well as those independent of the decreased BP [34,35].

A balance of sodium with other nutrients, especially potassium and magnesium, is important, not only in reducing and controlling BP, but also in decreasing cardiovascular and cerebrovascular events [3,32,33]. An increase in the sodium to potassium ratio is associated with significantly increased risk of CVD and all-cause mortality [34].

## Potassium

The average US dietary intake of potassium ($K^+$) is 45 mmol/day with a potassium-to-sodium ($K^+/Na^+$) ratio of less than 1:2 [8,35]. The recommended intake of $K^+$ is 4700 mg day (120 mmol) with a $K^+/Na^+$ ratio of about 4–5 to 1 [8,35]. Numerous epidemiologic, observational, and clinical trials have demonstrated a significant reduction in BP with increased dietary $K^+$ intake in both normotensive and hypertensive patients [8,35,36]. The average BP reduction with a $K^+$ supplementation of 60–120 mmol/day is 4.4/2.5 mmHg in hypertensive patients but may be as much as 8/4.1 mmHg with 120 mmol/day (4700 mg) [8,35,36]. The response depends on race (black>white) and sodium, magnesium, and calcium intake [8]. Those on a higher sodium intake have a greater reduction in BP with potassium [8]. Alteration of the $K^+/Na^+$ ratio to a higher level is important for both antihypertensive and cardiovascular and cerebrovascular effects [37]. High potassium intake reduces the incidence of cardiovascular (CHD, MI) and cerebrovascular accidents independent of the BP reduction [8,35–37]. There are also reductions in CHF, left ventricular hypertrophy (LVH), DM, and cardiac arrhythmias [8]. If the serum potassium is less than 4.0 meq/dL, there is an increased risk of CVD mortality, ventricular tachycardia, ventricular fibrillation, and CHF [8]. Red blood cell (RBC) potassium is a better indication of total body stores and thus CVD risk than serum potassium [8]. Gu et al. [37] found that potassium supplementation at 60 mmol of KCl/day for 12 weeks significantly reduced SBP −5.0 mmHg (range −2.13 to −7.88 mmHg) (p<0.001) in 150 Chinese men and women aged 35–64 years.

Potassium increases natriuresis, modulates baroreflex sensitivity, vasodilates, decreases the sensitivity to catecholamines and A-II, and increases sodium/potassium ATPase and DNA synthesis in the vascular smooth muscle cells and sympathetic nervous system (SNS) cells resulting in improved function [8]. In addition, potassium increases bradykinin and urinary kallikrein; decreases NADPH oxidase, which lowers oxidative stress and inflammation; improves insulin sensitivity; decreases asymmetric dimethylarginine (ADMA); reduces intracellular sodium; and lowers production of TGF-beta [8].

Each 1000 mg increase in potassium intake per day reduces all-cause mortality by approximately 20%. Each 1000 mg decrease in sodium intake per day will decrease all-cause mortality by 20% [34]. The recommended daily dietary intake for patients with hypertension is 4.7 g of potassium and less than

1500 mg of sodium [8]. Potassium in food or from supplementation should be reduced or used with caution in those patients with renal impairment or those on medications that increase renal potassium retention [8].

## Magnesium (Mg$^{\pm\pm}$)

A high dietary intake of magnesium of at least 500–1000 mg/day reduces BP in most of the reported epidemiologic, observational, and clinical trials, but the results are less consistent than those seen with Na$^+$ and K$^+$ [35,38]. In most epidemiologic studies, there is an inverse relationship between dietary magnesium intake and BP [35,38,39]. A study of 60 essential hypertensive subjects given magnesium supplements showed a significant reduction in BP over an 8 week period documented by 24 h ambulatory BP and home and office blood BP [35,38,39]. The maximum reduction in clinical trials has been 5.6/2.8 mmHg, but some studies have shown no change in BP [40]. The combination of high potassium and low sodium intake with increased magnesium intake had additive antihypertensive effects [40]. Magnesium also increases the effectiveness of all antihypertensive drug classes [40].

Magnesium competes with Na$^+$ for binding sites on vascular smooth muscle and acts like a CCB, increases prostaglandin E (PGE), increases NO, improves endothelial function, and binds in a necessary cooperative manner with potassium, inducing endothelium-dependent vasodilation (EDV) and BP reduction [35,38–40].

Magnesium is an essential cofactor for the delta-6-desaturase enzyme that is the rate-limiting step for conversion of linoleic acid (LA) to gamma linolenic acid (GLA) [35,38,39] needed for synthesis of the vasodilator and platelet inhibitor prostaglandin E$_1$ (PGE$_1$).

Intracellular level of magnesium (RBC) is more indicative of total body stores and should be measured in conjunction with serum and urinary magnesium [40]. Magnesium may be supplemented in doses of 500–1000 mg/day. Magnesium formulations chelated to an amino acid may improve absorption and decrease the incidence of diarrhea [40]. Adding taurine at 1000–2000 mg/day will enhance the antihypertensive effects of magnesium [40]. Magnesium supplements should be avoided or used with caution in patients with known renal insufficiency or in those taking medications that induce magnesium retention [40].

## Calcium (Ca$^{\pm\pm}$)

Population studies show a link between hypertension and calcium [41], but clinical trials that administer calcium supplements to patients have shown inconsistent effects on BP [35,41]. The heterogeneous responses to calcium supplementation have been explained by Resnick [42]. This is the "ionic hypothesis" [42] of hypertension, CVD, and associated metabolic, functional, and structural disorders. Calcium supplementation is not recommended at this time as an effective means to reduce BP.

## Zinc (Zn$^{++}$)

Low serum zinc levels in observational studies correlate with hypertension as well as CHD, type 2 DM, hyperlipidemia, elevated lipoprotein a (Lp(a)), 2 h postprandial plasma insulin levels, and insulin resistance [43]. Zinc is transported into cardiac and vascular muscle and other tissues by methothinein [44]. Genetic deficiencies of metallothionein with intramuscular zinc deficiencies may lead to increased oxidative stress, mitochondrial dysfunction, cardiomyocyte dysfunction and apoptosis with subsequent myocardial fibrosis, abnormal cardiac remodeling, heart disease, heart failure, or hypertension [44]. Intracellular calcium increases oxidative stress that is reduced by zinc [44].

Bergomi et al. [45] evaluated Zn$^{++}$ status in 60 hypertensive subjects compared to 60 normotensive control subjects. An inverse correlation of BP and serum Zn$^{++}$ was observed. The BP was also inversely correlated to a Zn$^{++}$-dependent enzyme-lysyl oxidase activity. Zn$^{++}$ inhibits gene expression and transcription through NF-kb and activated protein-1 (AP-1) [43,44]. These effects plus those on insulin resistance, membrane ion exchange, renin–angiotensin–aldosterone system (RAAS), and SNS effects may account for Zn$^{++}$ antihypertensive effects [43,44]. Zinc intake should be 50 mg/day [1].

## Protein

Observational and epidemiologic studies demonstrate a consistent association between a high protein intake and a reduction in BP and incident BP [46,47]. The protein source is an important factor in the BP effect, animal protein being less effective than nonanimal or plant protein, especially almonds [46–49]. In the Intersalt study of over 10,000 subjects, those with a dietary protein intake 30% above the mean had lower BP by 3.0/2.5 mmHg compared to those that were 30% below the mean (81 g versus 44 g/day) [46]. However, lean or wild animal protein with less saturated fat and more essential omega-3 fatty acids (FAs) may reduce BP, lipids, and CHD risk [46,49]. A meta-analysis confirmed these findings and also suggested that hypertensive patients and the elderly have the greatest BP reduction with protein intake [47]. A randomized crossover study in 352 adults with prehypertension and stage I hypertension found a significant reduction in SBP of 2.0 mmHg with soy protein and 2.3 mmHg with milk protein compared to a high-glycemic-index diet over each of the 8 week treatment periods [50]. There was a nonsignificant reduction in DBP. The daily recommended intake of protein from all sources is 1.0–1.5 g/kg body weight, varying with exercise level, age, renal function, and other factors [1,32,33].

Fermented milk supplemented with whey protein concentrate significantly reduces BP in human studies [51–55]. Administration of 20 g/day of hydrolyzed whey protein supplement rich in bioactive peptides significantly reduced BP over 6 weeks by 8.0±3.2 mmHg in SBP and 5.5±2.1 mm in DBP [52]. Milk peptides that contain both caseins and whey proteins are a rich source of ACEI peptides. Val-Pro-Pro and Ile-Pro-Pro given at 5–60 mg/day have variable reductions in BP with an average decrease in pooled studies of about 4.8/2.2 mmHg [33,50,53–55]. However, a recent meta-analysis did not show significant reductions in BP in humans [55]. Powdered fermented milk with *Lactobacillus helveticus* given at 12 g/day significantly lowered BP by 11.2/6.5 mmHg in 4 weeks in one study [53]. The clinical response is attributed to fermented milk's two peptides that inhibit ACE.

Pins and Keenan [56] administered 20 g of hydrolyzed whey protein to 30 hypertensive subjects and noted a BP reduction of 11/7 mmHg compared to controls at 1 week that was sustained throughout the study. These data indicate that the whey protein must be hydrolyzed in order to exhibit an antihypertensive effect and the maximum BP response is dose dependent.

Bovine casein-derived peptides and whey protein-derived peptides exhibit ACEI activity [51–56]. These components include B-caseins, B-lg fractions, B2-microglobulin, and serum albumin [51–53,56]. The enzymatic hydrolysis of whey protein isolates releases ACEI peptides.

Marine collagen peptides (MCPs) from deep-sea fish have antihypertensive activity [57–59]. A double-blind placebo-controlled trial in 100 hypertensive subjects with diabetes who received MCPs twice a day for 3 months had significant reductions in DBP and MAP [57]. Bonito protein (*Sarda orientalis*), from the tuna and mackerel family, has natural ACEI inhibitory peptides and reduces BP 10.2/7 mmHg at 1.5 g/day [58].

Sardine muscle protein, which contains valyl-tyrosine (VAL-TYR), significantly lowers BP in hypertensive subjects [60]. Kawasaki et al. treated 29 hypertensive subjects with 3 mg of valyl-tyrosine sardine muscle–concentrated extract for 4 weeks and lowered BP 9.7/5.3 mmHg (p < 0.05) [60]. Levels of A-I increased as serum A-II and aldosterone decreased indicating that valyl-tyrosine is a natural ACEI. A similar study with a vegetable drink with sardine protein hydrolysates significantly lowered BP by 8/5 mmHg in 13 weeks [61].

Soy protein intake was significantly and inversely associated with both SBP and DBP in 45,694 Chinese women consuming 25 g/day or more of soy protein over 3 years, and the association increased with age [62]. The SBP reduction was 1.9–4.9 mm lower and the DBP 0.9–2.2 mmHg lower [62]. However, randomized clinical trials and meta-analysis have shown mixed results on BP with no change in BP to reductions of 7%–10% for SBP and DBP [63–67]. Some studies suggest improvement in endothelial function, improved arterial compliance, reduction in HS-CRP and inflammation, ACEI activity, reduction in sympathetic tone, diuretic action, and reduction in both oxidative stress and aldosterone levels [66,68,69]. Fermented soy at about 25 g/day is recommended.

In addition to ACEI effects, protein intake may also alter catecholamine responses and induce natriuresis [60,61]. Low protein intake coupled with low omega-3 FA intake may contribute to hypertension in animal models [70]. The optimal protein intake, depending on level of activity, renal function, stress, and other factors, is about 1.0–1.5 g/kg/day [1].

## Amino acids and related compounds

### L-Arginine

L-Arginine and endogenous methylarginines are the primary precursors for the production of NO, which has numerous beneficial cardiovascular effects, mediated through conversion of L-arginine to NO by eNOS. Patients with hypertension, hyperlipidemia, DM, and atherosclerosis have increased levels of HS-CRP and inflammation, increased microalbumin, low levels of apelin (stimulates NO in the endothelium), increased levels of arginase (breaks down arginine), and elevated serum levels of ADMA, which inactivates NO [71–75].

Under normal physiological conditions, intracellular arginine levels far exceed the K(m) of eNOS. However, endogenous NO formation is dependent on extracellular arginine concentration. The NO production in endothelial cells is closely coupled to cellular arginine update indicating that arginine transport mechanisms play a major role in the regulation of NO-dependent function. Exogenous arginine can increase renal vascular and tubular NO bioavailability and influence renal perfusion, function, and BP [74].

Human studies in hypertensive and normotensive subjects of parenteral and oral administrations of L-arginine demonstrate an antihypertensive effect [71,76–80]. The BP decreased by 6.2/6.8 mmHg on 10 g/day of L-arginine when provided as a supplement or through natural foods to a group of hypertensive subjects [76]. Arginine produces a statistically and biologically significant decrease in BP and improved metabolic effect in normotensive and hypertensive humans that is similar in magnitude to that seen in the DASH-I diet [76]. Arginine given at 4 g/day also significantly lowered BP in women with gestational hypertension without proteinuria, reduced the need for antihypertensive therapy, decreased maternal and neonatal complications, and prolonged the pregnancy [77,78]. The combination of arginine (1200 mg/day) and N-acetylcysteine (600 mg bid) administered over 6 months to hypertensive patients with type 2 diabetes lowered SBP and DBP (p<0.05), increased HDL-C, decreased LDL-C and oxLDL, and reduced HS-CRP, ICAM, VCAM, PAI-I, fibrinogen, and IMT [79]. A study of 54 hypertensive subjects given arginine 4 g three times per day for 4 weeks had significant reductions in 24 h ambulatory BP monitoring (ABM) [80]. Although these doses of L-arginine appear to be safe, no long-term studies in humans have been published at this time, and there are concerns of a pro-oxidative effect or even an increase in mortality in patients who may have severely dysfunctional endothelium, advanced atherosclerosis, CHD, or MI [81].

### L-Carnitine

L-Carnitine is a nitrogenous constituent of muscle primarily involved in the oxidation of FAs in mammals. Human studies on the effects of L-carnitine are small, limited to older studies in which there is minimal to no change in BP [82–84]. Carnitine may be useful in the treatment of essential hypertension, type 2 DM with hypertension, hyperlipidemia, cardiac arrhythmias, CHF, and cardiac ischemic syndromes [1,82–84]. Doses of 2–3 g twice per day are recommended.

## Taurine

Taurine is a sulfonic beta-amino acid that is considered a conditionally essential amino acid, which is not utilized in protein synthesis but is found free or in simple peptides with its highest concentration in the brain, retina, and myocardium [85]. In cardiomyocytes, it represents about 50% of the free amino acids and has a role of an osmoregulator, inotropic factor, and antihypertensive agent [86].

Human studies have noted that essential hypertensive subjects have reduced urinary taurine as well as other sulfur amino acids [1,85,86]. Taurine lowers BP, SVR, and HR; decreases arrhythmias, CHF symptoms, and SNS activity; increases urinary sodium and water excretion; increases atrial natriuretic factor; improves insulin resistance; increases NO; improves endothelial function; and decreases A-II, plasma renin activity (PRA), aldosterone, plasma norepinephrine, and plasma and urinary epinephrine [1,85–87]. A study of 31 Japanese males with essential hypertension placed on an exercise program for 10 weeks showed a 26% increase in taurine levels and a 287% increase in cysteine levels. The BP reduction of 14.8/6.6 mmHg was proportional to increases in serum taurine and reductions in plasma norepinephrine [88]. Fujita et al. [86] demonstrated a reduction in BP of 9/4.1 mmHg (p<0.05) in 19 hypertension subjects given 6 g of taurine for 7 days. Taurine has numerous beneficial effects on the cardiovascular system and BP [87]. The recommended dose of taurine is 2–3 g/day at which no adverse effects are noted, but higher doses may be needed to reduce BP significantly [1,32,33,85–88].

## Omega-3 fats

The omega-3 FAs found in cold-water fish, fish oils, flax, flax seed, flax oil, and nuts lower BP in observational, in epidemiologic, and in some small prospective clinical trials [89–93]. The findings are strengthened by a dose-related response in hypertension as well as a relationship to the specific concomitant diseases associated with hypertension [89–93].

Studies indicate that docosahexaenoic acid (DHA) reduces BP and heart rate [89]. The average reduction in BP is 8/5 mmHg, and heart rate falls about 6 beats per minute [1,32,33,89,94–96]. However, formation of eicosapentaenoic acid (EPA) and ultimately DHA from alpha lipoic acid (ALA) is decreased in the presence of high LA (the essential omega-6 FA), saturated fats, *trans*-FAs, alcohol, several nutrient deficiencies, and aging, all of which inhibit the desaturase enzymes [89]. Eating cold-water fish three times per week may be as effective as high-dose fish oil in reducing BP in hypertensive patients, and the protein in the fish may also have antihypertensive effects [1,89]. In patients with chronic kidney disease, 4 g of omega-3 FAs reduced BP measured with 24 h ABM over 8 weeks by 3.3/2.9 mmHg compared to placebo ($p < 0.0001$) [97].

The ideal ratio of omega-6 FA to omega-3 FA is between 1:1 and 1:4 with a polyunsaturated-to-saturated (P/S) fat ratio greater than 1.5–2:0 [2]. The omega-6 FA family includes LA, GLA, dihomo-gamma linolenic acid (DGLA), and arachidonic acid (AA), which do not usually lower BP significantly but may prevent increases in BP induced by saturated fats [99]. GLA may block stress-induced hypertension by increasing $PGE_1$ and PGI 2, reducing aldosterone levels, and reducing adrenal AT1R density and affinity [95].

The omega-3 FA have a multitude of other cardiovascular consequences that modulates BP such as increases in eNOS and NO, improvement in ED, reduction in plasma norepinephrine and increased parasympathetic nervous system tone, suppression of ACE activity, and improvement in insulin resistance [99]. The recommended daily dose is 3000–5000 mg/day of combined DHA and EPA in a ratio of 3 parts EPA to 2 parts DHA and about 50% of this dose as GLA combined with gamma/delta tocopherol at 100 mg/g of DHA and EPA to get the omega-3 index of 8% or higher to reduce BP and provide optimal cardioprotection [100].

## Omega-9 fats

Olive oil is rich in omega-9 monounsaturated fat (MUFA) oleic acid, which has been associated with BP and lipid reduction in Mediterranean and other diets [101,102]. Olive oil and MUFAs have shown consistent reductions in BP in most clinical studies in humans [101,103–105]. In one study, the SBP fell 8 mmHg ($p \leq 0.05$) and the DBP fell 6 mmHg ($p \leq 0.01$) in both clinic and 24 h ABM in the MUFA-treated subjects compared to the PUFA-treated subjects [101]. In addition, the need for antihypertensive medications was reduced by 48% in the MUFA group versus 4% in the omega-6 PUFA group ($p < 0.005$). Extra-virgin olive oil was more effective than sunflower oil in lowering SBP in a group of 31 elderly hypertensive patients in a double-blind randomized crossover study [103]. The SBP was 136 mmHg in the extra-virgin olive oil–treated subjects versus 150 mmHg in the sunflower-treated group ($p < 0.01$). Olive oil also reduces BP in hypertensive diabetic subjects [104].

Extra-virgin olive oil also contains lipid-soluble phytonutrients such as polyphenol; approximately 5 mg of phenols are found in 10 g of oil [101,102]. About 4 tablespoons of extra-virgin olive oil is equal to 40 g, which is the amount required to get significant reductions in BP.

## Fiber

The clinical trials with various types of fiber to reduce BP have been inconsistent [106,107]. Soluble fiber, guar gum, guava, psyllium, and oat bran may reduce BP and reduce the need for antihypertensive medications in hypertensive subjects, diabetic subjects, and hypertensive–diabetic subjects [1,32,33,106,107]. The average reduction in BP is about 7.5/5.5 mmHg on 40–50 g/day of a mixed fiber. There is improvement in insulin sensitivity and endothelial function, reduction in SNS activity, and increase in renal sodium loss [1,32,33,106].

## Vitamin C

Vitamin C is a potent water-soluble electron donor. At physiologic levels, it is an antioxidant although at supraphysiologic doses such as those achieved with intravenous vitamin C, it donates electrons to different enzymes, which results in pro-oxidative effects. At physiologic doses, vitamin C recycles vitamin E, improves ED, and produces a diuresis [108]. Dietary intake of vitamin C and plasma ascorbate concentration in humans are inversely correlated to SBP, DBP, and heart rate [108–122].

An evaluation of published clinical trials indicates that vitamin C dosing at 250 mg twice daily will lower BP about 7/4 mmHg [108–122]. Vitamin C will induce a sodium–water diuresis, improve arterial compliance, improve endothelial function, increase NO and PGI2, decrease adrenal steroid production, improve sympathovagal balance, increase RBC Na/K ATPase, increase SOD, improve aortic elasticity and compliance, improve flow-mediated vasodilation, decrease pulse wave velocity and augmentation index, increase cyclic GMP, activate potassium channels, reduce cytosolic calcium, and reduce serum aldehydes [120]. Vitamin C enhances the efficacy of amlodipine, decreases the binding affinity of the AT 1 receptor for A-II by disrupting the ATR1 disulfide bridges, and enhances antihypertensive effects of medications in the elderly with refractory hypertension [1,32,33,112–117]. In elderly patients with refractory hypertension already on maximum pharmacologic therapy, 600 mg of vitamin C daily lowered the BP by 20/16 mmHg [117]. The lower the initial ascorbate serum level, the better is the BP response. A serum level of 100 μmol/L is recommended [1,32,33]. The SBP and 24 ABM show the most significant reductions with chronic oral administration of vitamin C [112–117]. Block et al. [118] in an elegant depletion–repletion study of vitamin C demonstrated an inverse correlation of plasma ascorbate levels, SBP, and DBP. In a meta-analysis of 13 clinical trials with 284 patients, vitamin C at 500 mg/day over 6 weeks reduced SBP 3.9 mmHg and DBP 2.1 mmHg [119]. Hypertensive subjects were found to have significantly lower plasma ascorbate levels compared to normotensive subjects (40 μmol/L and 57 μmol/L, respectively) [123], and plasma ascorbate is inversely correlated with BP even in healthy, normotensive individuals [118].

## Vitamin E

Most studies have not shown reductions in BP with most forms of tocopherols or tocotrienols [1,32,33]. Patients with type 2 DM and controlled hypertension (130/76 mmHg) on prescription medications with an average BP of 136/76 mmHg were administered mixed tocopherols containing 60% gamma, 25% delta, and 15% alpha tocopherols [124]. The BP actually increased by 6.8/3.6 mmHg in the study patients (p<0.0001) but was less compared to the increase with alpha tocopherol of 7/5.3 mmHg (p<0.0001). This may be a reflection of drug interactions with tocopherols via cytochrome P 450 (3A 4 and 4F2) and reduction in the serum levels of the pharmacologic treatments that were simultaneously being given [124]. Gamma tocopherol may have natriuretic effects by inhibition of the 70pS potassium channel in the thick ascending limb of the loop of Henle and lower BP [125]. Both alpha and gamma tocopherols improve insulin sensitivity and enhance adiponectin expression via PPAR gamma-dependent processes, which have the potential to lower BP and serum glucose [126]. If vitamin E has an antihypertensive effect, it is probably small and may be limited to untreated hypertensive patients or those with known vascular disease or other concomitant problems such as diabetes or hyperlipidemia.

## Vitamin D

Vitamin D3 may have an independent and direct role in the regulation of BP and insulin metabolism [127–133]. Vitamin D influences BP by its effects on calcium–phosphate metabolism, RAAS, immune system, control of endocrine glands, and ED [128]. If the Vitamin D level is below 30 ng/mL, the circulating PRA levels are higher, which increases A-II, increases BP, and blunts plasma renal blood flow [133]. The lower the level of vitamin D, the greater the risk of hypertension, with the lowest quartile of serum vitamin D having a 52% incidence of hypertension and the highest quartile having a 20% incidence [133]. Vitamin D3 markedly suppresses renin transcription by a VDR-mediated mechanism via the juxtaglomerular apparatus (JGA). Its role in electrolytes, volume, and BP homeostasis indicates that vitamin D3 is important in amelioration of hypertension. Vitamin D lowers ADMA, suppresses pro-inflammatory cytokines such as TNF alpha, increases NO, improves endothelial function and arterial elasticity, decreases vascular smooth muscle hypertrophy, regulates electrolytes and blood volume, increases insulin sensitivity, reduces free FA concentration, regulates the expression of the natriuretic peptide receptor, and lowers HS-CRP [129,130,132,133].

The hypotensive effect of vitamin D was inversely related to the pretreatment serum levels of $1,25$ $(OH)_2$ $D_3$ and additive to antihypertensive medications. Pfeifer et al. showed that short-term supplementation with vitamin $D_3$ and calcium is more effective in reducing SBP than calcium alone [215]. In a group of 148 women with low 25 $(OH)_2$ $D_3$ levels, the administration of 1200 mg calcium plus 800 IU of vitamin $D_3$ reduced SBP 9.3% more (p<0.02) compared to 1200 mg of calcium alone. The HR fell 5.4% (p=0.02), but DBP was not changed. The range in BP reduction was 3.6/3.1–13.1/7.2 mmHg. The reduction in BP is related to the pretreatment level of vitamin D3, the dose of vitamin D3, and serum level of vitamin D3, but BP is reduced only in hypertensive patients. A 25-hydroxyvitamin D level of 60 ng/mL is recommended.

## Vitamin B6 (pyridoxine)

Low serum vitamin B6 (pyridoxine) levels are associated with hypertension in humans [134]. One human study by Aybak et al. [135] proved that high-dose vitamin B6 significantly lowered BP. Pyridoxine (vitamin B6) is a cofactor in neurotransmitter and hormone synthesis in the central nervous system, increases cysteine synthesis to neutralize aldehydes, enhances the production of glutathione, blocks calcium channels, improves insulin resistance, decreases central sympathetic tone, and reduces end-organ responsiveness to glucocorticoids and mineralocorticoids [1,32,33,136,137]. Vitamin B6 is reduced with chronic diuretic therapy and heme pyrollactams (HPU). Vitamin B6, thus, has similar action to central alpha agonists, diuretics, and CCBs. The recommended dose is 200 mg/day orally.

## Flavonoids

Over 4000 naturally occurring flavonoids have been identified in such diverse substances as fruits, vegetables, red wine, tea, soy, and licorice [138]. Flavonoids (flavonols, flavones, and isoflavones) are potent free radical scavengers that inhibit lipid peroxidation, prevent atherosclerosis, promote vascular relaxation, and have antihypertensive properties [138]. In addition, they reduce stroke and provide cardioprotective effects that reduce CHD morbidity and mortality [139].

Resveratrol is a potent antioxidant and antihypertensive found in the skin of red grapes and in red wine. Resveratrol administration to humans reduces augmentation index, improves arterial compliance, and lowers central arterial pressure when administered as 250 mL of either regular or dealcoholized red wine [140]. There was a significant reduction in the aortic augmentation index of 6.1% with the dealcoholized red wine and 10.5% with regular red wine. The central arterial pressure was significantly reduced by dealcoholized red wine at 7.4 and 5.4 mmHg by regular red wine. Resveratrol increases flow-mediated vasodilation in a dose-related manner, improves ED, prevents uncoupling of eNOS, increases adiponectin, lowers HS-CRP, and blocks the effects of A-II [141–144]. The recommended dose is 250 mg/day of *trans*-resveratrol [142].

## Lycopene

Lycopene is a fat-soluble phytonutrient in the carotenoid family. Dietary sources include tomatoes, guava, pink grapefruit, watermelon, apricots, and papaya in high concentrations [145–149]. Lycopene has recently been shown to produce a significant reduction in BP, serum lipids, and oxidative stress markers [145–149]. Paran and Engelhard [149] evaluated 30 subjects with grade I hypertension, aged 40–65, taking no antihypertensive or antilipid medications treated with a tomato lycopene extract (10 mg lycopene) for 8 weeks. The SBP was reduced from 144 to 135 mmHg (9 mmHg reduction, $p < 0.01$), and DBP fell from 91 to 84 mmHg (7 mmHg reduction, $p < 0.01$).

Another study of 35 subjects with grade I hypertension showed similar results on SBP but not DBP [145]. Englehard gave a tomato extract to 31 hypertensive subjects over 12 weeks, demonstrating a significant BP reduction of 10/4 mmHg [146]. Patients on various antihypertensive agents including ACEI, CCB, and diuretics had a significant BP reduction of 5.4/3 mmHg over 6 weeks when administered a standardized tomato extract [147]. Other studies have not shown changes in BP with lycopene [148]. The recommended daily intake of lycopene is 10–20 mg in food or supplement form.

## Coenzyme Q10 (ubiquinone)

Coenzyme Q10 has consistent and significant antihypertensive effects in patients with essential hypertension [211]:

- Compared to normotensive patients, essential hypertensive patients have a higher incidence (sixfold) of coenzyme Q10 deficiency documented by serum levels.
- Doses of 120–225 mg/day of coenzyme Q10, depending on the delivery method or the concomitant ingestion with a fatty meal, are necessary to achieve a therapeutic level of 3 µg/mL. This dose is usually 3–5 mg/kg/day of coenzyme Q10. Oral dosing levels may become lower with nanoparticle and emulsion delivery systems intended to facilitate absorption. Adverse effects have not been characterized in the literature.
- Patients with the lowest coenzyme Q10 serum levels may have the best antihypertensive response to supplementation.
- The average reduction in BP is about 15/10 mmHg based on reported studies.
- The antihypertensive effect takes time to reach its peak level at 4 weeks. Then the BP remains stable during long-term treatment. The antihypertensive effect is gone within 2 weeks after discontinuation of coenzyme Q10.
- Approximately 50% of patients on antihypertensive drugs may be able to stop between one and three agents. Both total dose and frequency of administration may be reduced.

Other favorable effects on cardiovascular risk factors include improvement in the serum lipid profile and carbohydrate metabolism with reduced glucose and improved insulin sensitivity, reduced oxidative stress, reduced heart rate, improved myocardial LV function and oxygen delivery, and decreased catecholamine levels.

## Alpha lipoic acid

ALA is known as thioctic acid in Europe where it is a prescription medication. It is a sulfur-containing compound with antioxidant activity both in water and lipid phases [1,32,33]. Its use is well established in the treatment of certain forms of liver disease and in the delay of onset of peripheral neuropathy in patients with diabetes. Recent research has evaluated its potential role in the treatment of hypertension, especially as part of the MS [212–214]. Lipoic acid reduces oxidative stress and inflammation, reduces atherosclerosis in animal models, decreases serum aldehydes, and closes calcium channels, which improves vasodilation, improves endothelial function, and lowers BP. In addition, lipoic acid improves insulin sensitivity that lowers glucose, which improves BP control and lowers serum triglycerides. Morcos and others [150] showed stabilization of urinary albumin excretion in DM subjects given 600 mg of ALA compared to placebo for 18 months (p<0.05).

The recommended dose is 100–200 mg/day of R-lipoic acid with biotin 2–4 mg/day to prevent biotin depletion with long-term use of lipoic acid. R-lipoic acid is preferred to the L isomer because of its preferred use by the mitochondria [1,32,33].

## Pycnogenol

Pycnogenol, a bark extract from the French maritime pine, at doses of 200 mg/day resulted in a significant reduction in SBP from 139.9 to 132.7 mmHg (p<0.05) in 11 patients with mild hypertension over 8 weeks in a double-blind randomized placebo crossover trial. DBP fell from 93.8 to 92.0 mmHg. Pycnogenol acts as a natural ACEI, protects cell membranes from oxidative stress, increases NO, improves endothelial function, reduces serum thromboxane concentrations, decreases myeloperoxidase activity, improves renal cortical blood flow, reduces urinary albumin excretion, and decreases HS-CRP [151–155]. Other studies have shown reductions in BP and a decreased need for ACEI and CCB and reductions in endothelin-1, HgA1C, fasting glucose, and LDL-C [152,153,155].

## Garlic

Clinical trials utilizing the correct dose, type of garlic, and well-absorbed long-acting preparations have shown consistent reductions in BP in hypertensive patients with an average reduction in BP of 8.4/7.3 mmHg [156,157]. Not all garlic preparations are processed similarly and are not comparable in antihypertensive potency [1]. In addition, in cultivated garlic (*Allium sativum*), wild uncultivated garlic, or bear garlic (*Allium ursinum*), effects of aged, fresh, and long-acting garlic preparations differ [1,32,33,156,157]. Garlic is also effective in reducing BP in patients with uncontrolled hypertension already on antihypertensive medication [158]. In a double-blind parallel randomized placebo-controlled trial of 50 patients, 900 mg of aged garlic extract with 2.4 mg of S-allylcysteine was administered daily for 12 weeks and reduced SBP 10.2 mmHg (p=0.03) more than the control group in those with SBP over 140 mmHg (158).

Approximately 10,000 mcg of allicin (one of the active ingredients in garlic) per day, the amount contained in four cloves of garlic (5 g), is required to achieve a significant BP lowering effect [1,32–33,157,158]. Garlic has ACEI activity and calcium channel-blocking activity, reduces catecholamine sensitivity, improves arterial compliance, increases bradykinin and NO, and contains adenosine, magnesium, flavonoids, sulfur, allicin, phosphorous, and atoenes that reduce BP [1,32,33].

## Seaweed

Wakame seaweed (*Undaria pinnatifida*) is the most popular, edible seaweed in Japan [159]. In humans, 3.3 g of dried wakame for 4 weeks significantly reduced both the SBP $14 \pm 3$ mmHg and the DBP $5 \pm 2$ mmHg ($p < 0.01$) [160]. In a study of 62 middle-aged, male subjects with mild hypertension given a potassium-loaded, ion-exchanging, sodium-adsorbing, potassium-releasing seaweed preparation, significant BP reductions occurred at 4 weeks on 12 and 24 g/day of the seaweed preparation ($p < 0.01$) [161]. The MAP fell 11.2 mmHg ($p < 0.001$) in the sodium-sensitive subjects and 5.7 mmHg ($p < 0.05$) in the sodium-insensitive subjects, which correlated with PRA.

Seaweed and sea vegetables contain most all of the seawater's 77I minerals and rare earth elements, fiber, and alginate in a colloidal form [159–161]. The primary effect of wakame appears to be through its ACEI activity from at least four parent tetrapeptides and possibly their dipeptide and tripeptide metabolites, especially those containing the amino acid sequence Val-Tyr, Ile-Tyr, Phe-Tyr, and Ile-Try in some combination [159,162,163]. Its long-term use in Japan has demonstrated its safety. Other varieties of seaweed may reduce BP by reducing intestinal sodium absorption and increasing intestinal potassium absorption [161].

## Sesame

Sesame has been shown to reduce BP in a several small randomized, placebo-controlled human studies over 30–60 days [64–169]. Sesame lowers BP alone [165–169] or in combination with nifedipine [164,168] and with diuretics and beta blockers [165,169]. In a group of 13 mild hypertensive subjects, 60 mg of sesamin for 4 weeks lowered SBP 3.5 mmHg ($p < 0.044$) and DBP 1.9 mmHg ($p < 0.045$) [166]. Black sesame meal at 2.52 g/day over 4 weeks in 15 subjects reduced SBP by 8.3 mmHg ($p < 0.05$), but there was a nonsignificant reduction in DBP of 4.2 mmHg [167]. In addition, sesame lowers serum glucose, HgbAIC, and LDL-C; increases HDL; reduces oxidative stress markers; and increases glutathione, SOD, GPx, CAT, and vitamins C, E, and A [164,165,167–169]. The active ingredients are natural ACEIs, sesamin, sesamolin, sesaminol glucosides, furofuran lignans, and suppressors of NF-kb [170,171]. All of these effects lower inflammation and oxidative stress, improve oxidative defense, and reduce BP [170,171].

## Beverages: Tea, coffee, and cocoa

Green tea, black tea, and extracts of active components in both have demonstrated reduction in BP in humans [172–174].

Dark chocolate (100 g) and cocoa with a high content of polyphenols (30 mg or more) have been shown to significantly reduce BP in humans [175–184]. A meta-analysis of 173 hypertensive subjects given cocoa for a mean duration of 2 weeks had a significant reduction in BP 4.7/2.8 mmHg (p=0.002 for SBP and 0.006 for DBP) [175]. Fifteen subjects given 100 g of dark chocolate with 500 mg of polyphenols for 15 days had a 6.4 mmHg reduction in SBP (p<0.05) with a nonsignificant change in DBP [176]. Cocoa at 30 mg of polyphenols reduced BP in prehypertensive and stage I hypertensive patients by 2.9/1.9 mmHg at 18 weeks (p<0.001) [177]. Two more recent meta-analyses of 13 trials and 10 trials with 297 patients found a significant reduction in BP of 3.2/2.0 and 4.5/3.2 mmHg, respectively [179,182]. The BP reduction is the greatest in those with the highest baseline BP and those with a least 50%–70% cocoa at doses of 6–100 g/day [179,181]. Cocoa may also improve insulin resistance and endothelial function [176,183,184].

Polyphenols, chlorogenic acids (CGAs), the ferulic acid metabolite of CGAs, and di-hydro-caffeic acids decrease BP in a dose-dependent manner, increase eNOS, and improve endothelial function in humans [185–187]. CGAs in green coffee bean extract at doses of 140 mg/day significantly reduced SBP and DBP in 28 subjects in a placebo-controlled randomized clinical trial. A study of 122 male subjects demonstrated a dose–response in SBP and DBP with doses of CGA from 46 to 185 mg/day. The group that received the 185 mg dose had a significant reduction in BP of 5.6/3.9 mmHg (p<0.01) over 28 days. Hydroxyhydroquinone is another component of coffee beans, which reduces the efficacy of CGAs in a dose-dependent manner, which partially explains the conflicting results of coffee ingestion on BP [185,187]. Furthermore, there is genetic variation in the enzyme responsible for the metabolism of caffeine that modifies the association between coffee intake, amount of coffee ingested, and the risk of hypertension, heart rate, MI, arterial stiffness, arterial wave reflections, and urinary catecholamine levels [188]. Fifty-nine percent of the population has the IF/IA allele of the CYP1A2 genotype that confers slow metabolism of caffeine. Heavy coffee drinkers who are slow metabolizers had a 3.00 hazard ratio (HR) for developing hypertension. In contrast, fast metabolizers with the IA/IA allele had a 0.36 HR for incident hypertension [188].

## Additional compounds

Melatonin demonstrates significant antihypertensive effects in humans in a numerous double-blind randomized placebo-controlled clinical trials [189–199]. Beta blockers reduce melatonin secretion [200].

Hesperidin significantly lowered DBP 3–4 mmHg (p<0.02) and improved microvascular endothelial reactivity in 24 obese hypertensive male subjects in a randomized, controlled crossover study over 4 weeks for each of three treatment groups consuming 500 mL of orange juice, hesperidin, or placebo [201].

Pomegranate juice reduces SBP 5%–12%, reduces serum ACE activity by 36%, and has antiatherogenic, antioxidant, and anti-inflammatory effects [202,203]. Pomegranate juice at 50 mL/day reduced carotid IMT by 30% over one year, increased PON by 83%, decreased oxLDL by 59%–90%, decreased antibodies to oxLDL by 19%, increased total antioxidant status by 130%, reduced TGF-beta, and increased catalase, SOD, and GPx.

Grape seed extract (GSE) was administered to subjects in nine randomized trials, meta-analysis of 390 subjects, and demonstrated a significant reduction in SBP of 1.54 mmHg ($p < 0.02$) [204]. Significant reduction in BP of 11/8 mmHg ($p < 0.05$) was seen in another dose–response study with 150–300 mg/day of GSE over 4 weeks [205]. GSE has high phenolic content that activates the PI3K/Akt signaling pathway that phosphorylates eNOS and increases NO [205,206]. Hawthorn extract demonstrated borderline reductions on BP in patients not taking antihypertensive agents at 500 mg/day ($p < 0.081$) and significantly reduced DBP ($p < 0.035$) in diabetic patients taking antihypertensive drugs at doses of 1200 mg/day of hawthorn extract [207].

## Clinical considerations

### Combining food and nutrients with medications

Several of the strategic combinations of nutraceutical supplements with anti-hypertensive drugs mentioned in the previous section have been shown to lower BP more than medication alone. These are

- Sesame with beta blockers, diuretics, and nifedipine
- Pycnogenol with ACEI
- Lycopene with various antihypertensive medications
- ALA with ACEI
- Vitamin C with CCB
- N-acetylcysteine with arginine
- Garlic with ACEI, diuretics, and beta blockers
- Coenzyme Q10 with ACEI and CCB

Many antihypertensive drugs may cause nutrient depletions that can actually interfere with their antihypertensive action or cause other metabolic adverse effects manifest through the lab or with clinical symptoms [208,209]. Diuretics decrease potassium, magnesium, phosphorous, sodium, chloride, folate, vita-min B6, zinc, iodine, and coenzyme Q10; increase homocysteine, calcium, and creatinine; and elevate serum glucose by inducing insulin resistance. Beta blockers reduce coenzyme Q10 and ACEI, and ARBs reduce zinc.

**Table 9.4** Integrative Approach to the Treatment of Hypertension

| Intervention Category | Therapeutic Intervention | Daily Intake |
|---|---|---|
| Diet characteristics | DASH I, DASH II-Na⁺, or premier diet | Diet type |
| | Sodium restriction | 1500 mg |
| | Potassium | 4700 mg |
| | Potassium/sodium ratio | >3:1 |
| | Magnesium | 1000 mg |
| | Zinc | 50 mg |
| Macronutrients | *Protein* Total intake from nonanimal sources, organic lean or wild animal protein, or cold-water fish | 30% of total calories, which are 1.5–1.8 g/kg body weight |
| | Whey protein | 30 g |
| | Soy protein (fermented sources are preferred) | 30 g |
| | Sardine muscle concentrate extract | 3 g |
| | Milk peptides | 30–60 mg |
| | Fat | 30% of total calories |
| | Omega-3 FAs | 2–3 g |
| | Omega-6 FAs | 1 g |
| | Omega-9 FAs | 2–4 tablespoons of olive or nut oil or 10–20 olives |
| | Saturated FAs from wild game, bison, or other lean meat | <10% total calories |
| | Polyunsaturated-to-saturated fat ratio | >2.0 |
| | Omega-3-to-omega-6 ratio | 1.1–1.2 |
| | Synthetic *trans*-FAs | None (completely remove from diet) |
| | Nuts in variety | Ad libitum |
| | *Carbohydrates* as primarily complex carbohydrates and fiber | 40% of total calories |
| | Oatmeal | 60 g |
| | Oat bran | 40 g |
| | Beta-glucan | 3 g |
| | Psyllium | 7 g |
| Specific foods | Garlic as fresh cloves or aged | 4 fresh cloves (4 g) or 600 mg aged garlic taken twice daily |
| | Sea vegetables, specifically dried wakame | 3.0–3.5 g |
| | Lycopene as tomato products, guava, watermelon, apricots, pink grapefruit, papaya, or supplements | 10–20 mg |
| | Dark chocolate | 100 g |
| | Pomegranate juice | 8 oz |
| | Sesame | 60 mg sesamin or 2.5 g sesame meal |

**Table 9.4 (continued)** Integrative Approach to the Treatment of Hypertension

| Intervention Category | Therapeutic Intervention | Daily Intake |
|---|---|---|
| Exercise | Aerobic | 20 min daily at 4200 KJ/week |
| | Resistance | 40 min/day |
| Weight reduction | Body mass index <25 | Lose 1–2 lb per week and increasing the proportion of lean muscle |
| | Waist circumference | |
| | <35 in. for women | |
| | <40 in. for men | |
| | Total body fat | |
| | <22% for women | |
| | <16% for men | |
| Other lifestyle recommendations | Alcohol restriction | <20 g/day |
| | Among the choice of alcohol, red wine is preferred due to its vasoactive phytonutrients. | Wine <10 oz <br> Beer <24 oz <br> Liquor <2 oz |
| | Caffeine restriction or elimination depending on CYP 450 type | <100 mg/day |
| | Tobacco and smoking | Stop |
| Medical considerations | Medications that may increase BP | Minimize use when possible, such as by using disease-specific nutritional interventions |
| Supplemental foods and nutrients | ALA with biotin | 100–200 mg twice daily |
| | Amino acids <br> Arginine | 5 g twice daily |
| | Carnitine | 1–2 g twice daily |
| | Taurine | 1–3 g twice daily |
| | CGAs | 150–200 mg |
| | Coenzyme Q 10 | 100 mg once to twice daily |
| | GSE | 300 mg |
| | Melatonin | 2.5 mg |
| | Resveratrol (*trans*) | 250 mg |
| | Vitamin B6 | 100 mg once to twice daily |
| | Vitamin C | 250–500 mg twice daily |
| | Vitamin D3 | Dose to raise 25-hydroxyvitamin D serum level to 60 ng/mL |
| | Vitamin E as mixed tocopherols | 400 IU |

# Summary

- Vascular biology such as endothelial and VSMD plays a primary role in the initiation and perpetuation of hypertension, CVD, and TOD.
- Nutrient–gene interactions are a predominant factor in promoting beneficial or detrimental effects in cardiovascular health and hypertension.
- Food and nutrients can prevent, control, and treat hypertension through numerous vascular biology mechanisms.
- Oxidative stress, inflammation, and autoimmune dysfunction initiate and propagate hypertension and CVD.
- There is a role for the selected use of single and component nutraceutical supplements, vitamins, antioxidants, and minerals in the treatment of hypertension based on scientifically controlled studies as a complement to optimal nutritional, dietary intake from food and other lifestyle modifications [210].
- A clinical approach that incorporates diet, foods, nutrients, exercise, weight reduction, smoking cessation, alcohol and caffeine restriction, and other lifestyle strategies can be systematically and successfully incorporated into clinical practice (Table 9.4).

# References

1. Houston, MC. Treatment of hypertension with nutraceuticals. Vitamins, antioxidants and minerals. *Expert Rev Cardiovasc Ther* 2007; 5(4): 681–691.
2. Eaton SB, Eaton SB III, and Konner MJ. Paleolithic nutrition revisited: A twelve-year retrospective on its nature and implications. *Eur J Clin Nutr* 1997; 51: 207–216.
3. Houston MC and Harper KJ. Potassium, magnesium, and calcium: Their role in both the cause and treatment of hypertension. *J Clin Hypertens* 2008; 10(7 Supp 2): 3–11.
4. Layne J, Majkova Z, Smart EJ, Toborek M, and Hennig B. Caveolae: a regulatory platform for nutritional modulation of inflammatory diseases. *J Nutr Biochem* 2011; 22: 807–811.
5. Dandona P, Ghanim H, Chaudhuri A, Dhindsa S, and Kim SS. Macronutrient intake induces oxidative and inflammatory stress: Potential relevance to atherosclerosis and insulin resistance. *Exp Mol Med* 2010; 42(4): 245–253.
6. Berdanier CD. Nutrient-gene interactions. In: Ziegler EE, Filer LJ Jr, eds. *Present Knowledge in Nutrition*, 7th edn. Washington, DC: ILSI Press; 1996, pp. 574–580.
7. Talmud PJ and Waterworth DM. In-vivo and in-vitro nutrient-gene interactions. *Curr Opin Lipidol* 2000; 11: 31–36.
8. Houston MC. The importance of potassium in managing hypertension. *Curr Hypertens Rep* 2011; 13(4): 309–317.
9. Broadhurst CL. Balanced intakes of natural triglycerides for optimum nutrition: An evolutionary and phytochemical perspective. *Med Hypotheses* 1997; 49: 247–261.
10. Nayak DU, Karmen C, Frishman WH, and Vakili BA. Antioxidant vitamins and enzymatic and synthetic oxygen-derived free radical scavengers in the prevention and treatment of cardiovascular disease. *Heart Dis* 2001; 3: 28–45.
11. Kizhakekuttu TJ and Widlansky ME. Natural antioxidants and hypertension: promise and challenges. *Cardiovasc Ther* 2010; 28(4): e20–e32.
12. Kitiyakara C and Wilcox C. Antioxidants for hypertension. *Curr Opin Nephrol Hypertens* 1998; 7: 531–538.

13. Russo C, Olivieri O, Girelli D, Faccini G, Zenari ML, Lombardi S, and Corrocher R. Antioxidant status and lipid peroxidation in patients with essential hypertension. *J Hypertens* 1998; 16: 1267–1271.

14. Tse WY, Maxwell SR, Thomason H, Blann A, Thorpe GH, Waite M, and Holder R. Antioxidant status in controlled and uncontrolled hypertension and its relationship to endothelial damage. *J Hum Hypertens* 1994; 8: 843–849.

15. Mansego ML, Solar Gde M, Alonso MP, Martinez F, Saez GT, Escudero JC, Redon J, and Chaves FJ. Polymorphisms of antioxidant enzymes, blood pressure and risk of hypertension. *J Hypertens* 2011; 29(3): 492–500.

16. Galley HF, Thornton J, Howdle PD, Walker BE, and Webster NR. Combination oral antioxidant supplementation reduces blood pressure. *Clin Sci* 1997; 92: 361–365.

17. Dhalla NS, Temsah RM, and Netticadam T. The role of oxidative stress in cardiovascular diseases. *J Hypertens* 2000; 18: 655–673.

18. Saez G, Tormos MC, Giner V, Lorano JV, and Chaves FJ. Oxidative stress and enzymatic antioxidant mechanisms in essential hypertension. *Am J Hypertens* 2001; 14: 248A (Abstract P-653).

19. Ghanem FA and Movahed A. Inflammation in high blood pressure: A clinician perspective. *J Am Soc Hypertens* 2007; 1(2): 113–119.

20. Amer MS, Elawam AE, Khater MS, Omar OH, Mabrouk RA, and Taha HM. Association of high–sensitivity C reactive protein with carotid artery intima-media thickness in hypertensive older adults. *J Am Soc Hypertens* April 23, 2011; 5(5): 395–400.

21. Vongpatanasin W, Thomas GD, Schwartz R, Cassis LA, Osborne-Lawrence S, Hahner L, Gibson LL, Black S, Samois D, and Shaul PW. C-Reactive protein causes downregulation of vascular angiotensin subtype 2 receptors and systolic hypertension in mice. *Circulation* 2007; 115(8): 1020–1028.

22. Razzouk L, Munter P, Bansilal S, Kini AS, Aneja A, Mozes J, Ivan O, Jakkula M, Sharma S, and Farkouh ME. C reactive protein predicts long-term mortality independently of low-density lipoprotein cholesterol in patients undergoing percutaneous coronary intervention. *Am Heart J* 2009; 158(2): 277–283.

23. Kvakan H, Luft FC, and Muller DN. Role of the immune system in hypertensive target organ damage. *Trends Cardiovasc Med* 2009; 19(7): 242–246.

24. Rodriquez-Iturbe B, Franco M, Tapia E, Quiroz Y, and Johnson RJ. Renal inflammation, autoimmunity and salt-sensitive hypertension. *Clin Exp Pharmacol Physiol* 2012; 39(1): 96–103.

25. Tian N, Penman AD, Mawson AR, Manning RD Jr, and Flessner MF. Association between circulating specific leukocyte types and blood pressure: the atherosclerosis risk in communities (ARIC) study. *J Am Soc Hypertens* 2010; 4(6): 272–283.

26. Muller DN, Kvakan H, and Luft FC. Immune-related effects in hypertension on and target-organ damage. *Curr Opin Nephrol Hypertens* 2011; 20(2): 113–117.

27. Kotchen TA and McCarron DA. AHA science advisory. Dietary electrolytes and blood pressure. *Circulation* 1998; 98: 613–617.

28. Cutler JA, Follmann D, and Allender PS. Randomized trials of sodium reduction: An overview. *Am J Clin Nutr* 1997; 65: 643S–651S.

29. Svetkey LP, Sacks FM, Obarzanek E et al. The DASH diet, sodium intake and blood pressure (the DASH-Sodium Study): Rationale and design. *JADA* 1999; 99: S96–S104.

30. Weinberger MH. Salt sensitivity of blood pressure in humans. *Hypertension* 1996; 27: 481–490.

31. Morimoto A, Usu T, Fujii T et al. Sodium sensitivity and cardiovascular events in patients with essential hypertension. *Lancet* 1997; 350: 1734–1737.

32. Houston MC. Nutraceuticals, vitamins, antioxidants and mineral in the prevention and treatment of hypertension. *Prog Cardiovasc Dis* 2005; 47(6): 396–449.

33. Houston MC. Nutrition and nutraceutical supplements in the treatment of hypertension. *Expert Rev Cardiovasc Ther* 2010; 8(6): 821–833.

34. Messerli FH, Schmieder RE, Weir MR. Salt: a perpetrator of hypertensive target organ disease? *Arch Intern Med* 1997; 157: 2449–2452.

35. Kawasaki T, Delea CS, Bartter FC, Smith H. The effect of high-sodium and low-sodium intakes on blood pressure and other related variables in human subjects with idiopathic hypertension. *Am J Med* 1978; 64: 193–198.

36. Whelton PK and He J. Potassium in preventing and treating high blood pressure. *Semin Nephrol* 1999; 19: 494–499.

37. Gu D, He J, Xigui W, Duan X, Whelton PK. Effect of potassium supplementation on blood pressure in Chinese: A randomized, placebo-controlled trial. *J Hypertens* 2001; 19: 1325–1331.

38. Widman L, Wester PO, Stegmayr BG, and Wirell MP. The dose dependent reduction in blood pressure through administration of magnesium: A double blind placebo controlled cross-over trial. *Am J Hypertens* 1993; 6: 41–45.

39. Laurant P and Touyz RM. Physiological and pathophysiological role of magnesium in the cardiovascular system: implications in hypertension. *J Hypertens* 2000; 18: 1177–1191.

40. Houston MC. The role of magnesium in hypertension and cardiovascular disease. *J Clin Hypertens*. In press as of January 2012.

41. McCarron DA. Role of adequate dietary calcium intake in the prevention and management of salt sensitive hypertensive. *Am J Clin Nutr* 1997; 65: 712S–716S.

42. Resnick LM. Calcium metabolism in hypertension and allied metabolic disorders. *Diabetes Care* 1991; 14: 505–520.

43. Garcia Zozaya JL and Padilla Viloria M. Alterations of calcium, magnesium, and zinc in essential hypertension: Their relation to the renin-angiotensin-aldosterone system. *Invest Clin* 1997; 38: 27–40.

44. Shahbaz AU, Sun Y, Bhattacharya SK, Ahokas RA, Gerling IC, McGee JE, and Weber KT. Fibrosis in hypertensive heart disease: Molecular pathways and cardioprotective strategies. *J Hypetens* 2010; 28: S25–S32.

45. Bergomi M, Rovesti S, Vinceti M, Vivoli R, Caselgrandi E, and Vivoli G. Zinc and copper status and blood pressure. *J Trace Elem Med Biol* 1997; 11: 166–169.

46. Stamler J, Elliott P, Kesteloot H, Nichols R, Claeys G, Dyer AR, and Stamler R. Inverse relation of dietary protein markers with blood pressure. Findings for 10,020 men and women in the Intersalt Study. Intersalt Cooperative Research Group. International study of salt and blood pressure. *Circulation* 1996; 94: 1629–1634.

47. Altorf-van der Kuil W, Engberink MF, Brink EJ, van Baak MA, Bakker SJ, Navis G, van t'Veer P, and Geleijnse JM. Dietary protein and blood pressure: A systematic review. *PLoS One* 2010; 5(8): e12102–e12117.

48. Jenkins DJ, Kendall CW, Faulkner DA et al. Long-term effects of a plant-based dietary portfolio of cholesterol-lowering foods on blood pressure. *Eur J Clin Nutr* 2008; 62(6): 781–788.

49. Elliott P, Dennis B, Dyer AR et al. Relation of dietary protein (total, vegetable, animal) to blood pressure: INTERMAP epidemiologic study. *Presented at the 18th Scientific Meeting of the International Society of Hypertension*, Chicago, IL, August 20–24, 2000.

50. He J, Wofford MR, Reynolds K, Chen J, Chen CS, Myers L, Minor DL, Elmer PJ, Jones DW, and Whelton PK. Effect of dietary protein supplementation on blood pressure: A randomized controlled trial. *Circulation* 2011; 124(5): 589–595.

51. FitzGerald RJ, Murray BA, and Walsh DJ. Hypotensive peptides from milk proteins. *J Nutr* 2004: 134(4): 980S–988S.

52. Pins JJ and Keenan JM. Effects of whey peptides on cardiovascular disease risk factors. *J Clin Hypertens* 2006; 8(11): 775–782.

53. Aihara K, Kajimoto O, Takahashi R, and Nakamura Y. Effect of powdered fermented milk with *Lactobacillus helveticus* on subjects with high-normal blood pressure or mild hypertension. *J Am Coll Nutr* 2005; 24(4): 257–265.

54. Gemino FW, Neutel J, Nonaka M, and Hendler SS. The impact of lactotripeptides on blood pressure response in stage 1 and stage 2 hypertensives. *J Clin Hypertens* 2010; 12(3): 153–159.

55. Geleijnse JM and Engberink MF. Lactopeptides and human blood pressure. *Curr Opin Lipidol* 2010; 21(1): 58–63.

56. Pins J and Keenan J. The antihypertensive effects of a hydrolyzed whey protein supplement. *Cardiovasc Drugs Ther* 2002; 16(Suppl): 68.

57. Zhu CF, Li GZ, Peng HB, Zhang F, Chen Y, and Li Y. Therapeutic effects of marine collagen peptides on Chinese patients with type 2 diabetes mellitus and primary hypertension. *Am J Med Sci* 2010; 340(5): 360–366.

58. De Leo F, Panarese S, Gallerani R, and Ceci LR. Angiotensin converting enzyme (ACE) inhibitory peptides: production and implementation of functional food. *Curr Pharm Des* 2009; 15(31): 3622–3643.

59. Lordan S, Ross P, and Stanton C. Marine Bioactives as functional food ingredients: Potential to reduce the incidence of chronic disease. *Mar Drugs* 2011; 9(6): 1056–1100.

60. Kawasaki T, Seki E, Osajima K, Yoshida M, Asada K, Matsui T, Osajima Y. Antihypertensive effect of valyl-tyrosine, a short chain peptide derived from sardine muscle hydrolyzate, on mild hypertensive subjects. *J Hum Hypertens* 2000; 14: 519–523.

61. Kawasaki T, Jun CJ, Fukushima Y, and Seki E. Antihypertensive effect and safety evaluation of vegetable drink with peptides derived from sardine protein hydrolysates on mild hypertensive, high-normal and normal blood pressure subjects. *Fukuoka Igaku Zasshi* 2002: 93(10): 208–218.

62. Yang G, Shu XO, Jin F, Zhang X, Li HL, Li Q, Gao YT, and Zheng W. Longitudinal study of soy food intake and blood pressure among middle-aged and elderly Chinese women. *Am J Clin Nutr* 2005; 81(5): 1012–1017.

63. Teede HJ, Giannopoulos D, Dalais FS, Hodgson J, and McGrath BP. Randomised, controlled, cross-over trial of soy protein with isoflavones on blood pressure and arterial function in hypertensive patients. *J Am Coll Nutr* 2006; 25(6): 533–540.

64. Welty FK, Lee KS, Lew NS, and Zhou JR. Effect of soy nuts on blood pressure and lipid levels in hypertensive, prehypertensive and normotensive postmenopausal women. *Arch Inter Med* 2007; 167(10): 1060–1067.

65. Rosero Arenas MA, Roser Arenas E, Portaceli Arminana MA, and Garcia MA. Usefulness of phyto-oestrogens in reduction of blood pressure. Systematic review and meta-analysis. *Aten Primaria* 2008; 40(4): 177–186.

66. Nasca MM, Zhou JR, and Welty FK. Effect of soy nuts on adhesion molecules and markers of inflammation in hypertensive and normotensive postmenopausal women. *Am J Cardiol* 30008; 102(1): 84–86.

67. He J, Gu D, Wu X, Chen J, Duan X, Chen J, and Whelton PK. Effect of soybean protein on blood pressure: A randomized, controlled trial. *Ann Intern Med* 2005; 143(1): 1–9.

68. Hasler CM, Kundrat S, and Wool D. Functional foods and cardiovascular Disease. *Curr Atheroscler Rep* 2000; 2(6): 467–475.

69. Tikkanen MJ and Adlercreutz H. Dietary soy-derived isoflavone phytoestrogens. Could they have a role in coronary heart disease prevention? *Biochem Pharmacol* 2000; 60(1): 1–5.

70. Begg DP, Sinclair AJ, Stahl LA, Garg ML, Jois M, and Weisinger RS. Dietary protein level interacts with omega-3 polyunsaturated fatty acid deficiency to induce hypertension. *Am J Hypertens* 2010(Feb); 23(2):125–128.

71. Vallance P, Leone A, Calver A, Collier J, and Moncada S. Endogenous dimethyl-arginine as an inhibitor of nitric oxide synthesis. *J Cardiovasc Pharmacol* 1992; 20: S60–S62.

72. Sonmez A, Celebi G, Erdem G et al. Plasma apelin and ADMA levels in patients with essential hypertension. *Clin Exp Hypertens* 2010; 32(3): 179–183.

73. Michell DL, Andrews KL, and Chin-Dusting JP. Endothelial dysfunction in hypertension: The role of arginase. *Front Biosci (Schol Ed)* 2011; 3: 946–960.

74. Rajapakse NW and Mattson DL. Role of L-arginine in nitric oxide production in health and hypertension. *Clin Exp Pharmacol Physiol* 2009; 36(3): 249–255.

75. Tsioufis C, Dimitriadis K, Andrikou E, Thomopoulos C, Tsiachris D, Stefanadi E, Mihas C, Miliou A, Papademetriou V, and Stefanadis C. ADMA, C-reactive protein and albuminuria in untreated essential hypertension: A cross-sectional study. *Am J Kidney Dis* 2010; 55(6): 1050–1059.

76. Siani A, Pagano E, Iacone R, Iacoviell L, Scopacasa F, and Strazzullo P. Blood pressure and metabolic changes during dietary L-arginine supplementation in humans. *Am J Hypertens* 2000; 13: 547–551.
77. Facchinetti F, Saade GR, Neri I, Pizzi C, Longo M, and Volpe A. L-Arginine supplementation in patients with gestational hypertension: a pilot study. *Hypertens Pregnancy* 2007; 26(1): 121–130.
78. Neri I, Monari F, Sqarbi L, Berardi A, Masellis G, and Facchinetti F. L-Arginine supplementation in women with chronic hypertension: impact on blood pressure and maternal and neonatal complications. *J Matern Fetal Neonatal Med* 2010; 23(12): 1456–1460.
79. Martina V, Masha A, Gigliardi VR et al. Long-term N-acetylcysteine and L-arginine administration reduces endothelial activation and systolic blood pressure in hypertensive patients with type 2 diabetes. *Diabetes Care* 2008; 31(5): 940–944.
80. Ast J, Jablecka A, Bogdanski I, Krauss H, and Chmara E. Evaluation of the antihypertensive effect of L-arginine supplementation in patients with mild hypertension assessed with ambulatory blood pressure monitoring. *Med Sci Monit* 2010; 16(5): CR 266–271.
81. Schulman SP, Becker LC, Kass DA et al. L arginine therapy in acute myocardial infarction: The vascular interaction with age in myocardial infarction (VINTAGE MI) randomized clinical trial. *JAMA* 2006; 295(1): 58–64.
82. Digiesi V, Cantini F, Bisi G, Guarino G, and Brodbeck B. L-Carnitine adjuvant therapy in essential hypertension. *Clin Ter* 1994; 144: 391–395.
83. Ghidini O, Azzurro M, Vita G, and Sartori G. Evaluation of the therapeutic efficacy of L-carnitine in congestive heart failure. *Int J Clin Pharmacol Ther Toxicol* 1988; 26(4): 217–220.
84. Digiesi V, Palchetti R, and Cantini F. The benefits of L-carnitine therapy in essential arterial hypertension with diabetes mellitus type II. *Minerva Med* 1989; 80(3): 227–231.
85. Huxtable RJ. Physiologic actions of taurine. *Physiol Rev* 1992; 72: 101–163.
86. Fujita T, Ando K, Noda H, Ito Y, Sato Y. Effects of increased adrenomedullary activity and taurine in young patients with borderline hypertension. *Circulation* 1987; 75: 525–532.
87. Huxtable RJ and Sebring LA. Cardiovascular actions of taurine. *Prog Clin Biol Res* 1983; 125: 5–37.
88. Tanabe Y, Urata H, Kiyonaga A, Ikede M, Tanake H, Shindo M, and Arakawa K. Changes in serum concentrations of taurine and other amino acids in clinical antihypertensive exercise therapy. *Clin Exp Hypertens* 1989; 11: 149–165.
89. Mori TA, Bao DQ, Burke V, Puddey IB, and Beilin LJ. Docosahexaenoic acid but not eicosapentaenoic acid lowers ambulatory blood pressure and heart rate in humans. *Hypertension* 1999; 34: 253–260.
90. Bønaa KH, Bjerve KS, Straume B, Gram IT, Thelle D. Effect of eicosapentaenoic and docosahexaenoic acids on blood pressure in hypertension: A population-based intervention trial from the Tromso study. *N Engl J Med* 1990; 322: 795–801.
91. Mori TA, Burke V, Puddey I, and Irish A. The effects of omega 3 fatty acids and coenzyme Q 10 on blood pressure and heart rate in chronic kidney disease: A randomized controlled trial. *J Hypertens* 2009; 27(9): 1863–1872.
92. Ueshima H, Stamler J, Elliot B, Brown, CQ. Food omega 3 fatty acid intake of individuals (total, linolenic acid, long chain) and their blood pressure: INTERMAP study. *Hypertension* 2007; 50(20): 313–319.
93. Mon TA. Omega 3 fatty acids and hypertension in humans. *Clin Exp Pharmacol Physiol* 2006; 33(9): 842–846.
94. Liu JC, Conkin SM, Manuch SB, Yao JK, and Muldoon MF. Long-chain omega 3 fatty acids and blood pressure. *Am J Hypertens* 2011; 24(10): 1121–1126. July 14 E PUB.
95. Engler MM, Schambelan M, Engler MB, and Goodfriend TL. Effects of dietary gamma-linolenic acid on blood pressure and adrenal angiotensin receptors in hypertensive rats. *Proc Soc Exp Biol Med* 1998; 218(3): 234–237.

96. Sagara M, Njelekela M, Teramoto T, Taquchi T, Mori M, Armitage L, Birt N, Birt C, and Yamori Y. Effects of docosahexaenoic acid supplementation on blood pressure, heart rate, and serum lipid in Scottish men with hypertension and hypercholesterolemia. *Int J Hypertens* 2011; 8: 8091–8098.

97. Mori TA, Burke V, Puddey I, Irish A, Cowpland CA, Beilin L, Dogra G, and Watts GF. The effects of omega 3 fatty acids and coenzyme Q 10 on blood pressure and heart rate in chronic kidney disease: a randomized controlled trial. *J Hypertens* 2009; 27: 1863–1872.

99. Chin JP. Marine oils and cardiovascular reactivity. *Prostaglandins Leukot Essent Fatty Acids* 1994; 50: 211–222.

100. Saravanan P, Davidson NC, Schmidt EB, and Calder PC. Cardiovascular effects of marine omega-3 fatty acids. *Lancet* 2010; 376(9740): 540–550.

101. Ferrara LA, Raimondi S, and d'Episcopa I. Olive oil and reduced need for antihypertensive medications. *Arch Intern Med* 2000; 160: 837–842.

102. Thomsen C, Rasmussen OW, Hansen KW, Vesterlund M, and Hermansen K. Comparison of the effects on the diurnal blood pressure, glucose, and lipid levels of a diet rich in monounsaturated fatty acids with a diet rich in polyunsaturated fatty acids in type 2 diabetic subjects. *Diabet Med* 1995; 12: 600–606.

103. Perona JS, Canizares J, Montero E, Sanchez-Dominquez JM, Catala A, and Ruiz-Gutierrez V. Virgin olive oil reduces blood pressure in hypertensive elderly patients. *Clin Nutr* 2004; 23(5): 1113–1121.

104. Perona JS, Montero E, Sanchez-Dominquez JM, Canizares J, Garcia M, and Ruiz-Gutierrez V. Evaluation of the effect of dietary virgin olive oil on blood pressure and lipid composition of serum and low-density lipoprotein in elderly type 2 subjects. *J Agric Food Chem* 2009; 57(23): 11427–11433.

105. Lopez-Miranda J, Perez-Jimenez F, Ros E et al. Olive oil and health: Summary of the II international conference on olive oil and health consensus report, Jaen and Cordoba (Spain) 2008. *Nutr Metab Cardiovasc Dis* 2010; 20(4): 284–294.

106. He J and Whelton PK. Effect of dietary fiber and protein intake on blood pressure: A review of epidemiologic evidence. *Clin Exp Hypertens* 1999; 21: 785–796.

107. Pruijm M, Wuerzer G, Forni V, Bochud M, Pechere-Bertschi A, and Burnier M. Nutrition and hypertension: more than table salt. *Rev Med Suisse* 2010; 6(282): 1715–1720.

108. Sherman DL, Keaney JF, Biegelsen ES et al. Pharmacological concentrations of ascorbic acid are required for the beneficial effect on endothelial vasomotor function in hypertension. *Hypertension* 2000; 35: 936–941.

109. Ness AR, Khaw K-T, Bingham S, and Day NE. Vitamin C status and blood pressure. *J Hypertens* 1996; 14: 503–508.

110. Duffy SJ, Bokce N, and Holbrook M. Treatment of hypertension with ascorbic acid. *Lancet* 1999; 354: 2048–2049.

111. Enstrom JE, Kanim LE, and Klein M. Vitamin C intake and mortality among a sample of the United States population. *Epidemiology* 1992; 3:194–202.

112. Block G, Jensen CD, Norkus EP, Hudes M, and Crawford PB. Vitamin C in plasma is inversely related to blood pressure and change in blood pressure during the previous year in young black and white women. *Nutr J* 2008; 17(7): 35–46.

113. Hatzitolios A, Iliadis F, Katsiki N, and Baltatzi M. Is the antihypertensive effect of dietary supplements via aldehydes reduction evidence based: A systemic review. *Clin Exp Hypertens* 2008: 30(7): 628–639.

114. Mahajan AS, Babbar R, Kansai N, Agarwai, SK, and Ray PC. Antihypertensive and antioxidant action of amlodipine and vitamin C in patients of essential hypertension. *J Clin Biochem Nutr* 200740(2): 141–147.

115. Ledlerc PC, Proulx CD, Arquin G, and Belanger S. Ascorbic acid decreases the binding affinity of the AT! Receptor for angiotensin II. *Am J Hypertens* 2008: 21(1): 67–71.

116. Plantinga Y, Ghiadone L, Magagna A, and Biannarelli C. Supplementation with vitamins C and E improves arterial stiffness and endothelial function in essential hypertensive patients. *Am J Hypertens* 2007; 20(4): 392–397.

117. Sato K, Dohi Y, Kojima M, and Miyagawa K. Effects of ascorbic acid on ambulatory blood pressure in elderly patients with refractory hypertension. *Arzneimittelforschung* 2006; 56(7): 535–540.

118. Block G, Mangels AR, Norkus EP, Patterson BH, Levander OA, and Taylor PR. Ascorbic acid status and subsequent diastolic and systolic blood pressure. *Hypertension* 2001; 37: 261–267.

119. McRae MP. Is Vitamin C an effective antihypertensive supplement? A review and analysis of the literature. *J Chiropr Med* 2006; 5(2): 60–64.

120. Simon JA. Vitamin C and cardiovascular disease: A review. *J Am Coll Nutr* 1992; 11(2): 107–125.

121. Ness AR, Chee D, and Elliott P. Vitamin C and blood pressure—An overview. *J Hum Hypertens* 1997; 11(6): 343–350.

122. Trout DL. Vitamin C and cardiovascular risk factors. *Am J Clin Nutr* 1991; 53(1 Suppl): 322S–325S.

123. National Center for Health Statistics, Fulwood R, Johnson CL, and Bryner JD. Hematological and nutritional biochemistry reference data for persons 6 months-74 years of age: United States, 1976–80. Washington, DC, US Public Health Service, 1982 Vital and Health Statistics series 11, No. 232, DHHS publication No. (PHS) 83-1682.

124. Ward NC, Wu JH, Clarke MW, and Buddy IB. Vitamin E effects on the treatment of hypertension in type 2 diabetics. *J Hypertens* 2007; 227: 227–234.

125. Murray ED, Wechter WJ, Kantoci D, Wang WH, Pham T, Quiggle DD, Gibson KM, Leipold D, and Anner BM. Endogenous natriuretic factors 7: Biospecificity of a natriuretic gamma-tocopherol metabolite LLU alpha. *J Pharmacol Exp Ther* 1997; 282(2): 657–662.

126. Gray B, Swick J, and Ronnenberg AG. Vitamin E and adiponectin: Proposed mechanism for vitamin E-induced improvement in insulin sensitivity. *Nutr Rev* 2011; 69(3): 155–161.

127. Hanni LL, Huarfner LH, Sorensen OH, and Ljunghall S. Vitamin D is related to blood pressure and other cardiovascular risk factors in middle-aged men. *Am J Hypertens* 1995; 8: 894–901.

128. Bednarski R, Donderski R, and Manitius L. Role of vitamin D in arterial blood pressure control. *Pol Merkur Lekarski* 2007; 136: 307–310.

129. Ngo DT, Sverdlov AL, McNeil JJ, and Horowitz JD. Does vitamin D modulate asymmetric dimethylarginine and C-reactive protein concentrations? *Am J Med* 2010; 123(4): 335–341.

130. Rosen CJ. Clinical practice. Vitamin D insufficiency. *N Engl J Med* 2011; 364(3):248–254.

131. Pittas AG, Chung M, Trikalinos T, Mitri J, Brendel M, Patel K, Lichtenstein HA, Lau J and Balk EM. Systematic review: Vitamin D and cardiometabolic outcomes. *Ann Intern Med* 2010; 152(5): 307–314.

132. Motiwala SR and Want TJ. Vitamin D and cardiovascular disease. *Curr Opin Nephrol Hypertens* 2011; 20(4): 345–353.

133. Bhandari SK, Pashayan S, Liu IL, Rasgon SA, Kujubu DA, Tom TY, and Sim JJ. 25-Hydroxyvitamin D levels and hypertension rates. *J Clin Hypertens* 2011; 13(3): 170–177.

134. Keniston R and Enriquez JI Sr. Relationship between blood pressure and plasma vitamin B6 levels in healthy middle-aged adults. *Ann N Y Acad Sci* 1990; 585: 499–501.

135. Aybak M, Sermet A, Ayyildiz MO, and Karakilcik AZ. Effect of oral pyridoxine hydrochloride supplementation on arterial blood pressure in patients with essential hypertension. *Arzneimittelforschung* 1995; 45: 1271–1273.

136. Paulose CS, Dakshinamurti K, Packer S, and Stephens NL. Sympathetic stimulation and hypertension in the pyridoxine-deficient adult rat. *Hypertension* 1988; 11(4): 387–391.

137. Dakshinamurti K, Lal KJ, and Ganguly PK. Hypertension, calcium channel and pyridoxine (vitamin B 6). *Mol Cell Biochem* 1998; 188(1–2): 137–148.

138. Moline J, Bukharovich IF, Wolff MS, and Phillips R. Dietary flavonoids and hypertension: is there a link? *Med Hypotheses* 2000; 55: 306–309.

139. Knekt P, Reunanen A, Järvinen R, Seppänen R, Heliövaara M, and Aromaa A. Antioxidant vitamin intake and coronary mortality in a longitudinal population study. *Am J Epidemiol* 1994; 139: 1180–1189.

140. Karatzi KN, Papamichael CM, Karatizis EN, Papaioannou TG, Aznaouridis KA, Katsichti PP, and Stamatelopuolous KS. Red wine acutely induces favorable effects on wave reflections and central pressures in coronary artery disease patients. *Am J Hypertens* 2005; 18(9): 1161–1167.

141. Biala A, Tauriainen E, Siltanen A, Shi J, Merasto S, Louhelainen M, Martonen E, Finckenberg P, Muller DN, and Mervaala E. Resveratrol induces mitochondrial biogenesis and ameliorates Ang II-induced cardiac remodeling in transgenic rats harboring human renin and angiotensinogen genes. *Blood Press* 2010; 19(3): 196–205.

142. Wong RH, Howe PR, Buckley JD, Coates AM, Kunz L, and Berry NM. Acute resveratrol supplementation improves flow-mediated dilatation in overweight obese individuals with mildly elevated blood pressure. *Nutr Metab Cardiovasc Dis* November 2011; 21(11): 851–856.

143. Bhatt SR, Lokhandwala MF, and Banday AA. Resveratrol prevents endothelial nitric oxide synthase uncoupling and attenuates development of hypertension in spontaneously hypertensive rats. *Eur J Pharmacol* 2011; 667(1–3): 258–264.

144. Rivera L, Moron R, Zarzuelo A, and Galisteo M. Long-term resveratrol administration reduces metabolic disturbances and lowers blood pressure in obese Zucker rats. *Biochem Pharmacol* 2009; 77(6): 1053–1063.

145. Paran E and Engelhard YN. Effect of lycopene, an oral natural antioxidant on blood pressure. *J Hypertens* 2001; 19: S74. Abstract P 1.204.

146. Engelhard YN, Gazer B, and Paran E. Natural antioxidants from tomato extract reduce blood pressure in patients with grade-1 hypertension: A double blind placebo controlled pilot study. *Am Heart J* 2006; 151(1): 100.

147. Paran E, Novac C, Engelhard YN, Hazan-Halevy I. The effects of natural antioxidants form tomato extract in treated but uncontrolled hypertensive patients. *Cardiovasc Durgs Ther* 2009; 23(2): 145–151.

148. Reid K, Frank OR, and Stocks NP. Dark chocolate or tomato extract for prehypertension: A randomized controlled trial. *BMC Complement Altern Med* 2009; 9: 22.

149. Paran E and Engelhard Y. Effect of tomato's lycopene on blood pressure, serum lipoproteins, plasma homocysteine and oxidative stress markers in grade I hypertensive patients. *Am J Hypertens* 2001; 14: 141A. Abstract P-333.

150. Morcos M, Borcea V, Isermann B et al. Effect of alpha-lipoic acid on the progression of endothelial cell damage and albuminuria in patients with diabetes mellitus: an exploratory study. *Diabetes Res Clin Prac* 2001; 52(3): 175–183.

151. Hosseini S, Lee J, Sepulveda RT et al. A randomized, double-blind, placebo-controlled, prospective 16 week crossover study to determine the role of pycnogenol in modifying blood pressure in mildly hypertensive patients. *Nutr Res* 2001; 21: 1251–1260.

152. Zibadi S, Rohdewald PJ, Park D and Watson RR. Reduction of cardiovascular risk factors in subjects with type 2 diabetes by pycnogenol supplementation. *Nutr Res* 2008: 28(5): 315–320.

153. Liu X, Wei J, Tan F, Zhou S, Wurthwein G, and Rohdewald P. Pycnogenol French maritime pine bark extract improves endothelial function of hypertensive patients. *Life Sci* 2004; 74(7): 855–862.

154. Van der Zwan LP, Scheffer PG, and Teerlink T. Reduction of myeloperoxidase activity by melatonin and pycnogenol may contribute to their blood pressure lowering effect. *Hypertension* 2010; 56(3): e35.

155. Cesarone MR, Belcaro G, Stuard S, Schonlau F, Di Renzo A, Grossi MG Dugall M, Cornelli U, Cacchio M, Gizzi G, and Pellegrini L. Kidney low and function in hypertension: Protective effects of pycnogenol in hypertensive participants—A controlled study. *J Cardiovasc Pharmacol Ther* 2010; 15(1): 41–46.

156. Simons S, Wollersheim H, and Thien T. A systematic review on the influence of trial quality on the effects of garlic on blood pressure. *Neth J Med* 2009; 67(6): 212–219.

157. Reinhard KM, Coleman CI, Teevan C, and Vacchani P. Effects of garlic on blood pressure in patients with and without systolic hypertension: a meta-analysis. *Ann Pharmacother* 2008: 42(12): 1766–1771.

158. Reid K, Frank OR, and Stocks NP. Aged garlic extract lowers blood pressure in patients with treated but uncontrolled hypertension: A randomized controlled trial. *Maturitas* 2010; 67(2): 144–150.

159. Suetsuna K and Nakano T. Identification of an antihypertensive peptide from peptic digest of wakame (*Undaria pinnatifida*). *J Nutr Biochem* 2000; 11: 450–454.

160. Nakano T, Hidaka H, Uchida J, Nakajima K, and Hata Y. Hypotensive effects of wakame. *J Jpn Soc Clin Nutr* 1998; 20: 92.

161. Krotkiewski M, Aurell M, Holm G, Grimby G, and Szckepanik J. Effects of a sodium-potassium ion-exchanging seaweed preparation in mild hypertension. *Am J Hypertens* 1991; 4: 483–488.

162. Sato M, Oba T, Yamaguchi T, Nakano T, Kahara T, Funayama K, Kobayashi A, and Nakano T. Antihypertensive effects of hydrolysates of wakame (*Undaria pinnatifida*) and their angiotnesin-1-converting inhibitory activity. *Ann Nutr Metab* 2002; 46(6): 259–267.

163. Sato M, Hosokawa T, Yamaguchi T, Nakano T, Muramoto K, Kahara T, Funayama K, KobayashiA, and Nakano T. Angiotensin I converting enzyme inhibitory peptide derived from wakame (*Undaria pinnatifida*) and their antihypertensive effect in spontaneously hypertensive rats. *J Agric Food Chem* 2002; 50(21): 6245–6252.

164. Sankar D, Sambandam G, Ramskrishna Rao M, and Pugalendi KV. Modulation of blood pressure, lipid profiles and redox status in hypertensive patients taking different edible oils. *Clin Chim Acta* 2005; 355(1–2): 97–104.

165. Sankar D, Rao MR, Sambandam G, and Pugalendi KV. Effect of sesame oil on diuretics or beta-blockers in the modulation of blood pressure, anthropometry, lipid profile and redox status. *Yale J Biol Med* 2006; 79(1): 19–26.

166. Miyawaki T, Aono H, Toyoda-Ono Y, Maeda H, Kiso Y, and Moriyama K. Anti-hypertensive effects of sesamin in humans. *J Nutr Sci Vitaminol (Toyko)* 2009; 55(1): 87–91.

167. Wichitsranoi J, Weerapreeyakui N, Boonsiri P, Settasatian N, Komanasin N, Sirjaichingkul S, Teerajetgul Y, Rangkadilok N, and Leelayuwat N. Antihypertensive and antioxidant effects of dietary black sesame meal in pre-hypertensive humans. *Nutr J* 2011; 10(1): 82–88.

168. Sudhakar B, Kalaiarasi P, Al-Numair KS, Chandramohan G, Rao RK, and Pugalendi KV. Effect of combination of edible oils on blood pressure, lipid profile, lipid peroxidative markers, antioxidant status, and electrolytes in patients with hypertension on nifedipine treatment. *Saudi Med J* 2011; 32(4): 379–385.

169. Sankar D, Rao MR, Sambandam G, and Pugalendi KV. A pilot study of open label sesame oil in hypertensive diabetics. *J Med Food* 2006; 9(3): 408–412.

170. Harikumar KB, Sung B, Tharakan ST, Pandey MK, Joy B, Guha S, Krishnan S, and Aggarwai BB. Sesamin manifests chemopreventive effects through the suppression of NF-kappa-B-regulated cell survival, proliferation, invasion and angiogenic gene products. *Mol Cancer Res* 2010; 8(5): 751–761.

171. Nakano D, Ogura K, Miyakoshi M et al. Antihypertensive effect of angiotensin I-converting enzyme inhibitory peptides from a sesame protein hydrolysate in spontaneously hypertensive rats. *Biosci Biotechnol Biochem* 206; 70(5): 1118–1126.

172. Hodgson JM, Puddey IB, Burke V, Beilin LJ, and Jordan N. Effects on blood pressure of drinking green and black tea. *J Hypertens* 1999; 17: 457–463.

173. Kurita I, Maeda-Yamamoto M, Tachibana H, and Kamei M. Anti-hypertensive effect of Benifuuki tea containing O-methylated EGCG. *J Agric Food Chem* 2010; 58(3): 1903–1908.

174. McKay DL, Chen CY, Saltzman E, and Blumberg JB. *Hibiscus sabdariffa* L tea (tisane) lowers blood pressure in pre-hypertensive and mildly hypertensive adults. *J Nutr* 2010; 140(2): 298–303.

175. Taubert D, Roesen R, and Schomig E. Effect of cocoa and tea intake on blood pressure: A meta-analysis. *Arch Intern Med* 2007; 167(7): 626–634.

176. Grassi D, Lippi C, Necozione S, Desideri G, and Ferri C. Short-term administration of dark chocolate is followed by a significant increase in insulin sensitivity and a decrease in blood pressure in health persons. *Am J Clin Nutr* 2005; 81(3): 611–614.

177. Taubert D, Roesen R, Lehmann C, Jung N, and Schomig E. Effects of low habitual cocoa intake on blood pressure and bioactive nitric oxide: a randomized controlled trial. *JAMA* 2007; 298(1): 49–60.

178. Cohen DL and Townsend RR. Cocoa ingestion and hypertension-another cup please? *J Clin Hypertens* 2007; 9(8): 647–648.

179. Reid I, Sullivan T, Fakler P, Frank OR, and Stocks NP. Does chocolate reduce blood pressure? A meta-analysis. *BMC Med* 2010; 8: 39–46.

180. Egan BM, Laken MA, Donovan JL, and Woolson RF. Does dark chocolate have a role in the prevention and management of hypertension? Commentary on the evidence. *Hypertension* 2010; 55(6): 1289–1295.

181. Desch S, Kobler D, Schmidt J, Sonnabend M, Adams V, Sareben M, Eitel I, Bluher M, Shuler G, and Thiele H. Low vs higher-dose dark chocolate and blood pressure in cardiovascular high-risk patients. *Am J Hypertens* 2010; 23(6): 694–700.

182. Desch S, Schmidt J, Sonnabend M, Eitel I, Sareban M, Rahimi K, Schuler G, and Thiele H. Effect of cocoa products on blood pressure: Systematic review and meta-analysis. *Am J Hypertens* 2010; 23(1): 97–103.

183. Grassi D, Desideri G, Necozione S, Lippi C, Casale R, Properzi G, Blumberg JB, and Ferri C. Blood pressure is reduced and insulin sensitivity increased in glucose intolerant hypertensive subjects after 15 days of consuming high-polyphenol dark chocolate. *J Nutr* 2008; 138(9): 1671–1676.

184. Grassi D, Necozione S, Lippi C, Croce G, Valeri L, Pasqualetti P, Desideri G, Blumberg JB, and Ferri C. Cocoa reduces blood pressure and insulin resistance and improved endothelium-dependent vasodilation in hypertensives. *Hypertension* 2005; 46(2): 398–405.

185. Yamaquchi T, Chikama A, Mori K, Watanabe T, Shiova Y, Katsuraqi Y, and Tokimitsu I. Hydroxyhydroquinone-free coffee: A double-blind, randomized controlled dose-response study of blood pressure. *Nutr Metab Cardiovasc Dis* 2008; 18(6):408–414.

186. Chen ZY, Peng C, Jiao R, Wong YM, Yang N, and Huang Y. Anti-hypertensive nutraceuticals and functional foods. *J Agric Food Chem* 2009; 57(11): 4485–4499.

187. Ochiai R, Chikama A, Kataoka K, Tokimitsu I, Maekawa Y, Ohsihi M, Rakugi H, and MIkmai H. Effects of hydroxyhydroquinone-reduced coffee on vasoreactivity and blood pressure. *Hypertens Res* 2009; 32(11): 969–974.

188. Kozuma K, Tsuchiya S, Kohori J, Hase T, and Tokimitsu I. Anti-hypertensive effect of green coffee bean extract on mildly hypertensive subjects. *Hypertens Res* 2005; 28(9): 711–718.

188. Palatini P, Ceolotto G, Ragazzo F, Donigatti F, Saladini, F, Papparella I, Mos L, Zanata G, and Santonastaso M. CYP 1A2 genotype modifies the association between coffee intake and the risk of hypertension. *J Hypertens* 2008; 27(8): 1594–1601.

189. Scheer FA, Van Montfrans GA, van Someren EJ, Mairuhu G, and Buijs RM. Daily night-time melatonin reduces blood pressure in male patients with essential hypertension. *Hypertension* 2004; 43(2): 192–197.

190. Cavallo A, Daniels SR, Dolan LM, Khoury JC, and Bean JA. Blood pressure response to melatonin in type I diabetes. *Pediatr Diabetes* 2004; 5(1): 26–31.

191. Cavallo A, Daniels SR, Dolan LM, Bean JA, and Khoury JC. Blood pressure-lowering effect of melatonin in type 1 diabetes. *J Pineal Res* 2004; 36(4): 262–266.

192. Cagnacci A, Cannoletta M, Renzi A, Baldassari F, Arangino S, and Volpe A. Prolonged melatonin administration decreases nocturnal blood pressure in women. *Am J Hypertens* 2005; 18(12 Pt 1): 1614–1618.

193. Grossman E, Laudon M, Yalcin R, Zengil H Peleg E, Sharabi Y, Kamari Y, Shen-Orr Z, and Zisapel N. Melatonin reduces night blood pressure in patients with nocturnal hypertension. *Am J Med* 2006; 119(10): 898–902.

194. Rechcinski T, Kurpese M, Trzoa E, and Krzeminska-Pakula M. The influence of melatonin supplementation on circadian pattern of blood pressure in patients with coronary artery disease-preliminary report. *Pol Arch Med Wewn* 2006; 115(6): 520–528.

195. Merkureva GA and Ryzhak GA. Effect of the pineal gland peptide preparation on the diurnal profile of arterial pressure in middle-aged and elderly women with ischemic heart disease and arterial hypertension. *Adv Gerontol* 2008; 21(1): 132–142.

196. Zaslavskaia RM, Scherban EA, and Logvinenki SI. Melatonin in combined therapy of patients with stable angina and arterial hypertension. *Klin Med (Mosk)* 2009; 86: 64–67.

197. Zamotaev IuN, Enikeev AKh, and Kolomets NM. The use of melaxen in combined therapy of arterial hypertension in subjects occupied in assembly line production. *Klin Med (Mosk)* 2009; 87(6): 46–49.

198. Rechcinski T, Trzos E, Wierzbowski-Drabik K, Krzeminska-Pakute M, and Kurpesea M. Melatonin for nondippers with coronary artery disease: assessment of blood pressure profile and heart rate variability. *Hypertens Res* 2002; 33(1): 56–61.

199. Kozirog M, Poliwczak AR, Duchnowicz P, Koter-Michalak M, Sikora J, and Broncel M. Melatonin treatment improves blood pressure, lipid profile and parameters of oxidative stress in patients with metabolic syndrome. *J Pineal Res* 2011; 50(3): 261–266.

200. De-Leersnyder H, de Biois MC, Vekemans M, Sidi D, Villain E, Kindermans C, and Munnich A. Beta (1) adrenergic antagonists improve sleep and behavioural disturbances in a circadian disorder, Smith-Magenis syndrome. *J Med Genet* 2110; 38(9): 586–590.

201. Morand C, Dubray C, Milenkovic D, Lioger D, Martin JF, Scalber A, and Mazur A. Hesperidin contributes to the vascular protective effects of orange juice: A randomized crossover study in healthy volunteers. *Am J Clin Nutr* 2011; 93(1): 73–80.

202. Basu A and Penugonda K. Pomegranate juice: A heart-healthy fruit juice. *Nutr Rev* 2009; 67(1): 49–56.

203. Aviram M, Rosenblat M, Gaitine D et al. Pomegranate juice consumption for 3 years by patients with carotid artery stenosis reduces common carotid intima-media thickness, blood pressure and LDL oxidation. *Clin Nutr* 2004; 23(3):423–433.

204. Aviram M and Dornfeld L. Pomegranate juice inhibits serum angiotensin converting enzyme activity and reduces systolic blood pressure. *Atherosclerosis* 2001; 18(1): 195–198.

204. Feringa HH, Laskey DA, Dickson JE, and Coleman CI. The effect of grape seed extract on cardiovascular risk markers: A meta-analysis of randomized controlled trials. *J Am Diet Assoc* 2011; 111(8): 1173–1181.

205. Sivaprakasapillai B, Edirsinghe K, Randolph J, Steinberg F, and Kappagoda T. Effect of grape seed extract on blood pressure in subjects with the metabolic syndrome. *Metabolism* 2009: 58(12): 1743–1746.

206. Edirisinghe I, Burton-Freeman B, and Tissa Kappagoda C. Mechanism of the endothelium-dependent relaxation evoked by grape seed extract. *Clin Sci (Lond)* 2008; 114(4): 331–337.

207. Walker AF, Marakis G, Simpson E, Hope JL, Robinson PA, Hassanein M, and Simpson HC. Hypotensive effects of hawthorn for patients with diabetes taking prescription drugs: a randomized controlled trial. *Br J Gen Pract* 2006; 56(527): 437–443.

208. Trovato A, Nuhlicek DN, and Midtling JE. Drug-nutrient interactions. *Am Fam Physician* 1991; 44(5): 1651–1658.

209. McCabe BJ, Frankel EH, and Wolfe JJ. Eds. *Handbook of Food-Drug Interactions*. CRC Press, Boca Raton, FL, 2003.

210. Houston MC. The role of cellular micronutrient analysis and minerals in the prevention and treatment of hypertension and cardiovascular disease. *Therapeut Adv Cardiovasc Dis* 2010; 4: v165–v183.

211. Rosenfeldt FL, Haas SJ, Krum H, and Hadu A. Coenzyme Q 10 in the treatment of hypertension: A meta-analysis of the clinical trials. *J Hum Hypertens* 2007; 21(4): 297–306.

212. McMackin CJ, Widlansky ME, Hambury NM, and Haung, AL. Effect of combined treatment with alpha lipoic acid and acetyl carnitine on vascular function and blood pressure in patients with coronary artery disease. *J Clin Hypertens* 2007; 9: 249–255.

213. Salinthone S, Schillace RV, Tsang C, Regan JW, Burdette DN, and Carr DW. Lipoic acid stimulates cAMP production via G protein-coupled receptor-dependent and -independent mechanisms. *J Nutr Biochem* 2011; 22(7): 681–690.

214. Rahman ST, Merchant N, Hague T, Wahi J, Bhaheetharan S, Ferdinand KC, and Khan BV. The impact of lipoic acid on endothelial function and proteinuria in Quinapril-treated diabetic patients with stage I hypertension: Results from the quality study. *J Cardiovasc Pharmacol Ther* 2011 July 12.

215. Pfeifer M, Begerow B, Minne HW, Nachtigall D, and Hansen C. Effects of a short-term vitamin D(3) and calcium supplementation on blood pressure and parathyroid hormone levels in elderly women. *J Clin Endocrinol Metab* 2001; 86: 1633–1637.

# 10

# Nutraceuticals in the Treatment of Metabolic Syndrome

Francesco Visioli

## Contents

## Introduction

The definition of "metabolic syndrome" (MS) as a multifactorial metabolic disorder has been proposed by Reaven to explain the clustering of central obesity and of at least two of the following four additional factors: high triglycerides, low high-density lipoprotein cholesterol (HDL-C), high blood pressure, or raised fasting plasma glucose level. This notion has been discussed by the recent International Diabetes Federation definition (Alberti et al., 2009). In addition, the Adult Treatment Panel (ATP)-III definition (with which the general practitioner is more familiar) is that of the presence of any three of the five risk factors mentioned earlier (Gade et al., 2010). Within this context, it must be underlined that most plasma markers of inflammation, of which the most popular one is the C-reactive protein (CRP), are almost always increased in patients with the MS (Reaven, 2011). In this respect, the MS can be also defined as an inflammatory disorder. Therefore, nutritional or pharmacological interventions aimed at decreasing systemic inflammation would markedly affect prognosis. In synthesis and based on current knowledge, an appropriate approach to the MS (be it preventive or therapeutic) should target both inflammation and metabolic alterations.

This chapter reviews the available evidence of the role of specific food components, that is, those that can be exploited in nutraceutical preparations, in the treatment of the MS.

Given the multifaceted etiology of the MS, adjunct therapy with nutraceuticals should chiefly target both lipid disorders and inflammation. In particular, hypertriglyceridemia and an often concomitant low concentration of circulating HDL predispose MS patients to cardiovascular risk.

From a nutraceutical viewpoint, the principal therapeutic tool is omega-3 fatty acids, especially long-chain ones found in marine products such as fish or algal oils. Whereas appropriate dietary guidelines that incorporate biweekly fish consumption are sufficient for the general population and have been associated with better cardiovascular prognosis, MS patients often need pharmacological amounts of omega-3 fatty acids, namely, docosahexaenoic acids (DHA) and eicosapentaenoic acids (EPA). Such fatty acids should be provided in doses of >1 g/day and a clear distinction should be made between long- and medium-chain omega-3 fatty acids. Indeed, while alpha-linoleic acid might be endowed with biological activities, hypotriglyceridemic actions exclusively pertain to EPA and DHA. Alas, several manufacturers generically claim "omega-3 fatty acids" in their labels. The average consumer is not informed of the difference between the two classes, and close attention should be paid to avoid purchase and consumption of lipid inactive fatty acids.

In addition to triglycerides, omega-3 fatty acids affect other functional parameters linked to the MS, to more relevant and diversified degrees. One notable example is that of blood pressure (Mozaffarian, 2007) and of the production of adhesion molecules and pro-inflammatory cytokines by the arterial wall (Massaro et al., 2008). This latter phenomenon might decrease atherogenic risk by locally decreasing inflammation and the consequent recruitment of further pro-inflammatory cells. In turn, this local anti-inflammatory action would translate into a less inflamed and "reactive" arterial wall, resulting in better vasomotion and lower arterial narrowing.

Still in terms of inflammation, recent evidence demonstrates that DHA inhibits endothelial secreted phospholipase $A_2$ ($sPLA_2$; note that $sPLA_2$ is an independent risk factor for atherosclerosis (Rosenson and Gelb, 2009)) via antioxidant actions (Richard et al., 2008, 2009b) and PKC-mediated pathways (Le Guennec et al., 2010). In summary, consumption of adequate amounts of omega-3 fatty acids contributes to the amelioration of the MS via multiple pathways such as lowering triglycerides (Richard et al., 2009a), producing anti-inflammatory molecules (Chapkin et al., 2009; Levy, 2010), and inhibiting pro-inflammatory enzymes (Richard et al., 2009a). The near totality of these effects has been recorded by using pharmaceutical preparations of omega-3 fatty acids (usually given as ethylesters). Should we suggest patients to increase fish consumption or should we prescribe capsules? The former advice should certainly be part of the overall dietary strategy, but we should not forget that patients with elevated triglycerides need pharmacological treatment with at least 1 g/day of omega-3 capsules. Whether the levels of consumption recommended by international scientific bodies can be achieved by increasing fish consumption or by the use of supplements should be the subject of discussion between patient and physician. However, it has to be underlined that patients with severe hypertriglyceridemia need quantities of omega-3 fatty acids that cannot be provided by fish consumption; thus, they should be prescribed with adequate amounts of omega 3 via pharmaceutical preparations as integral part of their therapy. Alternatively, patients can be advised to consume appropriate functional foods, for example, omega-3-enriched milks or yoghurts. These products are gaining a remarkable share of the market; one caveat is that even though they provide highly bioavailable essential fatty acids (and, as an example, vitamin E), they do not contain such molecules in high concentrations to avoid unpleasant taste and development of off-flavors of rancidity. In brief, patients should verify the actual amount of omega-3 fatty acids that they are going to ingest via a daily dose of the functional food of choice and, eventually, discuss this issue with their primary care physician or specialist.

## Addressing inflammation with nutraceuticals

As mentioned, one of the major features shared by MS patients is systemic inflammation, often difficult to correctly diagnose and quantify due to the current limitations in our knowledge of biomarkers, for example, CRP. Therefore, nutritional or nutraceutical strategies aimed at lessening inflammation should be beneficial for a wide range of MS patients. In the past few years, several supplements have been investigated for their anti-inflammatory actions. The most prominent ones are—as mentioned earlier—omega-3 fatty acids. Yet, a plethora of molecules of vegetable origin, namely, polyphenols and other products of plants' secondary metabolism, appear to exert potent anti-inflammatory activities in various models. The widespread term *polyphenols* includes several classes of compounds that share a common structure; among polyphenols, flavonoids constitute the most important single group, including more than 5000 compounds that have been thus far identified. It is noteworthy that fruits and vegetables require a variety of compounds endowed with protective biological activities (Figure 10.1) to preserve their integrity, because they are constantly exposed to environmental stresses such as UV light radiation and relatively high temperatures. In brief, plants are subjected to exogenous oxidative stress and develop protective compounds accordingly.

One notable example is that of olive oil (OO) phenolics. The olive tree originated in Asia Minor, where meteorological conditions can be extreme and feature high heat and prolonged sunlight. Accordingly, olives are very rich in phenolic compounds, most of which bear the chemical moiety

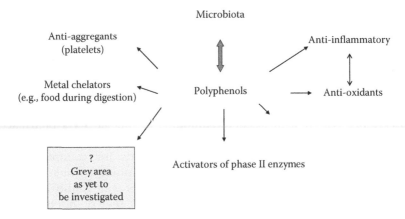

**Figure 10.1** *Multiple mechanisms of action of polyphenols, as related to the MS and its risk factors.*

of oleuropein (a very bitter phenol of large molecular weight). The anti-inflammatory effects of an extra-virgin olive oil (EVOO) extract, particularly rich in phenolic compounds, were investigated—in vitro—on NF-κB translocation in monocytes and monocyte-derived macrophages (MDM) isolated from healthy volunteers. In a concentration-dependent manner, the EVOO extract inhibited p50 and p65 NF-kB translocation in both unstimulated and phorbol myristate acetate (PMA)-challenged cells, being particularly effective on the p50 subunit (Carluccio et al., 2003). Interestingly, this effect occurred at concentrations found in human plasma after nutritional ingestion of virgin OO and was quantitatively similar to that exerted by ciglitazone, a peroxisome proliferator-activated receptors (PPAR)-γ ligand. However, the EVOO extract did not affect PPAR-γ expression in monocytes and MDM. These data provide further evidence of the beneficial effects of EVOO by indicating its ability to inhibit NF-κB activation in human monocyte/macrophages. The effects on endothelial activation mediated by adhesion molecules were confirmed by Dell'Agli et al. (2006) who also used physiological concentrations <1 μM. In monocytes, OO phenolics also inhibit matrix metalloproteinase (MMP)-9, an enzyme involved in plaque rupture (Bellosta et al., 2008; Bogani et al., 2007a), providing further evidence on the mechanisms by which OO reduces the inflammatory burden associated with cardiovascular disorders such as atherosclerosis (Dell'Agli et al., 2010). In vivo, Bogani et al. gave six subjects three oils (30 mL/day of OO, corn oil (CO), or EVOO, distributed among meals) in a Latin square design. The results demonstrate that EVOO is capable of reducing the postprandial events that associate with inflammation and oxidative stress and to increase serum antioxidant capacity (Bogani et al., 2007b). In addition, Bitler et al. demonstrated the anti-inflammatory activities of a wastewater extract in a mouse model of inflammation.

In synthesis, olive phenolics possess interesting activities that suggest their use in MS patients. Notably, their antioxidant activities, namely, toward low-density lipoprotein (LDL), have been recently acknowledged by the EFSA.

## Metabolic syndrome and dyslipidemia: Targeting HDL with nutraceuticals

The role of nutrition on HDL quality and functionality is currently poorly understood. Certain nutrients are known to increase HDL-C levels, but the current question is whether they can alter HDL functionality. Indeed, whereas one nutrient could increase the *concentration* of one HDL sub-population that is, however, noted efficient in RCT, another nutrient could also increase HDL *functionality*. Thus, total HDL-C levels could be increased by both diets, but only the latter would reduce the risk of cardiovascular disease. For this reason, it is imperative to assess which HDL populations are affected by nutritional interventions and whether these changes are ben-eficial in terms of RCT. For example, the ingestion of olive and palm oils is thought to be atheroprotective because—in humans—it decreases LDL and increases HDL levels. However, the expression of ABCA1 is decreased in vitro by oleic acid (Murthy et al., 2004; Wang and Oram, 2002), the main fatty acid of OO, suggesting that HDL should decrease after OO consump-tion. Then, why does OO increase HDL? Are the other components of OO modulating ATP-binding cassette transporters ABCA1, ABCG1, and ABCG4 expression, or is it that a more functional HDL particle is formed in the presence of OO? Berrougui et al. (2006) have shown that HDL preincubated with phenolic extracts of virgin argan oil increased cholesterol efflux from macrophages (THP-1), suggesting that polyphenolic extracts of OO might also be playing a role in the increase of HDL-C. Indeed, Covas et al. (2006) reported that the phenolic components of OO provided benefits for plasma lipid levels and oxidative damage.

Fish oil consumption is cardioprotective, largely because of the presence, in fish, of long-chain omega-3 fatty acids, namely, DHA, 22:6n-3, and EPA, 20:5n-3, acids (Holub, 2009). DHA consumption increases HDL, but the mech-anisms by which this occurs are not known (Wei and Jacobson, 2011). Indeed, DHA incorporates into phospholipids but is also found in free form (actually, as a fatty acid bound to albumin) in plasma.

It remains to be determined which of these forms, that is, phospholipid or free fatty acid, could be affecting the HDL rise and whether or not they affect HDL functionality.

Finally, moderate consumption of alcohol increases HDL and, thus, is con-sidered cardioprotective. The mechanisms by which alcohol increases HDL are not clear, but Berrougui et al. (2009) showed that resveratrol (which is contained in minute amounts in red wine) induces the efflux of cholesterol to apoA1. Nevertheless, the effects of alcohol on HDL functionality need in-depth investigation.

## Conclusions

The MS is a multifaceted disorder with several components that contribute to metabolic derangement. Notably, inflammation is common to most of the MS components and should be chiefly targeted by therapeutic approaches. The interest in nutraceuticals and adjunct pharmacological tools stems from the ability of several food components to lessen systemic inflammation and, therefore, decrease the risk for cardiovascular disease. In addition, some foods and extracts isolated from them, for example, soy protein, olive phenols, or green tea catechins, can positively modulate lipid profile.

In summary, whereas an appropriate pharmacological approach is paramount, lifestyle modifications, dietary advice, and the correct use of selected nutraceuticals should very often accompany drug therapy.

## References

Alberti, KG, Eckel, RH, Grundy, SM, Zimmet, PZ, Cleeman, JI, Donato, KA, Fruchart, JC, James, WP, Loria, CM, Smith, SC, Jr. (2009) Harmonizing the metabolic syndrome: a joint interim statement of the International Diabetes Federation Task Force on Epidemiology and Prevention; National Heart, Lung, and Blood Institute; American Heart Association; World Heart Federation; International Atherosclerosis Society; and International Association for the Study of Obesity. *Circulation* 120:1640–1645.

Bellosta, S, Bogani, P, Canavesi, M, Galli, C, Visioli, F. (2008) Mediterranean diet and cardioprotection: Wild artichoke inhibits metalloproteinase 9. *Mol Nutr Food Res*, 52:1147.

Berrougui, H, Cloutier, M, Isabelle, M, Khalil, A. (2006) Phenolic-extract from argan oil (*Argania spinosa* L.) inhibits human low-density lipoprotein (LDL) oxidation and enhances cholesterol efflux from human THP-1 macrophages. *Atherosclerosis* 184:389–396.

Berrougui, H, Grenier, G, Loued, S, Drouin, G, Khalil, A. (2009) A new insight into resveratrol as an atheroprotective compound: inhibition of lipid peroxidation and enhancement of cholesterol efflux. *Atherosclerosis* 207:420–427.

Bogani, P, Canavesi, M, Hagen, TM, Visioli, F, Bellosta, S. (2007a) Thiol supplementation inhibits metalloproteinase activity independent of glutathione status. *Biochem Biophys Res Commun* 363:651–655.

Bogani, P, Galli, C, Villa, M, Visioli, F. (2007b) Postprandial anti-inflammatory and antioxidant effects of extra virgin olive oil. *Atherosclerosis* 190:181–186.

Carluccio, MA, Siculella, L, Ancora, MA, Massaro, M, Scoditti, E, Storelli, C, Visioli, F, Distante, A, De Caterina, R. (2003) Olive oil and red wine antioxidant polyphenols inhibit endothelial activation: Antiatherogenic properties of Mediterranean diet phytochemicals. *Arterioscler Thromb Vasc Biol* 23:622–629.

Chapkin, RS, Kim, W, Lupton, JR, McMurray, DN. (2009) Dietary docosahexaenoic and eicosapentaenoic acid: Emerging mediators of inflammation. *Prostaglandins Leukot Essent Fatty Acids* 81:187–191.

Covas, MI, Nyyssonen, K, Poulsen, HE, Kaikkonen, J, Zunft, HJ, Kiesewetter, H, Gaddi, A. et al. and EUROLIVE Study Group. (2006) The effect of polyphenols in olive oil on heart disease risk factors: A randomized trial. *Ann Intern Med* 145:333–341.

Dell'Agli, M, Fagnani, R, Galli, GV, Maschi, O, Gilardi, F, Bellosta, S, Crestani, M, Bosisio, E, De Fabiani, E, Caruso, D. (2010) Olive oil phenols modulate the expression of metalloproteinase 9 in THP-1 cells by acting on nuclear factor-kappaB signaling. *J Agric Food Chem* 58:2246–2252.

Dell'Agli, M, Fagnani, R, Mitro, N, Scurati, S, Masciadri, M, Mussoni, L, Galli, GV et al.. (2006) Minor components of olive oil modulate proatherogenic adhesion molecules involved in endothelial activation. *J Agric Food Chem* 54:3259–3264.

Gade, W, Schmit, J, Collins, M, Gade, J. (2010) Beyond obesity: the diagnosis and pathophysiology of metabolic syndrome. *Clin Lab Sci* 23:51–61.

Holub, BJ. (2009) Docosahexaenoic acid (DHA) and cardiovascular disease risk factors. *Prostaglandins Leukot Essent Fatty Acids* 81:199–204.

Le Guennec, JY, Jude, S, Besson, P, Martel, E, Champeroux, P. (2010) Cardioprotection by omega-3 fatty acids: involvement of PKCs? *Prostaglandins Leukot Essent Fatty Acids* 82:173–177.

Levy, BD. (2010) Resolvins and protectins: natural pharmacophores for resolution biology. *Prostaglandins Leukot Essent Fatty Acids* 82:327–332.

Massaro, M, Scoditti, E, Carluccio, MA, De Caterina, R. (2008) Basic mechanisms behind the effects of n-3 fatty acids on cardiovascular disease. *Prostaglandins Leukot Essent Fatty Acids* 79:109–115.

Mozaffarian, D. (2007) Fish, n-3 fatty acids, and cardiovascular haemodynamics. *J Cardiovasc Med (Hagerstown)* 8(Suppl 1):S23–S26.

Murthy, S, Born, E, Mathur, SN, Field, FJ. (2004) Liver-X-receptor-mediated increase in ATP-binding cassette transporter A1 expression is attenuated by fatty acids in CaCo-2 cells: effect on cholesterol efflux to high-density lipoprotein. *Biochem J* 377:545–552.

Reaven, GM. (2011) The metabolic syndrome: Time to get off the merry-go-round? *J Intern Med* 269:127–136.

Richard, D, Bausero, P, Schneider, C, Visioli, F. (2009a) Polyunsaturated fatty acids and cardiovascular disease. *Cell Mol Life Sci* 66:3277–3288.

Richard, D, Kefi, K, Barbe, U, Bausero, P, Visioli, F. (2008) Polyunsaturated fatty acids as antioxidants. *Pharmacol Res* 57:451–455.

Richard, D, Wolf, C, Barbe, U, Kefi, K, Bausero, P, Visioli, F. (2009b) Docosahexaenoic acid down-regulates endothelial Nox 4 through a sPLA2 signalling pathway. *Biochem Biophys Res Commun* 389:516–522.

Rosenson, RS, Gelb, MH. (2009) Secretory phospholipase A2: A multifaceted family of proatherogenic enzymes. *Curr Cardiol Rep* 11:445–451.

Wang, Y, Oram, JF. (2002) Unsaturated fatty acids inhibit cholesterol efflux from macrophages by increasing degradation of ATP-binding cassette transporter A1. *J Biol Chem* 277:5692–5697.

Wei, MY, Jacobson, TA. (2011) Effects of eicosapentaenoic acid versus docosahexaenoic acid on serum lipids: a systematic review and meta-analysis. *Curr Atheroscler Rep* 13:474–483.

# Medicinal Herbs and Dietary Supplements in Diabetes Management

Poonam Tripathi and Awanish Kumar Pandey

## Contents

## Introduction

Diabetes mellitus is a metabolic disorder characterized by hyperglycemia, abnormal lipid, and protein metabolism along with specific long-term complications affecting the retina, kidney, and nervous system.[1] Diabetes mellitus has been recognized as a growing worldwide epidemic by many health advocacy group including WHO.[2] The WHO has estimated that diabetes will be one of the world's leading cause of death and disability in the next quarter century. The International Diabetes Federation (IDF) estimates the total number of diabetic subjects to be around 40.9 million in India, and this is estimated to rise to 69.9 million by the year 2025.[3]

## Need and scope of alternative medicine

Regardless of the type of diabetes, patients are required to control their blood glucose with medication and/or by adhering to an exercise program and a dietary plan. Due to the adoption of Western lifestyle, non-insulin-dependent diabetes mellitus is becoming a major health problem in developing countries. Patients with type-2 diabetes mellitus are usually placed on a restricted diet and are instructed to exercise, the purpose being weight control. If diet and exercise fail to control blood glucose at desired level, pharmacological treatment is prescribed.[4] These treatments have their own drawbacks ranging from development of resistance and adverse effects to lack of responsiveness in a large segment of the patient population. Moreover, none of the glucose-lowering agents adequately control hyperlipidemia that is frequently seen in diabetes.[5]

The limitation of currently available oral antidiabetic agents in terms of either efficacy or safety coupled with the emergence of the disease into global epidemic has encouraged alternative therapy that can manage diabetes more efficiently and safely.[6]

## Alternative approach

Complementary and alternative therapies (CAM) are treatments that are neither widely taught in medical schools nor widely practiced in hospitals. However, use of CAM is increasing worldwide. In 1997, 42% Americans had used an alternative medical therapy. Total visit to complementary practitioners (629 million visits) exceed total visit to U.S. primary care physicians (386 million).[7] In Canada, a recent survey found that 75% people with diabetes used non-prescribed supplements (herbal, vitamin, mineral, or others) and alternative medications.[8] Overall, research indicates that most people who use CAM therapies do so in addition to, rather than in place of, conventional medical treatment[8,9] although some do not receive any concurrent conventional medical care.[10] CAM for diabetes have been becoming increasingly popular the last several years, and there is an existing body of research on CAM and diabetes, but well-defined epidemiological studies are needed.

## Medicinal herbs

In ancient literature, more than 800 plants are reported to have antidiabetic properties.[11] Ethnopharmacological surveys indicate that more than 1200 plants are used in traditional medicine for hypoglycemic activity.[12]

Ancient Indian medicine has mentioned numerous *dravyas* (things that have biological functions and properties) that have been reported effective in *madhumeha* (diabetes).[13]

We believe that the search for a novel antidiabetic drug from plant materials should be advocated, since plants are well recognized as important sources of providing new drugs.[14] According to a review published by Newman and Cragg,[15] nearly 32 new chemical entities have been filed with FDA for treatment of diabetes, both types I and II in last 25 years. These drugs include a significant number of biologics, based upon varying modifications of insulin produced in general by biotechnological means. In 2005, the FDA approved exenatide, the first in a new class of therapeutic agents known as incretin mimetics, a natural product derivative. The drug exhibits glucose-lowering activity similar to the hormone glucagon-like peptide-1 (GLP-1) but is a 39-residue peptide based upon one of the peptide venoms of the Gila monster, *Heloderma suspectum*.[15] Metformin created by the Bristol–Myers Squibb Company is an oral antidiabetic drug from the biguanide class. It is the first-line drug of choice for the treatment of type-2 diabetes, particularly in overweight and obese people and those with normal kidney function, and it might be the best choice for the people with heart disease.[16-19] The biguanide class of antidiabetic drugs, which also includes the withdrawn agents phenformin and buformin, originates from the French lilac or goat's rue (*Galega officinalis*), a plant known for several centuries to reduce the symptoms of diabetes mellitus.[20-22]

To date, over 400 traditional plant treatments for diabetes have been reported,[23] although only a small number of these have received scientific and medicinal evaluation to assess their efficacy. The following is a summary of several of the most studied and commonly used medicinal herbs.

*Azadirachta indica*: This plant is commonly known as *neem (A. indica)* in India. It has been long used as a treatment for diabetes. Hydroalcoholic extract of *A. indica* showed hypoglycemic and antihyperglycemic effects in normal, glucose-fed, and streptozotocin (STZ)-diabetic rats.[24] The plant exerts its pharmacological activity independent of its time of administration, that is, either prior or after alloxan administration.[25] The plant blocks the action of epinephrine on glucose metabolism, thus, increasing peripheral glucose utilization.[26] It also increased glucose uptake and glycogen deposition in isolated rat hemidiaphragm.[27]

*Beta vulgaris*: It is commonly known as garden beet and is used traditionally in the management of diabetes in different parts of India. Various glycosides (beta-vulgarosides I–IV) isolated from the root of this plant were investigated for hypoglycemic activity. The extract of the plant was also found to be effective in inhibiting nonenzymatic glycolization of skin proteins in STZ-induced diabetic rats.[28]

*Cajanus cajan*: It is used in Panamanian folk medicine for the treatment of diabetes. A single dose of unroasted seeds of *C. cajan* administration as a 60%

and 80% diet to normal and alloxanized mice caused a significant reduction in the serum glucose levels after 1–2 h and a significant rise at 3 h. However, roasted seeds administration caused a significant increase in the serum glucose levels during a 3 h experimental period. Roasting of seeds at high temperature for 30 min resulted in the total loss of hypoglycemic principle but not the hyperglycemic principle present in the seeds.[29] Aqueous fraction of the leaves and stems of *C. cajan* (500 and 1000 mg/kg) lacked hypoglycemic effect in normal mice. However, it significantly increased glucose tolerance at 1 and 2 h in OGTT.[30] Hypolipidemic effect has also been reported earlier.[31,32] Cooked diet of *C. cajan* has also shown significant hypoglycemic effect in healthy human volunteers.[33]

*Gymnema sylvestre*: It is commonly known as Gurmar in India and has long been used as a treatment for diabetes. It appeared on the U.S. market several years ago, known as a "sugar blocker." In a study of type-2 diabetes, 22 patients were given 400 mg *G. sylvestre* extract daily along with their oral hypoglycemic drugs. All patients demonstrated improved blood sugar control. Out of these 22, 21 were able to discontinue oral medication and maintain blood sugar control with the Gymnema extract alone.[34] It was postulated that G. sylvestre enhance the production of endogenous insulin.[35]

*Momordica charantia*: *M. charantia*, also known as bitter melon, has been used extensively in folk medicine as a remedy for diabetes. The blood sugar–lowering action of fresh juice or unripe fruit has been established in animal experimental models as well as human clinical trials.[36–41] It is composed of several compounds with confirmed antidiabetic activity. Alcohol-extracted charantin from *M. charantia* consists of mixed steroids and was found to be more potent than the oral hypoglycemic agent tolbutamide in an animal study.[42]

*Ocimum sanctum*: It is a herb found throughout India, up to an altitude of 1800 m in the Himalayas and is cultivated in temples and gardens. Dhar et al. reported a hypoglycemic effect of the ethanolic extract (50%) of the leaves.[43] The ethanol (70%) leaf extract of *O. sanctum* has been associated with significant reduction of blood glucose level in normal, glucose-fed hyperglycemic and STZ (50 mg/kg IP)-induced diabetic rats. This effect was 91.55% and 70.43% of that of tolbutamide in normal and diabetic rats, respectively.[44] A diet containing leaf powder (1%) fed to normal and diabetic rats for 1 month significantly reduced fasting blood sugar, uronic acid, total amino acids, total cholesterol, triglycerides, and total lipids.[45] The plant has also demonstrated antioxidant[46] and hypolipidemic effects.[47] Results of a randomized, placebo-controlled, crossover, single-blind clinical trial of leaf extract of *Ocimum album* showed significant decrease in fasting and postprandial blood levels and mean total cholesterol levels in treated subjects ($n=40$) as compared to controls.[48]

*Pterocarpus marsupium*: It is a well-known plant commonly known as *vijaysar*, found throughout India. In folk medicine, the plant is used as a hypoglycemic agent and studies validated the hypoglycemic potential of plant.[49-51] Different parts of the plant such as the bark and latex were investigated and reported to have hypoglycemic activity.[52-56] Various active components like (–)-epicatechin, marsupsin, pterosupin, and pterostilbene, isolated from the bark and heartwood of the plant, were also found to possess blood sugar–lowering activity.[57-61]

*Scoparia dulcis*: This plant is commonly known as "sweet broomweed." Various extracts of the plant have been reported to increase the activities of insulin and to reduce the blood glucose level in STZ-diabetic rats. The extracts also produced significant antioxidant activity in the liver, kidney, and brain of diabetic rats.[62] A beneficial effect of the extract on glycoproteins of diabetic rats[63] and insulin-receptor-binding effect, resulting in significant increase in plasma insulin, has been reported.[64]

*Trigonella foenum-graecum*: It is commonly known as *fenugreek*, popular for its aromatic properties, and often used as a flavoring agent. It has also been used as a remedy for diabetes, particularly in India.[65] The active principle is in the defatted portion of the seed, which contains the alkaloid gonelline, nicotinic acid, and coumarin. Several animal experimental studies have confirmed the antidiabetic potential of *T. foenum-graecum*.[4,66-68] Human studies have confirmed glucose and lipid-lowering effects of *T. foenum-graecum*.[69] The seeds contain at least 50% fiber and may constitute another potential mechanism of fenugreek's beneficial effect in diabetic patients. In type-2 diabetes patients, the ingestion of 15 g of powder of fenugreek seed soaked in water significantly reduced postprandial glucose levels during the glucose tolerance test.[70]

*Zingiber officinale*: It is a dietary component commonly known as ginger. The juice of *Z. officinale* administered at a dose of 4 mL/kg p.o. daily for 6 weeks significantly prevented hyperglycemia and hypoinsulinemia in STZ-induced type-I diabetic rats. It also produced a significant increase in insulin levels and a decrease in fasting glucose levels in diabetic rats. In an oral glucose tolerance test, the extract was found to decrease the glucose level and to increase the insulin level significantly in STZ-diabetic rats, suggesting that hypoglycemic activity of the juice of *Z. officinale* in type-I diabetic rats possibly involved 5-HT receptors.[71]

## Other plants

Other plants that are most effective and the most commonly used in treatment of diabetes and their complications are *Allium cepa, Allium sativum, Aloe vera, Catharanthus roseus, Coccinia indica, Caesalpinia bonducella, Cucurbita ficifolia, Ficus benghalensis, Ginkgo biloba, Polygala senega, Swertia chirayita, Syzygium cumini, Tinospora cordifolia, Eugenia jambolana, Mucuna pruriens, Murraya koenigii, and Brassica juncea*. All of these plants have shown varying degrees of hypoglycemic and antihyperglycemic activities.[72,73]

## Beverages

Some beverages may also be important in the prevention of diabetes. We briefly consider the potential roles of tea and red wine diabetes control and treatment.

## Black tea

The hot water extract of black tea significantly reduced blood glucose level and was found to possess both preventive and curative effects on STZ-induced diabetes in rats.[74] A study reported hypoglycemic effect of Sri Lankan BOPF black tea in rats,[75] while a double-blind randomized study did not find a hypoglycemic effect of extract of black tea in type-2 diabetes mellitus subjects.[76]

## Green tea

Green tea polyphenols (GTP) especially epigallocatechin gallate injected IP into rats significantly reduced food intake, body weight, blood levels of insulin, glucose, cholesterol, and triglyceride.[77] Epicatechin gallate showed the highest inhibition of glucose uptake by human intestinal epithelial Caco-2-cells suggesting tea catechins could play a role in controlling the dietary glucose uptake at the intestinal tract and possibly contribute to blood glucose homeostasis.[78] Daily administration of an aqueous solution of GTP at 50 and 100 mg/kg body weight for 15 days to alloxan-diabetic rats produced 29% and 44% reduction in serum glucose level. Significant increases in hepatic and renal enzymes as well as the serum lipid peroxide levels were restored by GTP (100 mg/kg bw). In addition, there was a significant increase in liver glycogen and improvements in superoxide dismutase and glutathione, antioxidant enzymes.[79] In STZ rats administered with epicatechin (30 mg/kg) twice a day for 6 days, hyperglycemia and weight loss were not observed and islet morphology was well preserved compared with the STZ-treated rats.[80] Green tea extract was found to have hypoglycemic effect in diabetic rats, which may be associated with the inhibitory effects on α-glucosidase activity and glucose transport ability in the small intestinal mucosa.[81] A double-blind randomized study did not find a hypoglycemic effect of green tea extract in type-2 diabetes mellitus subjects.[76]

## Red wine

Red wine consumption (300 mL) during a meal was associated with significant preservation of plasma antioxidant defenses and reduction of both LDL oxidation and thrombotic activation in type-2 diabetics, demonstrating a potential role in cardiovascular diseases control in these patients.[82] A polyphenol extract from red wine (200 mg/kg) administered for 6 weeks reduced glycemia and decreased food intake and body growth in STZ-diabetic and nondiabetic animals. Ethanol (1 mL/kg) administered alone or in combination with polyphenols (extract from red wine) corrected the diabetic state.[83] Resveratrol is a phytoalexin present in the skin of grapes and red wine. Antihyperglycemic action of resveratrol was demonstrated in obese rodents[84,85] and in two animal models of diabetes: in rats with STZ-induced diabetes or with STZ–nicotinamide-induced diabetes.[86–92]

Some studies also revealed that administration of resveratrol to diabetic rats resulted in diminished levels of glycosylated hemoglobin (HbA1C), which reflects the prolonged reduction of glycemia.[87,92]

## Dietary supplement

Vitamins and minerals are micronutrients that humans require in small quantities for specific biological functions. They function as essential coenzymes and cofactors for metabolic reactions. Some micronutrients have been investigated as potential preventive and treatment agents for both type-1 and type-2 diabetes and for common complications of diabetes.[93,94]

## Chromium

The trace element trivalent chromium ($Cr^{+3}$) is an essential micronutrient for human. It is required for the maintenance of normal glucose metabolism.[95] Effects of chromium on glycemic control, dislipidemia, weight loss, body composition, and bone density have all been studied.[96] Considerable experimental and epidemiological evidence now indicates that chromium levels (in the blood) are a major determinant of insulin sensitivity, as it functions as a cofactor in all insulin-regulating activities.[97] Although a low recommended daily allowance (RDA) has been established for chromium, approximately 200 mg/day appears necessary for optimal blood sugar regulation. A good supply of chromium is assured by supplemental chromium.[98] Among the larger trials, one using organic chromium in brewer's yeast ($n=78$) and another using chromium chloride ($n=180$) reported decreases in fasting and postprandial glucose.[99,100]

One large noncontrolled open-label trial of chromium picolinate followed 833 type-2 diabetic patients in China for up to 10 months. Investigators reported a decrease in fasting and postprandial glucose and a decrease in fatigue, excessive thirst, and frequent urination.[101]

## Vanadium

The trace element vanadium has not been established as an essential nutrient and human deficiency has not been documented.[96,102] Vanadium exists in natural valence states with vanadate (+4) and vanadyl (+5) forms, most common in biological system.

Small trials[103-106] have evaluated the use of oral vanadium supplements in diabetes, mostly on type-2 diabetes,[103] although animal study suggests that vanadium also has potential benefits in type-1 diabetes.[107] In subjects with type-2 diabetes, vanadium increased insulin sensitivity, which is assessed by euglycemic hyperinsulinemic clamp studies in some[103-105] but not all[106] trials. Two small studies have confirmed the potential effectiveness of vanadyl sulfate at a dose of 100 mg/day in improving insulin sensitivity.[104-105]

## Magnesium

The mineral magnesium functions as an essential cofactor for more than 300 enzymes. Magnesium is one of the more common micronutrient deficiencies in diabetes.[93,94,108,109] Low dietary magnesium intake has been associated with increased incidence of type-2 diabetes in some[110] but not in all[111] studies. Magnesium deficiency has been associated with complication of diabetes, retinopathy in particular. One study found that patients with the most severe retinopathy had the lowest level magnesium in the circulation.[112] Diets low in magnesium are associated with increased insulin level[113] and clinical magnesium deficiency is strongly associated with insulin resistance.[108,109]

## Nicotinamide

Niacin (vitamin B3) occurs in two forms, nicotinic acid and nicotinamide. The active coenzyme form (nicotinamide adenine dinucleotide NAD and NAD phosphate) is essential for the functions of hundreds of enzymes and for the metabolism of carbohydrate, lipid, and protein.[93,96] The effects of nicotinamide supplementation have been studied in several trials focusing on the development[114-117] and progression[118-120] of type-I diabetes, which includes a meta-analysis[121] and one small trial in type-2 diabetes.[122]

Nicotinamide appears to be effective against newly diagnosed diabetes and in subjects with positive islets cell antibodies but not diabetes. People who develop type-I diabetes after puberty appear to be more responsive to nicotinamide treatment.[118-121] There is also some evidence to support the idea that nicotinamide helps to preserve β-cell function[119] than for its possible role in diabetes prevention.[123]

## Vitamin E

This essential fat-soluble vitamin functions primarily as an antioxidant.[124] Low levels of vitamin E are associated with increased incidence of diabetes[125] and some research suggest that people with diabetes have decreased levels of antioxidants.[126] People with diabetes may also have greater antioxidant requirement because of increased free radical production with hyperglycemia.[127,128]

Increased levels of oxidative stress markers have been documented in people with diabetes.[129,130] Improvement in glycemic control decreases markers of oxidative stress as does vitamin supplementation.[127–132] Clinical trials involving people with diabetes have investigated the effect of vitamin E on diabetes prevention[133], insulin sensitivity[134,135], glycemic control,[136–138] protein glycation,[139] microvascular complication of diabetes[140,141], and cardiovascular disease and its risk factor.[131,132,142,143] Various clinical trials suggested that vitamin E has shown an important role in the management of diabetes and its associated complications.

## Alpha-Lipoic acid

α-Lipoic acid (ALA) sometimes called lipoic acid or thioctic acid, a naturally occurring compound and a radical scavenger, was shown to enhance glucose transport and utilization in different experimental and animal models. Oral administration of ALA can improve insulin sensitivity in patients with type-2 diabetes.[144] A study from Egypt reported the hypoglycemic effect of ALA in diabetic rats.[145] ALA has the potential in improving glucose levels and dyslipidemia and may exert some protective effect on atherosclerosis vascular changes in diabetic rats.[146]

Plants and dietary supplements having their mechanism of action to lower blood glucose level are shown in Table 11.1 while antidiabetic activity of plants proved in clinical trials are given in Table 11.2.

**Table 11.1** Plants, Dietary Supplements and Their Possible Mechanism of Action

| Plant | Possible Mechanism |
|---|---|
| A. indica (neem) | Inhibits action of epinephrine on glucose metabolism, resulting in increased utilization of peripheral glucose.[26,27] |
| G. sylvestre (Gurmar) | This is attributed to the ability of gymnemic acids to delay the glucose absorption in the blood.[147] |
| M. charantia (bitter gourd) | Not known (in diabetic rabbit models, it possesses a direct action similar to insulin).[148] |
| O. sanctum (tulsi) | Potentiated the action of exogenous insulin in normal rats.[44] |
| Punica granatum (pomegranate) | Inhibits intestinal alpha-glucosidase activity, leading to antihyperglycemic property.[149] |
| S. dulcis (sweet broomweed) | Causes significant increase in insulin and insulin-receptor-binding effect.[64] |
| T. foenum-graecum (fenugreek) | Hypoglycemic effect may be mediated through stimulating insulin synthesis and/or secretion from the beta pancreatic cells of Langerhans.[150] |
| **Dietary Supplements** | **Possible Mechanism** |
| Nicotinamide | Act by protecting pancreatic β-cells.[123,151] |
| L-Carnitine | Effect insulin sensitivity and enhance glucose uptake and storage.[152] |
| Vanadium | Insulin mimetic with upgradation of insulin receptors.[153] |
| Chromium | Facilitates insulin binding and subsequent uptake of glucose into cell.[94] |
| Vitamin E | Potent lipophilic antioxidant activity with possible influences on protein glycation lipid oxidation and insulin sensitivity and secretion.[94,153] |

**Table 11.2** Plants and Their Human Evidences

| Plant | Human Evidence |
|---|---|
| Asteracantha longifolia (Hygrophila) | Oral administration of extract significantly improves glucose tolerance in healthy human subjects and diabetic patients.[154] |
| C. cajan (pigeon pea) | Cooked diet of C. cajan has shown significant hypoglycemic activity in human healthy volunteers.[155] |
| Co. indica (ivy gourd) | Dried extract of plant (500 mg/kg for 6 weeks) has shown antihyperglycemic activity in human volunteers.[156] |
| G. sylvestre (Gurmar) | Extract of leaves (400 mg/day) enhances endogenous insulin release possibly by regeneration of β-cell in 27 patients with type-I diabetes.[157] |
| M. charantia (bitter gourd) | Water-soluble extract of fruits of M. charantia significantly reduces blood glucose concentration in the 9 NIDDM diabetes on OGTT (50 g).[158] |
| Mu. koenigii (curry leaf) | Powder supplementation (12 g providing 2.5 g fiber) for 1 month in 30 NIDDM patients has shown reduction in fasting and postprandial blood sugar.[159] |
| O. album (holy basil) | Leaf extract of plant has shown significant decreases in fasting and postprandial blood glucose levels in a randomized placebo-controlled crossover single-blind clinical trial.[48] |
| Withania somnifera (winter cherry) | Powder of W. somnifera roots produced a decrease in blood glucose levels in six mild NIDDM subjects that were comparable to oral hypoglycemic agent.[160] |
| T. foenum-graecum (fenugreek) | Administration of fenugreek seed powder (50 g with lunch and dinner) in insulin-dependent (type I) diabetic patients for 10 days significantly reduced fasting blood sugar and improved OGTT.[161] |

## Conclusion and recommendations

Diabetes affects 5% of the global population and the management of diabetes without side effect is still a challenge to the healthcare system. Various alternative therapies with antihyperglycemic effect like dietary supplements, herbs, and beverages are increasingly sought by patients with diabetes. Herbal medications are most commonly used as an alternative therapy by patients with diabetes along with their medication. Indian plant species have proved the efficacy of botanicals in reducing the sugar levels. Various studies suggest that the herbs have a major role in the management of diabetes, which needs further exploration for necessary development of herbal drugs and nutraceuticals from natural sources. However, their safety and efficacy need to be further evaluated by well-designed controlled clinical studies because various standardized forms of herbs and nutraceuticals are urgently needed.

## References

1. David, M.N., James, M., and Daniel, E.S. The epidemiology of cardiovascular disease in type 2 diabetes mellitus, how sweet it is… or is it? *Lancet* 1997; 350(Suppl. 1): S14–S19.
2. Prevention of Diabetes Mellitus, World Health Organization Technical Report Series 1994, No. 844.

3. Sicree, R., Shaw, J., and Zimmet, P. Diabetes and impaired glucose tolerance. In: Gan D, editor. *Diabetes Atlas. International Diabetes Federation.* 3rd edn. Belgium: International Diabetes Federation; 2006 pp. 15–103.

4. Khosla, P., Gupta, D.D., and Nagpal, R.K. Effect of *Trigonella foenum graecum* (Fenugreek) on blood glucose in normal and diabetic rats. *Indian Journal of Physiology and Pharmacology* 1995; 39(2): 173–174.

5. Zia, T., Hasnain, S.N., and Hassan, S.K. Evaluation of the oral hypoglycemic effect of *Trigonella foenum-graecum* in normal mice. *Journal of Ethnopharmacology* 2001; 75: 191–195.

6. Ranjan, C. and Ramanujam, R. Diabetes and insulin resistance associated disorders: disease and the therapy. *Current Science* 2002; 83: 1533–1538.

7. Eisenberg, D.M., Davis, R.B., Ettner, S.L., Appel, S., Wilkey, S., Van Rompay, M., and Kessler, R.C. Trends in alternative medicine use in the United States, 1990–1997: results of a follow-up national survey. *JAMA* 1998; 280: 1569–1575.

8. Ryan, E.A., Pick, M.E., and Marceau, C. Use of alternative medicines in diabetes mellitus. *Diabetic Medicine* 2001; 18: 242–245.

9. Astin, J.A. Why patients use alternative medicine: results of a national survey. *JAMA* 1998; 279: 1548–1553.

10. Eisenberg, D.M., Kessler, R.C., Foster, C., Norlock, F.E., Calkins, D.R., and Delbanco, T.L. Unconventional medicine in United States: prevalence, cost and pattern use. *New England Journal of Medicine* 1993; 328: 246–252.

11. Eddouks, M. and Maghrani, M. Phlorizin-like effect of Fraxinus excelsior in normal and diabetic rats. *Journal of Ethnopharmacology* 2004; 9: 149–154.

12. Kesari, A.N., Kesari, S., Santosh, K.S., Rajesh, K.G., and Geeta, W. Studies on the glycemic and lipidemic effect of *Murraya koenigii* in experimental animals. *Journal of Ethnopharmacology* 2007; 112(2): 305–311.

13. Sabu, M.C. and Subburaju, T. Effect of *Cassia auriculata* Linn. On serum glucose level, glucose utilization by isolated rat hemidiaphragm. *Journal of Ethnopharmacology* 2002; 80(2–3): 203–206.

14. Harvey, A.L. (1993). *Drugs from Natural Products-Pharmaceuticals and Agrochemicals*, Ellis Horwood Limited, Chichester, England.

15. Newman, D.J. and Cragg, G.M. Natural products as sources of new drugs over the last 25 years. *Journal of Natural Products* 2007; 70: 461–477.

16. Clinical Guidelines Task Force, International Diabetes Federation. Glucose Control: Oral Therapy. In: *Global Guideline for Type 2 Diabetes*. Brussels, Belgium: International Diabetes Federation, 2005, pp. 35–38.

17. Clinical guideline 66: Diabetes—type 2 (update). National Institute for Health and Clinical Excellence, London, 2008.

18. American Diabetes Association, Standards of medical care in diabetes—2007. *Diabetes Care* 2007; 30(Suppl. 1): S4–S41.

19. Eurich, D.T., McAlister, F.A., Blackburn, D.F., Majumdar, S.R., Tsuyuki, R.T., Varney, J., and Johnson, J. A. Benefits and harms of antidiabetic agents in patients with diabetes and heart failure: systematic review. *BMJ* 2007; 335(7618): 497–506.

20. Witters, L. The blooming of the French lilac. *Journal of Clinical Investigation* 2001; 108(8): 1105–1107.

21. Pandey, V.N., Rajagopalan, S.S., and Chowdhary, D.P. An effective Ayurvedic hypoglycemic formulation. *Journal of Research in Ayurveda and Siddha* 1995; XVI(1–2): 1–14.

22. Oubre, A.Y., Carlson, T.J., King, S.R., and Reaven, G.M. From plant to patient: an ethnomedical approach to the identification of new drugs for the treatment of NIDDM. *Diabetologia* 1997; 40(5): 614–617.

23. Bailey, C.J. and Day, C. Traditional plant medicines as treatments for diabetes. *Diabetes Care* 1989; 12: 553–564.

24. Chattopadhyay, R.R., Chattopadhyay, R.N., Nandy, A.K., Poddar, G., and Maitra, S.K. Preliminary report on antihyperglycemic effect of a fraction of fresh leaves of *Azadirachta indica* (Beng. Neem). *Bulletin of the Calcutta School of Tropical Medicine* 1987; 35: 29–33.

25. Khosla, P., Bhanwra, S., Singh, J., Seth, S., and Srivastava, R.K. A study of hypoglycemic effects of *Azadirachta indica* (Neem) in normal and alloxan diabetic rabbits. *Indian Journal of Physiology and Pharmacology* 2000; 44(1): 69–74.

26. Chattopadhyay, R.R. Possible mechanism of antihyperglycemic effect of Azadirachta indica leaf extract. Part IV. *General Pharmacology* 1996; 27(3): 431–434.

27. Chattopadhyay, R.R., Chattopadhyay, R.N., Nandy, A.K., Poddar, G., and Maitra, S.K. The effect of fresh leaves of *Azadirachta indica* on glucose uptake and glycogen content in the isolated rat hemi diaphragm. *Bulletin of the Calcutta School of Tropical Medicine* 1987; 35: 8–12.

28. Tunali, T., Yarat, A., Yanardag, R., Ozcelik, F., Ozsoy, O., Ergenekon, G., and Emekli, N. The effect of chard (*Beta vulgaris* L. var. cicla) on the skin of streptozotocin induced diabetic rats. *Die Pharmazie* 1998; 53: 638–640.

29. Amalraj, T. and Ignacimuthu, S. Hypoglycemic activity of *Cajanus cajan* (seeds) in mice. *Indian Journal of Experimental Biology* 1998a; 36: 1032–1033.

30. Esposito Avella, M., Diaz, A., de Gracia, I., de Tello, R., and Gupta, M.P. Evaluation of traditional medicine: effects of *Cajanus cajan* L. and of *Cassia fistula* L. on carbohydrate metabolism in mice. *Revista Medica de Panama* 1991; 16: 39–45.

31. Prema, L. and Kurup, P.A. Effect of protein fractions from *Cajanus cajan* (redgram) and *Dolichos biflorus* (horsegram) on the serum, liver and aortic lipid levels in rats fed a high-fat cholesterol diet. *Atherosclerosis* 19973a; 18: 369–377.

32. Prema, L. and Kurup, P.A. Hypolipidemic activity of the protein isolated from *Cajanus cajan* in high fat-cholesterol diet fed rats. *Indian Journal of Biochemistry and Biophysics* 19973b; 10: 293–296.

33. Panlasigui, L.N., Panlilio, L.M., and Madrid, J.C. Glycemic response in normal subjects to five different legumes commonly used in the Philippines. *International Journal of Food Science and Nutrition* 1995; 46(2): 155–160.

34. Baskaran, K., Kizar Ahamath, B., Radha Shanmugasundaram, K., and Shanmugasundaram, E.R. Antidiabetic effect of a leaf extract from *Gymnema sylvestre* in non-insulin dependent diabetes mellitus patients. *Journal of Ethnopharmacology* 1990; 30: 295–300.

35. Shanmugasundaram, E.R., Rajeswari, G., Baskaran, K. et al. Use of *Gymnema sylvestre* leaf extract in the control of blood glucose in insulin-dependent diabetes mellitus. *Journal of Ethnopharmacology* 1990; 30: 281–294.

36. Karunanayake, E.H., Jeevathayaparan, S., and Tennekoon, K.H. Effect of *Momordica charantia* fruit juice on Streptozotocin induced diabetes in rats. *Journal of Ethnopharmacology* 1990; 30(2): 199–204.

37. Khanna, P., Jain, S.C., Panagariya, A., and Dixit, V.P. Hypoglycemic activity of polypeptide-p from a plant source. *Journal of Natural Products* 1981; 44(6): 648–655.

38. Bailey, C.J., Day, C., Turner, S.L., and Leatherdale, B.A. Cerasee, a traditional treatment for diabetes. Studies in normal and streptozotocin diabetic mice. *Diabetes Research* 1985; 2(2): 81–84.

39. Singh, N., Tyagi, S.D., and Agarwal, S.C. Effects of long term feeding of acetone extract of *Momordica charantia* (whole fruit powder) on alloxan diabetic albino rats. *Indian Journal of Physiology and Pharmacology* 1989; 33(2): 97–100.

40. Jayasooriya, A.P., Sakono, M., Yukizaki, C., Kawano, M., Yamamoto, K., and Fukuda, N. Effects of *Momordica charantia* powder on serum glucose levels and various lipid parameters in rats fed with cholesterol-free and cholesterol-enriched diets. *Journal of Ethnopharmacology* 2000; 72(1/2): 331–336.

41. Ahmad, N., Hassan, M.R., Halder, H., and Bennoor, K.S. Effect of *Momordica charantia* (Karela) extracts on fasting and postprandial serum glucose levels in NIDDM patients. *Bangladesh Medical Research Council Bulletin* 1999; 25(1): 11–13.

42. Sarkar, S., Pranava, M., and Marita, R. Demonstration of the hypoglycemic action of *Momordica charantia* in a validated animal model of diabetes. *Pharmacological Research* 1996; 33: 1–4.

43. Dhar, M.L., Dhar, M.M., Dhawan, B.N., Mehrotra, B.N., and Ray, C. Screening of Indian plants for biological activity. *Indian Journal of Experimental Biology* 1968; 6(4): 232–247.

44. Chattopadhyay, R.R. Hypoglycemic effect of *Ocimum sanctum* leaf extract in normal and streptozotocin diabetic rats. *Indian Journal of Experimental Biology* 1993; 31(11): 891–893.

45. Rai, V., Iyer, U., and Mani, U.V. Effect of Tulsi (*Ocimum sanctum*) leaf powder supplementation on blood sugar levels, serum lipids and tissue lipids in diabetic rats. *Plant Foods and Human Nutrition* 1997; 50(1): 9–16.

46. Kelm, M.A., Nair, M.G., Strasburg, G.M., and DeWitt, D.L. Antioxidant and cyclooxygenase inhibitory phenolic compounds from *Ocimum sanctum* Linn. *Phytomedicine* 2000; 7(1): 7–13.

47. Sarkar, A., Lavania, S.C., Pandey, D.N., and Pant, M.C. Changes in the blood lipid profile after administration of *Ocimum sanctum* (Tulsi) leaves in the normal albino rabbits. *Indian Journal of Physiology and Pharmacology* 1994; 38(4): 311–312.

48. Agrawal, P., Rai, V., and Singh, R.B. Randomized placebo controlled, single blind trial of holy basil leaves in-patients with non insulin-dependent diabetes mellitus. *International Journal of Clinical Pharmacology and Therapy* 1996; 34(9): 406–409.

49. Bose, S.N. and Sepaha, G.C. Clinical observations on the hypoglycemic properties of *Pterocarpus marsupium* and *Eugenia jambolana*. *Journal of the Indiana State Medical Association* 1956; 27: 388–391.

50. Gupta, S.S. Effect of *Gymnema sylvestre* and *Pterocarpus marsupium* on glucose tolerance in albino rats. *Indian Journal of Medical Research* 1963; 51: 716–724.

51. Shah, D.S. A preliminary study of the hypoglycemic action of heartwood of *Pterocarpus marsupium* roxb. *Indian Journal of Medical Research* 1967; 55(2): 166–168.

52. Vats, V., Grover, J.K., and Rathi, S.S. Evaluation of antihyperglycemic and hypoglycemic effect of *Trigonella foenum graecum*, *Ocimum sanctum* and *Pterocarpus marsupium* in normal and alloxanized diabetic rats. *Journal of Ethnopharmacology* 2002; 79; 95–100.

53. Vats, V., Yadav, S.P., Biswas, N.R., and Grover, J.K., Anti-cataract activity of *Pterocarpus marsupium* bark and *Trigonella foenum-graecum* seeds extract in alloxan diabetic rats. *Journal of Ethnopharmacology* 2004b; 93: 289–294.

54. Vats, V., Yadav, S.P., and Grover, J.K. Ethanolic extract of *Ocimum sanctum* leaves partially attenuates streptozotocin-induced alterations in glycogen content and carbohydrate metabolism in rats. *Journal of Ethnopharmacology* 2004a; 90: 155–160.

55. Kar, A., Choudhary, B.K., and Bandyopadhyay, N.G. Comparative evaluation of hypoglycaemic activity of some Indian medicinal plants in alloxan diabetic rats. *Journal of Ethnopharmacology* 2003; 84: 105–108.

56. Abesundara, K.J., Matsui, T., and Matsumoto, K. Alpha-glucosidase inhibitory activity of some Sri Lanka plant extracts, one of which, *Cassia auriculata*, exerts a strong antihyperglycemic effect in rats comparable to the therapeutic drug acarbose. *Journal of Agricultural and Food Chemistry* 2004; 52: 2541–2545.

57. Sheehan, E.W., Zemaitis, M.A., Slatkin, D.J., and Schiff Jr., P.L. A constituent of *Pterocarpus marsupium*, (−)-epicatechin, as a potential hypoglycemic agent. *Journal of Natural Products* 1983; 46: 232–234.

58. Ahmad, F., Khalid, P., Khan, M.M., Rastogi, A.K., and Kidwai, J.R. Insulin like activity in (−)-epicatechin. *Acta Diabetologica Latina* 1989; 26: 291–300.

59. Ahmad, F., Khan, M.M., Rastogi, A.K., Chaubey, M., and Kidwai, J.R. Effect of (−)-epicatechin on cAMP content, insulin release and conversion of proinsulin to insulin in immature and mature rat islets in vitro. *Indian Journal of Experimental Biology* 1991; 29: 516–520.

60. Rizvi, S.I., Abu Zaid, M., and Suhail, M. Insulin-mimetic effect of (−)-epicatechin on osmotic fragility of human erythrocytes. *Indian Journal of Experimental Biology* 1995; 33: 791–792.

61. Manickam, M., Ramanathan, M., Jahromi, M.A., Chansouria, J.P., and Ray, A.B. Antihyperglycemic activity of phenolics from *Pterocarpus marsupium*. *Journal of Natural Products* 1997; 60: 609–610.

62. Latha, M. and Pari, L. Modulatory effect of *Scoparia dulcis* in oxidative stress-induced lipid peroxidation in streptozotocin diabetic rats. *Journal of Medicinal Food* 2003a; 6: 379–386.

63. Latha, M. and Pari, L. Effect of an aqueous extract of *Scoparia dulcis* on plasma and tissue glycoproteins in streptozotocin induced diabetic rats. *Pharmazie* 2005; 60: 151–154.

64. Pari, L. and Latha, M. Effect of *Scoparia dulcis* (Sweet Broomweed) plant extract on plasma antioxidants in streptozotocin-induced experimental diabetes in male albino Wistar rats. *Pharmazie* 2004; 59: 557–560.

65. Miller, L.G. Herbal medications, nutraceuticals, and diabetes. In: Miller, L.G. and Murray, W.J., eds. *Herbal Medicinals, A Clinician's Guide*. Binghamton, NY: Pharmaceutical Products Press, Imprint of the Haworth Press, Inc.; 1998, pp. 115–133.

66. Ribes, G., Sauvaire, Y., Da Costa, C., Baccou, J.C., and Loubatieres- Mariani, M.M. Antidiabetic effects of subfractions from fenugreek seeds in diabetic dogs. *Proceedings of the Society for Experimental Biology and Medicine* 1986; 182(2): 159–166.

67. Abdel-Barry, J.A., Abdel-Hassan, I.A., and Al-Hakiem, M.H. Hypoglycemic and antihyperglycemic effects of *Trigonella foenum-graecum* leaf in normal and alloxan induced diabetic rats. *Journal of Ethnopharmacology* 1997; 58(3): 149–155.

68. Ribes, G., Sauvaire, Y., Baccou, J.C. et al. Effects of fenugreek seeds on endocrine pancreatic secretions in dogs. *Annals of Nutrition and Metabolism* 1984; 28: 37–43.

69. Sharma, R.D., Raghuram, T.C., and Rao, N.S. Effect of fenugreek seeds on blood glucose and serum lipids in type I diabetes. *European Journal of Clinical Nutrition* 1990; 44(4): 301–306.

70. Madar, Z., Abel, R., Samish, S., and Arad, J. Glucose-lowering effect of fenugreek in non-insulin dependent diabetics. *European Journal of Clinical Nutrition* 1988; 42: 51–54.

71. Akhani, S.P., Vishwakarma, S.L., and Goyal, R.K. Anti-diabetic activity of *Zingiber officinale* in streptozotocin-induced type I diabetic rats. *Journal of Pharmacy and Pharmacology* 2004; 56: 101–105.

72. Grover, J.K., Yadav, S., and Vats, V. Medicinal plants of India with anti-diabetic potential. *Journal of Ethnopharmacology* 2002; 81: 81–100.

73. Bnouham, M., Ziyyat, A., Mekhfi, H., Tahri, A., and Legssyer, A. Medicinal plants with potential antidiabetic activity—a review of ten years of herbal medicine research (1990–2000). *International Journal of Diabetes and Metabolism* 2006; 14: 1–25.

74. Gomes, J.R., Vedasiromoni, M., Das, R., Sharma, M., and Ganguly, D.K. Anti-hyperglycemic effect of black tea (*Camellia sinensis*) in rat. *Journal of Ethnopharmacology* 1995; 45(3): 223–226.

75. Abeywickrama, K.R.W., Ratnasooriya, W.D., and Amarakoon, A.M.T. Oral hypogly-caemic, antihyperglycaemic and antidiabetic activities of Sri Lankan Broken Orange Pekoe Fannings (BOPF) grade black tea (*Camellia sinensis* L.) in rats. *Journal of Ethnopharmacology* 2011; 135: 278–282.

76. MacKenzie, T., Leary, L., and Blair Brooks, W. The effect of an extract of green and black tea on glucose control in adults with type 2 diabetes mellitus: double-blind randomized study. *Metabolism Clinical and Experimental* 2007; 56: 1340–1344.

77. Kao, Y., Hiipakka, R.A., and Liao, S. Modulation of endocrine systems and food intake by green tea epigallocatechin gallate. *Endocrinology* 2000; 141(3): 980–987.

78. Shimizu, M., Kobayashi, Y., Suzuki, M., Satsu, H., and Miyamoto, Y. Regulation of intestinal glucose transport by tea catechins. *Biofactors* 2000; 13(1–4): 61–65.

79. Sabu, M.C., Smitha, K., and Kuttan, R. Anti-diabetic activity of green tea polyphenols and their role in reducing oxidative stress in experimental diabetes. *Journal of Ethnopharmacology* 2002; 83(1–2): 109–116.

80. Kim, M.J., Ryu, G.R., Chung, J.S. et al. Protective effects of epicatechin against the toxic effects of streptozotocin on rat pancreatic islets: in vivo and in vitro. *Pancreas* 2003; 26(3): 292–299.

81. Quan, J., Yin, X., Liu, M., Shen, M., and Jin, M. Effects of green tea extract on aglucosidase and glucose transport in small intestinal mucosa. *Zhongcaoyao* 2005; 36(3): 411–412.

82. Ceriello, A., Bortolotti, N., Motz, E. et al. Red wine protects diabetic patients from meal-induced oxidative stress and thrombosis activation: a pleasant approach to the prevention of cardiovascular disease in diabetes. *European Journal of Clinical Investigation* 2001; 31(4): 322–328.

83. Al-Awwadi, N., Azay, J., Poucheret, P. et al. Antidiabetic activity of red wine polyphenolic extract, ethanol, or both in streptozotocin-treated rats. *Journal of Agricultural and Food Chemistry* 2004; 52(4): 1008–1016.

84. Lekli, I., Szabo, G., Juhasz, B. et al. Protective mechanisms of resveratrol against ischemia-reperfusion-induced damage in hearts obtained from Zucker obese rats: the role of GLUT-4 and endothelin. *American Journal of Physiology—Heart and Circulatory Physiology* 2008; 294: H859–H866.

85. Sharma, S., Misra, C.S., Arumugam, S. et al. Antidiabetic activity of resveratrol, a known SIRT1 activator in agenetic model for type-2 diabetes. *Phytotherapy Research* 2010; 25(1): 67–73.

86. Su, H.C., Hung, L.M., and Cheng, J.K. Resveratrol, a red wine antioxidant, possesses an insulin-like effect in streptozotocin-induced diabetic rats. *American Journal of Physiology—Endocrinology and Metabolism* 2006; 290: 1339–1346.

87. Palsamy, P. and Subramanian, S. Resveratrol, a natural phytoalexin, normalizes hyperglycemia in streptozotocin-nicotinamide induced experimental diabetic rats. *Biomedicine and Pharmacotherapy.* 2008; 62: 598–605.

88. Palsamy, P. and Subramanian, S. Modulatory effects of resveratrol on attenuating the key enzymes activities of carbohydrate metabolism in streptozotocin-nicotinamide-induced diabetic rats. *Chemico-Biological Interactions* 2009; 179: 356–362.

89. Silan, C. The effects of chronic resveratrol treatment on vascular responsiveness of streptozotocin-induced diabetic rats. *Biological and Pharmaceutical Bulletin* 2008; 31: 897–902.

90. Thirunavukkarasu, M., Penumathsa, S.V., Koneru, S. et al. Resveratrol alleviates cardiac dysfunction in streptozotocin-induced diabetes: role of nitric oxide, thiore-doxin, and heme oxygenase. *Free Radical Biology and Medicine* 2007; 43: 720–729.

91. Penumathsa, S.V., Thirunavukkarasu, M., Zhan, L. et al. Resveratrol enhances GLUT-4 translocation to the caveolar lipid raft fractions through AMPK/Akt/eNOS signalling pathway in diabetic myocardium. *Journal of Cellular and Molecular Medicine* 2008; 12: 2350–2361.

92. Palsamy, P. and Subramanian, S. Ameliorative potential of resveratrol on proinflammatory cytokines, hyperglycemia mediated oxidative stress, and pancreatic beta-cell dysfunction in streptozotocin-nicotinamide-induced diabetic rats. *Journal of Cellular Physiology* 2010; 224: 423–432.

93. Franz, M.J. and Bantle, J.P., Eds. *American Diabetes Association Guide to Medical Nutrition Therapy for Diabetes.* Alexandria, VA: American Diabetes Association, 1999.

94. Mooradian, A.D., Failla, M., Hoogwerf, B., Marynuik, M., and Wylie-Rosett, J. Selected vitamins and minerals in diabetes. *Diabetes Care* 1994; 17: 464–479.

95. Anderson, R.A. Chromium, glucose intolerance and diabetes. *Journal of the American College of Nutrition* 1998; 17: 548–555.

96. Sarubin, A. *The Health Professional's Guide to Popular Dietary Supplements.* Chicago, IL: The American Dietetic Association, 2000.

97. Offenbacher, E.G. and Pi-Sunyer, F.X. Beneficial effect of chromium-rich yeast on glucose tolerance and blood lipids in elderly subjects. *Diabetes* 1980; 29: 919–925.

98. Anderson, R.A., Bryden, N.A., and Polansky, M.M. Dietary chromium intake. Freely chosen diets, institutional diet, and individual foods. *Biological Trace Element Research* 1992; 32: 117–121.

99. Anderson, R.A., Cheng, N., Bryden, N.A., Polansky, M.M., Cheng, N., Chi, J., and Feng, J: Elevated intakes of supplemental chromium improve glucose and insulin variables in individuals with type 2 diabetes. *Diabetes* 1997; 46: 1786–1791.

100. Bahijiri, S.M., Mira, S.A., Mufti, A.M., and Ajabnoor, M.A. The effects of inorganic chromium and brewer's yeast supplementation on glucose tolerance, serum lipids and drug dosage in individuals with type 2 diabetes. *Saudi Medical Journal* 2000; 21: 831–837.

101. Cheng, N., Zhu, X., and Shi, H. Follow-up survey of people in China with type 2 diabetes mellitus consuming supplemental chromium. *Journal of Trace Elements in Experimental Medicine* 1999; 12: 55–60.

102. Harland, B.F. and Harden-Williams, B.A. Is vanadium of nutritional importance yet? *Journal of the American Dietetic Association* 1994; 94: 891–894.

103. Goldfine, A., Simonson, D., Folli, F., Patti, M.E., and Kahn, R. Metabolic effects of sodium metavanadate in humans with insulin-dependent and noninsulin dependent diabetes mellitus in vivo and in vitro studies. *Journal of Clinical Endocrinology and Metabolism* 1995; 80: 3311–3320.

104. Halberstam, M., Cohen, N., Shlimovich, P., Rossetti, L., and Shamoon, H. Oral vanadyl sulfate improves insulin sensitivity in NIDDM but not obese nondiabetic subjects. *Diabetes* 1996; 45: 659–666.

105. Cohen, N., Halberstam, M., Schilmovich, P., Chang, C.J., Shamoon, H., and Rosetti, L. Oral vanadyl sulfate improves hepatic and peripheral insulin sensitivity in patients with non-insulin dependent diabetes mellitus. *Journal of Clinical Investigation* 1995; 95: 2501–2509.

106. Boden, G., Chen, X., Ruiz, J., van Rossum, G.D., and Turco, S. Effects of vanadyl sulfate on carbohydrate and lipid metabolism in patients with noninsulin dependent diabetes mellitus. *Metabolism* 1996; 45: 1130–1135.

107. Poucheret, P., Verma, S., Grynpas, M.D., and McNeil, J.H. Vanadium and diabetes. *Molecular and Cellular Biology* 1998; 188: 73–80.

108. de Valk, H. Magnesium in diabetes mellitus. *Netherlands Journal of Medicine* 1999; 54: 139–146.

109. American Diabetes Association. Magnesium supplementation in the treatment of diabetes (Consensus statement). *Diabetes Care* 1992; 15: 1065–1067.

110. Meyer, K.A., Kushi, L.H., Jacobs, D.R., Slavin, J., Sellers, T.A., and Folsom, A.R. Carbohydrates, dietary fiber, and incident type 2 diabetes in older women. *The American Journal of Clinical Nutrition* 2000; 71: 921–930.

111. Kao, W.H.L., Folsom, A.R., Nieto, F.J., Mo, J.-P., Watson, R.L., and Brancati, F.L. Serum and dietary magnesium and the risk of type 2 diabetes mellitus: the Atherosclerosis Risk in Communities (ARIC) Study. *Archives of Internal Medicine* 1999; 159: 2151–2159.

112. McNair, P., Christiansen, C., Madsbad, S. et al. Hypomagnesemia, a risk factor in diabetic retinopathy. *Diabetes* 1978; 27: 1075–1077.

113. Ma, J., Folsom, A.R., Melnick, S.L., Eckfeldt, J.H., Sharrett, A.R., and Nabuulsi, A.A., Hutchinson, R.G., and Metcalf, P.A. Associated of serum and dietary magnesium with cardiovascular disease, hypertension, diabetes, insulin and carotid arterial wall thickness in the ARIC study. *Journal of Clinical Epidemiology* 1995; 48: 927–940.

114. Greenbaum, C.J., Kahn, S.E., and Palmer, J.P. Nicotinamides effects on glucose metabolism in subjects at risk for IDDM. *Diabetes* 1996; 45: 1631–1634.

115. Lampeter, E.F., Klinghammer, A., Scherbaum, W.A., Heinze, E., Haastert, B., Giani, G., Kolb, H., and the DENIS group. The Deutsch Nicotinamide Intervention Study: an attempt to prevent type 1 diabetes. *Diabetes* 1998; 47: 980–984.

116. Elliott, R.B., Pilcher, C.C., Fergusson, D.M., and Stewart, A.W. A population-based strategy to prevent insulin-dependent diabetes using nicotinamide. *Journal of Pediatric Endocrinology and Metabolism* 1996; 9: 501–509.

117. Gale, E.A.M. Nicotinamide: potential for the prevention of type 1 diabetes? *Hormone and Metabolic Research* 1996; 28: 361–364.

118. Pozzilli, P., Vissali, N., Girhlanda, G., Manna, R., and Andreani, D. Nicotinamide increases C-peptide secretion in patients with recent onset type 1 diabetes. *Diabetic Medicine* 1989; 6: 568–572.

119. Pozzilli, P., Vissali, N., Signore, A. et al. Double-blind trial of nicotinamide in recent onset IDDM (the IMDIAB III study). *Diabetologia* 1995; 38: 848–852.
120. Vissali, N., Cavallo, M.G., Signore, A. et al. A multi-center, randomized trial of two different doses of nicotinamide in patients with recent-onset type 1 diabetes (the IMDIAB VI). *Diabetes/Metabolism Research and Reviews* 1999; 15: 181–185.
121. Pozzolli, P., Browne, P.D., and Kolb, H. The Nicotinamide Trialists: meta-analysis of nicotinamide treatment in patients with recent-onset IDDM. *Diabetes Care* 1996; 19: 1357–1363.
122. Polo, V., Saibene, A., and Pontiroli, A.E. Nicotinamide improves insulin secretion and metabolic control in lean type 2 patients with secondary failure to sulphonylureas. *Acta Diabetologica* 1998; 35: 61–64.
123. Kolb, H. and Volker, B. Nicotinamide in type 1 diabetes mechanism of action revisited. *Diabetes Care* 1999; 22(Suppl. 2): B16–B20.
124. Shils, M.E., Olson, J.A., Shike, M., and Ross, A.C., Eds. *Modern Nutrition in Health and Disease.* 9th edn. Philadelphia, PA: Lea & Febiger, 1999.
125. Salonen, J.T., Nyyssonen, K., Tuomainen, T.P., Maenpaa, P.H., Korpela, H., Kaplan, G.A., Lynch, J., Helmrich, S.P., and Salonen, R. Increased risk of noninsulin-dependent diabetes mellitus at low plasma vitamin E concentrations: a four-year study in men. *BMJ* 1995; 311: 1124–1127.
126. Polidori, M.C., Mecocci, P., Stahl, W., Parente, B., Cecchetti, R., Cherubini, A., Cao, P., Sies, H., and Senin, U. Plasma levels of lipophilic antioxidants in very old patients with type 2 diabetes. *Diabetes/Metabolism Research and Reviews* 2000; 16: 15–19.
127. Sharma, A., Kharb, S., Chugh, S.N., Kakkar, R., and Singh, G.P. Evaluation of oxidative stress before and after control of glycemia and after vitamin E supplementation in diabetic patients. *Metabolism* 2000; 49: 160–162.
128. Ceriello, A., Bortolotti, N., Motz, E., Crescentini, A., Lizzio, S., Russo, A., Tonutti, L., and Taboga, C. Meal-generated oxidative stress in type 2 patients. *Diabetes Care* 1998; 21: 1529–1533.
129. Santini, S.A., Marra, G., Giardina, B., Cottroneo, P., Mordente, A., Martorana, G.E., Manto, A., and Ghirlanda, G. Defective plasma antioxidant defenses and enhanced susceptibility to lipid peroxidation in uncomplicated IDDM. *Diabetes* 1997; 46: 1853–1858.
130. Ceriello, A., Bortolotti, N., Falleti, E., Taboga, C., Tonutti, L., Crescentini, A., Motz, E., Lizzio, S., Russo, A., and Bartoli, E. Total radical-trapping antioxidant parameter in non-insulin dependent diabetic patients. *Diabetes Care* 1997; 20: 194–197.
131. Reaven, P.D., Herold, D.A., Barnett, J., and Edelman, S. Effects of vitamin E on susceptibility of low-density lipoprotein and low-density lipoprotein subfractions to oxidation and on protein glycation in NIDDM. *Diabetes Care* 1995; 18: 807–816.
132. Devaraj, S. and Jialal, I. Low-density lipoprotein postsecretory modification, monocyte function, and circulating adhesion molecules in type 2 diabetic patients with and without macrovascular complications. *Circulation* 2000; 102:191–196.
133. Pozzolli, P., Vissali, N., Cavallo, M.G. et al. Vitamin E and nicotinamide have similar effects in maintaining residual β cell function in recent onset insulin-dependent diabetes (The IMDIAB IV study). *European Journal of Endocrinology* 1997; 137: 234–239.
134. Paolisso, G., D'Amore, A., Giugliano, D., Cereillo, A., Varricchio, M., and D'Onofrio, F. Pharmacological doses of vitamin E improve insulin action in healthy subjects and non-insulin dependent diabetic patients. *The American Journal of Clinical Nutrition* 1993; 57: 650–656.
135. Skrha, J., Sindelka, G., Kvasnicka, J., and Hilgertova, J. Insulin action and fibrinolysis influenced by vitamin E in obese type 2 diabetes mellitus. *Diabetes Research and Clinical Practice* 1999; 4: 27–33.
136. Gomez-Perez, F.J., Valles-Sanchez, V.E., Lopez-Alvarenga, J.C., Pascuali, J.J.I., Orellana, R.G., Padilla, O.B.P., Salinas, E.G.R., and Rull, J.A. Vitamin E modifies neither fructosamine nor HbA1c levels in poorly controlled diabetes. *Revista de Investigación Clínica* 1996; 48: 421–424.

137. Jain, S.K., McVie, R., Jaramillo, J.J., Palmer, M., and Smith, T. Effect of modest vitamin E supplementation on blood glycated hemoglobin and triglyceride levels and red cell indices in type 1 diabetic patients. *Journal of the American College of Nutrition* 1996; 15: 458–461.
138. Paolisso, G., D'Amore, A., Galzerano, D., Balbi, V., Giugliano, D., Varriccho, M., and D'Onofrio, F. Daily vitamin E supplements improve metabolic control but not insulin secretion in elderly type 2 diabetic patients. *Diabetes Care* 1993; 16: 1433–1437.
139. Cerillo, A., Giugliano, D., Quataro, A., Donzella, C., Dipalo, G., and Lefebrve, P.J. Vitamin E reduction of protein glycosylation in diabetes. *Diabetes Care* 1991; 14: 68–72.
140. Tutuncu, N.B., Bayraktar, N., and Varli, K. Reversal of defective nerve conduction with vitamin E supplementation in type 2 diabetes. *Diabetes Care* 1998; 21: 1915–1918.
141. Bursell, S.E., Clermont, A.C., Aiello, L.P., Aiello, L.M., Schlossman, D.K., Feener, E.P., Laffel, L., and King, G.L. High-dose vitamin E supplementation normalizes retinal blood flow and creatinine clearance in patients with type 1 diabetes. *Diabetes Care* 1999; 22: 1245–1251.
142. Yusuf, S., Dagenais, G., Pogue, J., Bosch, J., and Sleight, P. Vitamin E supplementation and cardiovascular events in high-risk patients. The Heart Outcomes Prevention Evaluation Study. *New England Journal of Medicine* 2000; 342: 154–160.
143. Andrew, R., Skyrme-Jones, P., O'Brien, R.C., Berry, K.L., and Meredith, I.T. Vitamin E supplementation improves endothelial function in type 1 diabetes mellitus: a randomized, placebo-controlled study. *Journal of the American College of Cardiology* 2000; 36: 94–102.
144. Jacob, S., Ruus, P., Heramann, R., Tritchler, H.J., Maeker, E., Renn, W., Augustin, H.J., Dietze, G.J., and Rett, K. Oral administration of rac-α-lipoic acid modulates insulin sensitivity in patients with type-2 diabetes mellitus: A placebo-controlled pilot trial. *Free Radical Biology and Medicine* 1999; 27(3–4): 309–314.
145. Salama, R.H.M. Effect of lipoic acid, carnitine and nigella sativa on diabetic rat model. *International Journal of Health Sciences, Qassim University* 2011; 5(2): 126–134.
146. Siti Balkis, B., Khairul, O., Wan Nazaimoon, W.M., Faizah, O., Santhana, R.L., Mokhtar, A.B., Megat, R., and Jamaludin, M. Alpha lipoic acid reduces plasma glucose and lipid and the ultra-microscopic vascular changes in streptozotocin-induced diabetic rats. *Annal of Microscopy* 2008; 8: 58–65.
147. Kanetkar, P., Singhal, R., and Kamat, M. *Gymnema sylvestre*: a Memoir. *Journal of Clinical Biochemistry and Nutrition* 2007; 41: 77–81.
148. Akhtar, M.S., Athar, M.A., and Yaqub, M. Effect of *Momordica charantia* on blood glucose level of normal and alloxan diabetic rabbits. *Planta Medica* 1981; 42: 205–212.
149. Li, Y., Wen, S., Kota, B.P., Peng, G., Li, G.Q., Yamahara, J., and Roufogalis, B.D. *Punica granatum* flower extract, a potent alpha-glucosidase inhibitor, improves postprandial hyperglycemia in Zucker diabetic fatty rats. *Journal of Ethnopharmacology* 2005; 99: 239–244.
150. Puri, D., Prabhu, K.M., and Murthy, P.S. Mechanism of action of a hypoglycemic principle isolated from fenugreek seeds. *Indian Journal of Physiology and Pharmacology* 2002; 46: 457–462.
151. Knip, M., Douek, I.F., Moore, W.P.T., Gillmor, H.A., Mclean, A.E.M., Bingley, P.J., and Gale, E.A.M. For the ENDIT group. Safety of high dose nicotinamide: A review. *Diabetologia* 2000; 43: 1337–1345.
152. Giancaterini, A., De Gaetano, A., Mingrone, G. et al. Acetyl-L-Carnitine infusion increases glucose disposal in type-2 diabetes patients. *Metabolism* 2000; 49: 704–707.
153. O'Connell, B. Select vitamins and minerals in the management of diabetes. *Diabetes Spectrum* 2001; 14: 133–148.
154. Fernando, M.R., Wickramasinghe, N., Thabrew, M.I., Ariyananda, P.L., and Karunanayake, E.H. Effect of *Artocarpus heterophyllus* and *Asteracanthus longifolia* on glucose tolerance in normal human subjects and in maturity-onset diabetic patients. *Journal of Ethnopharmacology* 1991; 31: 277–282.

155. Panlasigui, L.N., Panlilio, L.M., and Madrid, J.C. Glycemic response in normal subjects to five different legumes commonly used in the Philippines. *International Journal of Food Science and Nutrition* 1995; 46(2): 155–160.

156. Kamble, S.M., Kamlakar, P.L., Vaidya, S., and Bambole, V.D. Influence of *Coccinia indica* on certain enzymes in glycolytic and lipolytic pathway in human diabetes. *Indian Journal of Medical Science* 1998; 52: 143–146.

157. Shanmugasundaram, E.R., Rajeswari, G., Baskaran, K., Rajesh Kumar, B.R., Radha hanmugasundaran, K., and Kizar Ahmath, B. Use of *Gymnema sylvestre* leaf extract in the control of blood glucose in insulin-dependent diabetes mellitus. *Journal of Ethnopharmacology* 1990; 30: 281–294.

158. Leatherdale, B.A., Panesar, R.K., Singh, G., Atkins, T.W., Bailey, C.J., and Bignell, A.H. Improvement in glucose tolerance due to *Momordica charantia* (karela). *British Medical Journal (Clinical Research Edition)* 1981; 282(6279): 1823–1824.

159. Iyer, U.M. and Mani, U.V. Studies on the effect of curry leaves supplementation (*Murraya Koenigii*) on lipid profile, glycated proteins and amino acids in non-insulin-dependent diabetic patients. *Plant Foods for Human Nutrition* 1990; 40(4): 275–282.

160. Andallu, B. and Radhika, B. Hypoglycaemic, diuretic and hypocholesterolemic effect of winter cherry (*Withania somnifera*, Dunal) root. *Indian Journal of Experimental Biology* 2000; 38: 607–609.

161. Sharma, R.D., Raghuram, T.C., and Rao, N.S. Effect of fenugreek seeds on blood glucose and serum lipids in type I diabetes. *European Journal of Clinical Nutrition* 1990; 44(4): 301–306.

# V

# Respiratory Health

# 12

# Nutrients for Optimal Lung Function

Susan Ettinger

## Contents

## Structure and function of the normal lung

The lung is designed for the exchange of gases, predominately carbon dioxide and oxygen. Air entering the airways passes through progressively smaller bronchi and bronchioles until it reaches the acinus, a roughly spherical structure with a diameter of about 7 mm. Gas exchange occurs at the blind ends of the respiratory airways, in the thin-walled alveoli. Several alveolar ducts branch off each acinus and terminate in alveolar sacs. At this point, acinar basement

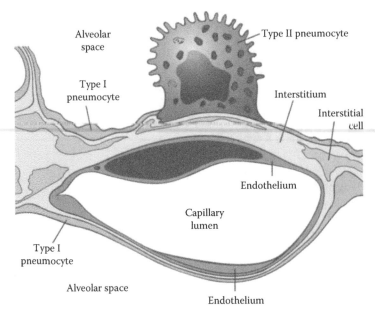

**Figure 12.1** *A section of the alveolar wall, depicting types I and II pneumocytes, interstitial cells, and the endothelial cells lining the capillary lumen. The basement membrane is thin on one side and widened where it is continuous with the interstitial space. (Reproduced from Kumar, V. et al., eds.,* Robbins and Cotran Pathological Basis of Disease, *8thedn, Saunders Elsevier, Philadelphia, PA, 2010.)*

membrane is intimately annexed to an intertwining network of capillaries and delicate interstitial tissue. Thin sections of the alveolar septum contain only fused endothelium and epithelium, while thick sections, the pulmonary interstitia, contain fine elastic fibers, small collagen bundles, a few fibroblast-like cells, smooth muscle cells, mast cells, and a rare lymphocyte and monocyte.

Histologically, the alveolar epithelium is a continuous layer of two cell types: flattened, platelike cells that cover 95% of the alveolar surface (type I pneumocytes) and rounded cells (type II pneumocytes) that synthesize and store surfactant in lamellar bodies. Type II cells differentiate into type I cells and are, thus, essential to alveolar repair. Loosely attached to the alveolar epithelium are alveolar macrophages that ingest carbon and alveolar debris. Numerous *pores of Kohn* perforate the epithelium, allowing passage of bacteria and exudate between adjacent alveoli (Figure 12.1). Surfactant produced by type II cells forms a thin layer over the alveolar membrane (Kumar et al. 2010).

## Lung injury and host defense

The lung is a major target of oxidant stress. The pathogenesis of several pulmonary diseases, including adult respiratory distress syndrome (ARDS) and asthma, involves an imbalance in the prooxidant vs. antioxidant state of lung tissue. The delicate pulmonary systems that facilitate gas exchange can also be damaged by a multitude of agents. For example, sheer stresses exerted by conditions that increase the pressure of blood entering the lung can injure the delicate capillaries. Imbalance of blood or tissue oncotic pressure across the interstitium can perturb fluid flux across the alveolar membranes, resulting in accumulation of fluid in the air sacs (pulmonary edema). Any limitation of blood flow to the lung results in hypoxic injury and cell death. Drugs, pollutants, and inhaled chemicals such as tobacco smoke cause cell injury, stimulating a vigorous immune response and cascades of inflammatory mediators. Thus, host defense sets in motion compensatory mechanisms that also damage the lung and stimulate fibrotic changes that limit lung expansion and compromise gas exchange. Microbes and their toxins also injure the membranes and stimulate host response. Asthma, cystic fibrosis (CF), and chronic obstructive pulmonary disease (COPD) are characterized by persistent airway inflammation and airway remodeling. In asthma, the airways are narrowed with smooth muscle hypertrophy, while CF and COPD exhibit localized, irreversible dilation and destruction of the airways.

## Neoplastic change in the lung

Carcinomas of the lung, as in cancers at other sites, are initiated by DNA damage and progress to the neoplastic phenotype through a series of genetic mutations. It has been estimated that between 10 and 20, stepwise changes have occurred by the time the tumor is clinically diagnosed. Lung cancers can be divided clinically into two subgroups: small-cell carcinoma and non-small-cell carcinoma. Common oncogenes expressed in lung tumors include cMYC, KRAS, EGFR, cMET, and cKit. Lung tumors also have commonly deleted or inactivated tumor suppressor genes including p53 and RB1 and multiple loci on chromosome 3p. Lung tumor subgroups have unique characteristic molecular changes and respond differently to therapy. Precursor epithelial lesions include (1) squamous dysplasia and carcinoma in situ, (2) atypical adenomatous hyperplasia, and (3) diffuse idiopathic pulmonary neuroendocrine cell hyperplasia, although not all "precursor" lesions progress to cancer (Kumar et al. 2010).

## Diet components that increase susceptibility to lung injury
### Alcohol

A history of chronic ethanol consumption, while not directly damaging pulmonary tissues, increases the risk for acute lung injury upon exposure to damaging agents (Brown et al. 2004). While the lung does not use cytoplasmic alcohol dehydrogenase (ADH) to convert alcohol to acetaldehyde, its toxic metabolite, elevated concentrations of alcohol are readily metabolized by microsomal and mitochondrial cytochrome P450 enzymes in the lung (Lieber 1997). These reactions generate damaging oxidants that must be detoxified to avoid cell injury. Glutathione (GSH), a thiol antioxidant predominantly synthesized in the liver, is maintained in high concentrations in alveolar fluid and cells, where it detoxifies oxidants and modulates inflammatory cytokine cascades. Chronic alcohol abuse reduces hepatic GSH synthesis and the enzymes that use GSH for peroxide detoxification. Thus, the reduction of GSH observed in alveolar fluid of chronic alcoholics may be insufficient to protect alveolar cells from inhaled oxidants.

While ethanol is not itself a carcinogen, its metabolite, acetaldehyde, is both carcinogenic and mutagenic. Ethanol also induces the microsomal enzyme oxidizing system (MEOS) CYT2E1 that also metabolizes xenobiotics and dietary procarcinogens to ultimate carcinogens (Lieber 1997). MEOS induction may explain, in part, the increased cancer risk from supplemental beta-carotene tested in heavy smokers as a cancer-preventive agent (Omenn et al. 1994). Alcohol-induced changes in the microbiota bacterial flora also increase acetaldehyde production, thereby increasing cancer risk. Cysteine, a sulfur-containing amino acid synthesized from dietary methionine, provides tissue protection by binding acetaldehyde into a stable, nonreactive complex (Salaspuro 2007).

Alcohol consumption also compromises nutrient bioavailability, especially folate, vitamin $B_6$, thiamin, zinc, and selenium (Pöschl and Seitz 2004). Folate is an essential cofactor for one-carbon metabolism and for production of the universal methyl donor, S-adenosylmethionine (SAM). Homocysteine is methylated to form methionine, the substrate for SAM. Following donation of its methyl group to an acceptor molecule, SAM becomes S-adenosylhomocysteine (SAH) and reenters the homocysteine–methionine (H-M) cycle. Decreased hepatic SAM and increased SAH levels have been observed in folate-depleted animal models, suggesting that folate and other cofactors for the H-M cycle may be limiting in chronic alcoholics.

DNA methylation and histone modification are major epigenetic changes known to regulate gene transcription, underlie phenotypic changes, and modify risk for disease. Epigenetic DNA modification has been directly linked to environmental influences and nutritional status (Choi and Friso 2010). DNA methylation patterns are altered during tumor progression. Herceg et al. (Vaissière et al. 2009) reported that tobacco smoking, sex, and alcohol intake had a strong influence on methylation levels of distinct genes associated with lung cancer. Ulrich et al. (Jung et al. 2008) used case–control methodology to evaluate polymorphisms in methyltransferase (DNMT) and ADH genes in colon cancer and adenoma patients. The authors concluded that gene–diet interactions that regulate DNA methylation or alcohol metabolism may play a role in cancer risk.

## Simple carbohydrates and high glycemic index

Although associations between simple carbohydrates and inflammatory lung diseases have been observed, few mechanistic studies have assessed a causal relationship between dietary carbohydrates and lung inflammation. A recent epidemiologic study explored the associations between dietary glycemic index (GI), dietary fiber (DF), and consumption of carbohydrate-containing food groups and mortality attributable to noncardiovascular, noncancer inflammatory disease (Buyken et al. 2010). Analysis of validated food-frequency questionnaires from 1490 postmenopausal women and 1245 men over a 13 year period revealed that women in the highest GI tertile had a 2.9-fold increased risk for inflammatory death compared with women in the lowest GI tertile, adjusted for age, smoking, diabetes, and alcohol and fiber consumption. Diet patterns that included foods high in refined sugars and limited bread, cereals, or vegetables other than potatoes also independently predicted a greater mortality risk. Further studies are needed to study the influence of metabolic changes incurred by this diet pattern, such as over-production of inflammatory cytokines and reactive species seen with insulin resistance and recurrent postprandial hyperglycemia.

There is also scant evidence that human lung cancer risk is associated with dietary sucrose and simple carbohydrates. However, the link between high refined sugar intake, hyperinsulinemia, and insulin-like growth factor (IGF) aberration is of interest. Recent data were presented that IGF and IGF receptors exert mitogenic action on lung cancer cells and reduce apoptosis (Pavelić et al. 2005) together with observations that IGF is overexpressed in obesity and its comorbid conditions (Giovannucci 2001), suggesting that further studies in this area are warranted.

## Fatty acids and high-fat diets

Dietary fatty acids have differential effects on airway inflammation and hyper-responsiveness, possibly because of their contribution to membrane viscosity and of their ability to induce changes in membrane protein localization, with impact on T cell signaling and proliferation (Shaikh and Edidin 2006). An alternate possibility is that dietary n-6 fatty acids from grain oils are converted to arachidonic acid and provide substrate for proinflammatory prostaglandins and leukotrienes through the cyclooxygenase (COX) and lipoxygenase (LOX) pathways. Saturated fatty acids such as palmitic acid (C16:0) initiate Toll-like receptor (TLR) activation and stimulate the innate immune response, activating the nuclear factor kB (NFkB) and other inflammatory cascades (Wood et al. 2011; Huang et al. 2012).

Taken together, these observations suggest that chronic intake of excess fatty acids with potential to amplify an inflammatory response can increase susceptibility to lung injury. Further, excess calories consumed in a "high-fat diet" can induce metabolic syndrome with dyslipidemia. Recent work (Madenspacher et al. 2010) revealed that while dyslipidemia in an obese population increases systemic neutrophil trafficking and secretion of proinflammatory mediators, neutrophil influx is attenuated in the air spaces. Consequent reduction in chemokine secretion and NFkB activation could compromise microbial clearance, placing the obese patient at greater risk for pneumonia.

Despite higher lung cancer rates in countries with the highest fat intake and several case–control studies that reported a positive association between fat and/or cholesterol and lung cancer, a direct relationship between human dietary fat consumption and lung cancer has been difficult to establish. Possible reasons include confounding variables such as excess energy intake, type of fat and oil (Chang et al. 2007), type of meat (red meat or fowl), mutagens or heme iron content in meat, cooking preferences (Yang et al. 2012), dilution of micronutrients in the diet, fatty foods contaminated with lipid oxidation products, abnormal metabolic flux in obesity and its comorbidities, and inaccuracies in diet assessment. To explore this question further, data from 280,419 female and 149,862 male participants in eight prospective cohort studies that met predefined criteria were analyzed. After adjusting for smoking and other potential risk factors, the authors reported no associations between intakes of total or specific types of fat and lung cancer risk among never, past, or current smokers (Smith-Warner et al. 2002).

## Aging processes, inadequate protein, and compromised micronutrient availability

Chronological aging reduces erythrocyte GSH concentrations in the face of increased age-associated oxidative stress (Sekhar et al. 2011). It is unclear whether this reduction is due primarily to inadequate protein intake, reduced protein turnover, or increased protein degradation; however, the authors demonstrated that supplementation of aging subject with glycine and cysteine precursors increased their GSH concentrations to the levels observed in young subjects. Rojas et al. (Iyer et al. 2009) infused intraperitoneal endotoxin into murine lungs and observed severe oxidative damage accompanied by reduced GSH concentrations, supporting the hypothesis that compromised, age-associated GSH availability increases the risk for acute lung injury in aging individuals. While GSH and its disulfide dimer are the primary intracellular redox couple in the body, cysteine and its disulfide, cystine, constitute the most abundant, low molecular weight thiol/disulfide redox couple in the plasma. Since the reduction in cysteine concentration is tightly linked to the progression and severity of lung injury following endotoxin exposure, the authors propose that cysteine availability may be essential to host defense following oxidative injury. Cysteine is synthesized from dietary methionine via pathways that require folate, pyridoxine, and vitamin $B_{12}$. Further, intracellular

GSH functions as an endogenous antioxidant by cycling between the reduced and dimer forms by means of enzymes that require riboflavin, niacin, and selenium cofactors. Taken together, these lines of evidence suggest that inadequate dietary protein and micronutrient availability may increase susceptibility to oxidative stress and lung injury, especially in an aging population.

## Food allergens and asthma

Asthma is a chronic inflammation of the airways resulting in recurrent bouts of wheezing, chest tightness, shortness of breath, and coughing. The condition is characterized by increased airway responsiveness to stimuli, inflammation of the bronchial walls, and increased secretion of mucus. Lung epithelium is infiltrated with lymphocytes, eosinophils, mast cells, macrophages, and neutrophils. Asthma is also accompanied by significant airway remodeling. The basement membrane becomes thickened with connective tissue elements not found in normal basement membrane. Smooth muscle cells proliferate and hypertrophy; the tissue is infiltrated by fibroblasts and myofibroblasts that secrete fibrous matrix into the area. Bronchoconstriction contributes to airflow limitation; severe constriction can remit spontaneously or with treatment.

A subset of patients with asthma mount this hypersensitivity response to allergens in the diet. The most common allergens in the diet are cow's milk, eggs, peanuts, tree nuts, wheat, soy, fish, and shellfish. Allergens in these foods cross the gut mucosa, enter the blood stream, and bind IgE on mast cells, setting off the allergic cascade as described earlier. Reactions can be localized or result in system-wide anaphylaxis. Common food patterns associated with food allergy include (1) inadequate food/oral beverage intake; (2) poor food choices, which can be secondary to food insecurity, inadequate cooking facilities, and lack of time; and (3) disordered eating/meal patterns, reliance on fast food, and inadequate knowledge of healthy eating. The rapid "Westernization" of the diet is thought to play a key role in the complex genetics and developmental pathophysiology of asthma (Kim et al. 2009). Hypotheses to explain increased incidence of allergy include changes in hygiene (reduced early exposure to allergens to induce tolerance); reduced consumption of dietary antioxidants, lipids, and other anti-inflammatory nutrients; increased fast-food intake and unhealthy dietary patterns; as well as changes in breastfeeding and modification in intestinal microbiota.

Early mechanistic studies focused on major dietary factors such as food allergens, antioxidants, and lipids, while more recent investigations have explored dietary patterns, prevalence of obesity, fast-food consumption, and the Mediterranean diet. Several surveys suggest that a diet pattern rich in fruits and vegetables, as well as nuts, is inversely related to asthma symptoms. However, evidence about the role of specific nutrients, antioxidant supplementation, food types, or dietary patterns past early childhood on asthma prevalence is inconclusive. The influence of diet on airway hypersensitivity has also been linked to the fetal environment; the maternal diet may be a significant factor in the development of the fetal airway and immune system and may play an epigenetic role in sensitizing fetal airways to respond abnormally to environmental insults.

## Excess calories, obesity, and metabolic syndrome

Agrawal et al. have recently reviewed epidemiologic data demonstrating associations between asthma, lung inflammation, and metabolic syndrome complicated by abdominal adiposity, hypertension, dyslipidemia, and insulin resistance (Agrawal et al. 2011). Several mechanistic links have been explored, including the association of obesity with altered immune response, specifically Th2 inflammation or atopic tendencies. Growing evidence has implicated metabolic abnormalities such as hyperglycemia, hyperinsulinemia, and excess IGF secretion associated with obesity. Experimental animal models of metabolic syndrome have been used to demonstrate abnormal lung nitric oxide–arginine response and oxonitrosative stress in response to injury.

Evidence from human studies is sparse. Mustajoki et al. intervened with a very low calorie diet (~450 kcal) for 8 weeks followed by a low calorie diet and reported weight loss and symptom reduction, but no change in medication in obese patients with asthma (Stenius-Aarniala et al. 2000). It is not clear from this and similar studies whether airway hyperreactivity was secondary to increased fat storage or to associated metabolic abnormalities and comorbidities. Conversely, reduced physical activity, mechanical stress induced from breathing difficulties, or side effects from asthma medications could have increased the risk for obesity and metabolic changes. Indeed, it is possible that weight loss simply modified the therapeutic threshold of their drugs. Finally, part of the improvement in airway hyperreactivity could be due to substitution of the subject's usual nutrient poor obesogenic diet with a diet that improved protein and micronutrient status. Nonetheless, several potential mechanisms relating obesity and lung inflammation are possible and require further investigation (McGinley and Punjabi 2011):

- Obesity can impose a mechanical load on the diaphragm and decrease expiratory residual volume and functional residual capacity, reducing airway diameter and smooth muscle function and compromising clearance of pathogenic debris.
- Obesity-associated inflammation may increase visceral adipocyte secretion of proinflammatory cytokines such as interleukin-6, leptin, tumor necrosis factor, tumor growth factor-$\beta$1, and eotaxin. The altered mediator milieu can modify atopy and the Th1-Th2 imbalance (common in obesity), increasing airway hyperreactivity.
- Obesity is associated with a cadre of metabolic derangements; there is growing evidence that hyperglycemia, hyperinsulinemia, and IGFs exert negative effects on airway structure and function.

On the other hand, improved clinical outcomes and lower levels of inflammatory cytokines have been observed in obese patients with acute lung injury. Suratt et al. investigated the effects of obesity in leptin-resistant db/db obese and diet-induced obese mice using an inhaled lipopolysaccharide (LPS) model of acute lung injury (Kordonowy et al. 2012). Neutrophil chemoattractant response was attenuated in obese mice, with diminished chemotaxis to the chemokine keratinocyte cytokine, together with a reduced neutrophil migration into the air spaces.

Thus, despite theoretical considerations that lung inflammation and lung cancer are associated with consumption of a high glycemic diet rich in simple carbohydrate, variables related to the type of obesity (visceral or subcutaneous), diabetes, and body composition are complex and confounded. Rohan et al. (Kabat et al. 2008) analyzed eight years of data from the Women's Health Study and reported that after mutually adjusting for BMI and waist circumference, BMI was inversely associated with lung cancer risk in both current smokers and former smokers, and waist circumference was positively associated with risk.

## Diet components that enhance lung defense

Recently, several investigators have reasoned that human inflammatory lung conditions, including adult respiratory distress syndrome, might respond to diet that modifies inflammatory processes. Singer et al. (Pontes-Arruda et al. 2008) conducted a meta-analysis of three long-term clinical trials on 411 mechanically ventilated patients with acute lung injury. Patients placed on diets using eicosapentaenoic acid (EPA), γ-linolenic acid (GLA), and elevated antioxidants exhibited significantly reduced risk for mortality, risk of developing new organ failures, time on mechanical ventilation, and stay in the intensive care unit.

## n-3 and n-6 Polyunsaturated fatty acids (PUFA: Eicosanoids)

Inflammatory leukotrienes and prostaglandins of the two series are generated when arachidonic acid (20:4 n-6) is cleaved from the membrane and metabolized through the COX or LOX pathways. In contrast, dietary n-3 fatty acids from marine and flaxseed provide substrate for anti-inflammatory eicosanoids (Simopoulos 2008). Inflammatory eicosanoids play major roles in the cascades that occur early in asthmatic airways, sometimes within 2–12 months after diagnosis. Inflammatory leukotrienes are potent inducers of bronchospasm, airway edema, mucus secretion, and inflammatory cell migration, all of which are important to the asthmatic symptomatology (Simopoulos 2008). In contrast, membrane n-3 fatty acids, EPA (20:5), and docosahexaenoic acid (DHA 22:6) produce eicosanoids with lesser inflammatory and bronchoconstrictive actions. Several studies have shown that low dietary availability of n-3 polyunsaturated fatty acid (PUFA) is associated with increased respiratory distress upon asthmatic stimulation. Conversely, leukotrienes with less inflammatory potential were increased in urine after dietary EPA and DHA ingestion, associated with improved respiratory capacity and treatment efficacy. Other inflammatory eicosanoid mediators include cytokines and growth factors (peptide mediators) that also play roles in the underlying inflammatory mechanisms of asthma, but leukotrienes and prostaglandins appear to have the greatest relevance to the pathogenesis of asthma.

In addition to their well-documented competition with proinflammatory eicosanoid precursors (e.g., arachidonic acid), immunosuppressive EPA and DHA incorporated into the membrane modulate lipid–protein lateral organization and inhibit downstream signaling mediated by T cell receptors, suppressing T cell activation and proliferation. These n-3 fatty acids can also alter the surface expression and trafficking of class I and II major histocompatibility complexes, with consequences for recognition by effector T cells (Shaikh and Edidin 2006). This activity modifies antigen presentation, possibly by altering its conformation, orientation, and lateral organization, thereby reducing asthmatic response. Finally, all PUFA acyl chains modify membrane lateral organization and protein function and modulate membrane structure and function, producing differential actions on both the innate and adaptive immune responses.

Both n-3 and n-6 PUFAs are reduced in several types of cancer cells. This reduction is thought to contribute to cancer cell resistance to chemotherapy and to generation of lipid peroxides. Addition of either arachidonic acid or DHA to human lung cancer (A549) cells reduced growth in a dose-dependent manner. In both models, the mechanism appeared to involve biomembrane changes, production of lipid-peroxide-derived cytostatic aldehydes and induction of peroxisomal proliferator-activated receptors (PPAR) (Trombetta et al. 2007). Both n-6 and n-3 PUFAs are ligands for PPARγ, a member of the family of ligand-inducible nuclear transcription factors. PPARs heterodimerize with retinoid X receptors and bind to peroxisomal proliferator response elements located in the promoter region of PPAR target genes. Roman et al. (Han et al. 2009) reported that fish oil inhibits non-small-cell lung carcinoma by suppressing integrin-linked kinase (ILK) expression. ILK overexpression facilitates neoplastic changes by linking cell adhesion receptors, integrins, and growth factors to the actin cytoskeleton and to signaling pathways implicated in the regulation of anchorage-dependent cell growth/survival, cell cycle progression, invasion and migration, and tumor angiogenesis.

## Conjugated linoleic acid

Conjugated linoleic acid (CLA) refers to a class of positional and geometric conjugated dienoic isomers of linoleic acid (LA), of which *cis*-9, *trans*-11 (c9,t11) and *trans*-10, *cis*-12 (t10,c12) CLA predominate. In nature, most of LA (18:2, n-6) is conjugated to $c9,t11$ (~90%) by bacterial action in the rumen of cows, goats, and other ruminants. Thus, milk fat from ruminants contains CLA and full-fat milk consumption from early childhood onward has been found to be inversely correlated with allergic sensitization and the onset of bronchial asthma. Similar results were shown with consumption of butter and grass-fed beef, also rich in naturally occurring CLA (O'Shea et al. 2004).

The concentration of CLA in meat and dairy products depends on the availability of LA substrate and bacterial activity in the rumen. Increasingly in the United States, commercial beef cattle are "finished" on feedlots where they are prevented from exercise and fed a high-grain diet to maximize marbling, deposition of fat in the muscle. While this regimen maximizes weight gain and produces a tender meat product, feedlot finishing dramatically lowers rumen pH, changes the bacterial composition, and can make the animals very sick. Additionally, prophylactic antibiotics are given to prevent infection; thus, rumen bacteria are less able to conjugate LA. Meat and dairy products from animals finished on grass and pasture contain substantially more CLA than grain-fed meat (Mir et al. 2004).

Mechanisms that reduce risk of airway hypersensitivity associated with the consumption of CLA are unclear. Recent studies in mice demonstrated that CLA reduced the Th2 cytokine, IL-5, in bronchoalveolar lavage fluid, reduced IgE and eosinophil production, and modulated allergen-induced in vivo airway hyperresponsiveness. Less mucous plugging of segmental bronchi and membrane remodeling were also seen, most likely via a PPARγ-related mechanism and by reducing eicosanoid precursors (O'Shea et al. 2004; Jaudszus et al. 2008). Because CLA is incorporated into membranes and is also a ligand for PPARγ, it has potential to modify cancer risk. A comprehensive evaluation of CLA mechanisms and its efficacy in cancer prevention was recently conducted. The authors concluded that further research is required to fully evaluate the causal relationship between CLA and risk for cancers (Gebauer et al. 2011).

## Dietary fiber and resistant starch

In contrast to the deleterious effects of diets rich in sugars and readily available carbohydrates, consumption of DF and resistant starch (RS) has been associated with protection from inflammation. DF polysaccharides are not cleaved by salivary or pancreatic amylase and pass into the colon undigested. RS has a 3D structure that also prevents its digestion by amylase. These undigested carbohydrates are fermented by the gastrointestinal microbiota. Whole grains, fruits, and vegetables contain fermentable RS and DF as pectins, gums, mucilages, and nonstructural hemicelluloses that can be variably viscous and fermentable depending on their structures. Several mechanisms have been advanced to explain the anti-inflammatory influence of DF and RS ingestion, including effects on transit time and maintenance of specific species of gut flora (Maslowski et al. 2009):

- DF and RS increase bacterial populations that compete with dangerous pathogens capable of stimulating the gut immune system by a variety of pattern recognition molecules such as the TLRs.
- Commensal bacterial species may also positively influence immune responses, in part by modulating antibody production regulated by CD4+ helper T cells.
- Bacteria of the *Bacteroidetes* phylum produce high levels of acetate and propionate, whereas bacteria of the *Firmicutes* phylum produce high amounts of butyrate. These anti-inflammatory short-chain fatty acids (SCFA), especially acetate (C2) and propionate (C3), have been shown to bind and activate the G-protein coupled receptor, GPR43. Subsequent signaling between commensal bacteria and SCFA regulate immune and inflammatory responses.

Butyrate consumed as food or synthesized by the microbiota is a histone deacetylase inhibitor and exerts cell cycle and transcriptional control (Choi and Friso 2010). Recent work suggests that histone deacetylase inhibitors enhance expression of Thy-1, a glycoprotein present on the surface of normal lung fibroblasts. This protein is absent in fibroblastic foci of idiopathic pulmonary fibrosis and its expression correlates inversely with fibrogenic phenotypic characteristics and functions and has been characterized as a "fibrosis suppressor" by modulating histone acetylation (Sanders et al. 2011). Butyrate has been studied for its synergistic effect on inducing apoptosis in lung cancer cells (Denlinger et al. 2004).

# Vitamin D

Vitamin D is a steroid-derived vitamin obtained through a limited number of food sources (saltwater fish, liver, egg yolks) and is added to milk and other food staples as a supplement. The vitamin is also synthesized from cholesterol in the skin by ultraviolet rays in sunlight. Although it has long been known as an essential nutrient for calcium metabolism and bone integrity, it has been increasingly recognized as an important immunomodulator and as a regulator of pro- and anti-inflammatory mediators. Following absorption, vitamin D is hydroxylated in the liver to $25(OH)D_3$, the inactive precursor. Several factors contribute to low serum levels of $25(OH)D_3$. Inadequate dietary intake of the few foods that contain vitamin D and infrequent exposure to the sun, especially in urban dwellers and home-bound elderly, increase risk for depletion. Melanin acts as a sunscreen to prevent vitamin D biosynthesis in the skin. Obese patients are at risk, presumably because fat-soluble vitamin D is sequestered in adipocytes. An age-associated decline in 7-dehydrocholesterol precursor reduces the capacity for vitamin D biosynthesis in the skin. Time of day, season, latitude, and use of sunscreen also modulate vitamin D biosynthesis (Holick 2004).

Active vitamin D (1–25 $D_3$) is a steroid hormone synthesized from $25(OH)D_3$ by 1-$\alpha$ hydroxylase produced by the kidney and other tissues (Finklea et al. 2011). The vitamin D axis includes 1–25 $D_3$, vitamin D receptor (VDR), and vitamin D-binding protein (VDBP). Serum VDBP action in the lung relates to macrophage activation and neutrophil chemotaxis, and gene variants have been associated with airway disease, especially COPD. VDBP clears extracellular G-actin released from necrotic cells and may be important in severe lung infections and possibly cancer risk (Chishimba et al. 2010). The macrophage VDR and activating enzyme, $25(OH)D_3$-1$\alpha$ hydroxylase, expressions are upregulated in response to infection. Active $1,25(OH)_2$ $D_3$ increases antimicrobial cathelicidin and induces destruction of the infectious agent. $1,25(OH)_2$ $D_3$ regulates cytokine and immunoglobulin synthesis, modulating the immune response. It also regulates genes involved in proliferation, angiogenesis, differentiation, and apoptosis. Locally produced $1,25(OH)_2$ $D_3$ also inhibits the expression and synthesis of parathyroid hormone. $1,25(OH)_2$ $D_3$ produced in the kidney downregulates renin production and stimulates pancreatic $\beta$-cell insulin secretion.

VDRs are present on many cells, including lung epithelial and peripheral blood mononuclear cells. Intensive research is ongoing to understand the observation that adequate maternal vitamin D status is associated with reduced childhood wheezing and asthma. On the other hand, some studies showed that vitamin D supplements increased the risk for childhood asthma, suggesting that the therapeutic window for vitamin D is narrow. Levels of the inactive $25(OH)D_3$ are associated with upper respiratory tract infection (URI) (Ginde et al. 2009). Secondary analysis of data from 18,883 subjects with recent URI revealed that serum $25(OH)D_3$ levels had an independent, inverse association with URI throughout the year. The authors postulated that individuals with diseases such as asthma, who have low serum $25(OH)D_3$ levels, may be even more susceptible to infection due to inability of vitamin D-depleted macrophages to kill infectious agents.

Vitamin D is also effective in increasing differentiation and decreasing proliferation of transformed cells (Nithya et al. 2011). $1,25(OH)_2$ $D_3$ inhibited the growth of lung cancer cell lines, possibly mediated by VDR regulation of the cell cycle. $1,25(OH)_2D_3$ has also been shown to inhibit lung tumor growth and metastases in mouse models. While the mechanisms are still unclear, it is known that normal tracheobronchial cells contain high levels of CYP27B1 (the enzyme that converts inactive $25(OH)D_3$ to active $1,25(OH)_2$ $D_3$) and low levels of CYP24A1 (the enzyme that catabolizes $1,25(OH)_2$ $D_3$ to an inactive form, $1,24,25(OH)_3$ $D_3$). The regulation of enzyme activity ensures optimal intracellular concentrations of active vitamin D. In lung cancer cells, enzyme activities are reversed, potentially reducing the availability of active vitamin

D in the cells. Additionally, vitamin D is synergistic with lung cancer chemo-therapy, including platinum analogues and taxanes.

## Vitamin E

Vitamin E consists of a group of eight structurally related compounds: α-, β-, γ-, and δ-tocopherols (α-, β-, γ-, and δ-T) and α-, β-, γ-, and δ-tocotrienols (α-, β-, γ-, and δ-TTs). Of the tocopherols plentiful in vegetable oils (e.g., soy, corn, sesame, and cotton seed oils) and nuts, γ-T is estimated to be found at sever-alfold higher levels than α-T. Despite the low concentrations of α-T in the food supply, this form has been used for supplements because of its higher serum levels and efficacy in restoring experimentally reduced fertility (Ju et al. 2010). TTs have additional double bonds and are less available in Northern climates; they are present in trace amounts in oils derived from rice bran, barley, wheat germ, and rye but are high in tropical oils such as palm oil.

Despite observations that vitamin E functions as an effective membrane anti-oxidant and anti-inflammatory agent in animal models (Rocksén et al. 2003), there is a paucity of clinical trials demonstrating conclusive benefit in human subjects (Chow et al. 2003). Causes for this discrepancy have been suggested, including species differences and the methods used to induce experimental lung damage. It is also possible that the tocopherol most often tested (α-T) is not the active form of the vitamin. Actually, high supplements of α-T intake are known to decrease blood and tissue levels of γ-T; thus, it is possible that supplemental α-T is actually deleterious (Ju et al. 2010).

Inconclusive results for vitamin E as an anticancer agent were reported by Yang et al. (Ju et al. 2010) who analyzed case–control and cohort studies published since 1986. Two of the three cohort studies found a significant inverse asso-ciation between dietary intake of vitamin E and risk of lung cancer in current smokers, suggesting that dietary vitamin E may protect against smoke-related lung injury. The authors also cited evidence suggesting that other forms of vitamin E have greater potential to prevent cancer than α-T. Khanna et al. (Ling et al. 2012) reviewed evidence suggesting that TT should actually be considered the anticancer form of vitamin E. In addition to its antioxidant and proapop-totic functions, TT inhibited epithelial to mesenchymal transitions, suppressed tumor angiogenic growth factors, and induced anticancer immunity. Since prior clinical trials with α-T did not show a robust anticancer activity, the authors called for clinical trials using alternate vitamin E forms, especially the TTs.

## Immunomodulatory nonnutritive components in the diet

Diets rich in fruits and vegetables have been repeatedly associated with reduced lung inflammation injury and neoplastic lung diseases. The disease-protective properties observed with vegetable and fruit consumption are likely to be mediated through a spectrum of "bioactive compounds" in each individual plant that induce a variety of physiologic functions. Specific bioactive compounds have been shown to act as direct or indirect antioxidants, modulate enzymes, control apoptosis, and regulate the cell cycle (Finley 2005). Because they are pharmacologically active, the safety of large quantities of bioactive compounds extracted from plants is uncertain. Since whole food contains many different compounds, including vitamins, minerals, and fibers, many of which are known to have synergistic effects, it is difficult to separate the actions of each bioactive component. For these reasons, consumption of a wide variety of plant foods is prudent.

Difficulties in conducting clinical research using whole foods or even extracted compounds include variability in concentrations and/or structure with environmental conditions, plant species, attack by diseases or pests, and growing season. Further, when compounds are ingested, they are known to be extensively modified in the gastrointestinal tract, especially by the microbiota, or in the body before they reach their target tissues. Since the substrate preferences and products of the human microbiota can vary markedly in response to diet and environmental influences (Zoetendal et al. 2012), the form and extent to which ingested bioactive compounds reach the blood stream are uncertain in a given individual. Thus, studies conducted using a pure extracted bioactive compound in vitro and in animal models may produce results not reproducible in human trials (Egert and Rimbach 2011).

## Flavonoids

The flavonoid class of bioactive components in fruits and vegetables shares a common chemical structure consisting of two phenol rings linked through three carbons. These compounds are known to prevent cytotoxic effects of oxidative stress in lung tissue, quenching free radicals by means of their conjugated ring structure. They also modify gene transcription via induction or inhibition of specific transcription factors. Antioxidant function is especially important in the lung, since inhaled oxidants (e.g., cigarette smoke [CS]) first come in contact with the lung epithelium at the epithelial lining fluid interface. Endogenous and diet-derived antioxidants in the epithelial lining fluid neutralize the oxidants and prevent epithelial damage (Egert and Rimbach 2011). To date, over 5000 naturally occurring flavonoids are known, most of which have anti-inflammatory and anticancer properties. This diversity precludes an exhaustive discussion; thus, selected flavonoids are described in the following.

*Resveratrol* is a phytoalexin that is found in the skin and seeds of grapes and produced by grapevines in response to injury or fungal attack. It has been reported to have antioxidant, anti-inflammatory, and anticarcinogenic properties. Resveratrol is a more effective antioxidant than vitamins E and C; it has been shown to scavenge free radicals such as lipid hydroperoxyl, hydroxyl (OH), and superoxide anion ($O_2 \cdot$) radicals. Resveratrol inhibits LPS-induced airway neutrophilia and inflammation through an NF-κB-independent mechanism. A recent report demonstrated that oxidants (cigarette smoke [CS]) depleted and resveratrol restored GSH lung epithelial lining fluid by upregulation of glutamate cysteine ligase via activation of Nrf2 and also quenched CS-induced release of reactive oxygen species (Kode et al. 2008). Resveratrol is also known to activate sirtuins, a conserved family of NAD⁺-dependent protein deacetylases that are involved in gene silencing processes related to aging, blockade of apoptosis, and promotion of cell survival (Bishayee 2009). In vitro and animal studies have shown that resveratrol acts on multiple molecular targets in cancer progression (Chen et al. 2008); several clinical trials are underway to determine the optimal therapeutic window for use of pure resveratrol or in whole foods as a cancer-preventive agent (Bishayee 2009).

*Curcumin* is a yellow polyphenol found in the spice turmeric. It has antioxidant, anti-inflammatory, and anticancer effects. Despite the low bioavailability of curcumin after oral application, it is now in phase I clinical studies for several respiratory diseases. Curcumin also restores oxidative stress–impaired histone deacetylase-2 activity and corticosteroid efficacy in monocytes. Hence, curcumin has potential to reverse corticosteroid resistance, which is common in patients with COPD and severe asthma (Meja et al. 2008). Curcumin is also able to suppress cancer proliferation, invasion, angiogenesis, and metastasis through a number of molecular mechanisms (Chen et al. 2008).

*Quercetin* is the major dietary flavonoid in the human diet; we consume as much as 25 mg/day. It is found in many plants including onions, broccoli, apples, berries, and tea. By combining with free radical species, it forms less reactive phenoxy radicals. It also inhibits several kinases, thus mediating many of its potent antiproliferative, proapoptotic, and anti-inflammatory effects on biological systems. Quercetin inhibited cytokine and chemokine production in cultured cells and inhibited mast cell activation and histamine release. It attenuated LPS-induced nitric oxide production via inducible nitric oxide synthase expression, inhibited release of TNFα and IL6 from cultured macrophages, and reduced activation of mitogen-activated protein kinases and NFkB (Nanua et al. 2006). Recent studies have investigated the gene–nutrient relationships among quercetin, smoking, and lung cancer risk (Lam et al. 2010).

## Evidence linking nutraceuticals to reduction of human lung disease risk

Meta-analysis of cohort and case–control studies suggested that dietary vegetable and fruit consumption reduced risk for lung cancer (Riboli and Norat 2003). More recent epidemiologic studies revealed that lung function is improved (Bentley et al. 2012) and risk for lung cancer is reduced by specific phytoestrogens in fruits and vegetables (Schabath et al. 2005). Prospective analysis of dietary questionnaires from the NIH-AARP Diet and Health Study identified total fruit and vegetable intake as well as intake of specific fruit and vegetable groups as cancer protective. After adjusting for smoking, alcohol, and other risk factors, significant inverse associations between lung cancer and total plant sources were seen for both men and women. Intakes of Rosaceae (apples, peaches, nectarines, plums, pears, and strawberries), Convolvulaceae (sweet potatoes and yams), and Umbelliferae (carrots) reduced risk for men; borderline inverse relations are also apparent for Compositae and Cruciferae (Wright et al. 2008). The authors speculated that the inverse relationship in men between lung cancer risk and consumption of carrots, sweet potatoes, and yams could be due to their provitamin A carotenoids (beta-carotene) and ascorbate content. Both components are potent antioxidants, and vitamin A regulates epithelial cell division, growth, differentiation, and proliferation. Several prospective studies have shown an inverse association between flavonoid intake (particularly quercetin, which is concentrated in apples) and lung cancer risk (Arts and Hollman 2005).

On the other hand, clinical interventions have been largely inconclusive and sometimes negative. The Beta-Carotene and Retinol Efficacy Trial (CARET) and Alpha-Tocopherol, Beta-Carotene Cancer Prevention (ATBC) studies, two large, randomized, double-blinded, placebo-controlled chemoprevention trials designed to study the effect of beta-carotene and vitamins A and E on lung cancer in high-risk populations (Albanes et al. 1996; Omenn et al. 1996), actually identified increased lung cancer in groups with supplemental beta-carotene. Several conclusions to be drawn from these studies include the probability that the spectrum of bioactive components in each plant exert synergistic or inhibitory influences. Pure compounds extracted from plants are not so modified and can have unexpected pharmacologic effects. Several authors have offered caveats on study designs appropriate to evaluate the complex links between diet and human health (Finley 2005; Moiseeva and Manson 2009; Traka and Mithen 2011). Until these links are clearly established, recommendations to prevent and control lung disease with diet must counsel moderation. Taken together, current evidence suggests consumption of a nutrient-dense diet, adequate to maintain optimal weight; rich in varied plant sources, RS, and fibers; and optimal in protein and micronutrients.

# References

Agrawal, A., U. Mabalirajan et al. (2011). Emerging interface between metabolic syndrome and asthma. *American Journal of Respiratory Cell and Molecular Biology* **44**(3): 270–275.

Albanes, D., O.P. Heinonen, et al. (1996). α-Tocopherol and β-Carotene supplements and lung cancer incidence in the Alpha-Tocopherol, Beta-Carotene Cancer Prevention Study: Effects of base-line characterstics and study compliance. *Journal of the National Cancer Institute* **88**(21): 1560–1570.

Arts, I. C. and P. C. Hollman (2005). Polyphenols and disease risk in epidemiologic studies. *The American Journal of Clinical Nutrition* **81**(1): 317S–325S.

Bentley, A. R., S. B. Kritchevsky et al. (2012). Dietary antioxidants and forced expiratory volume in 1 s decline: the Health, Aging and Body Composition study. *European Respiratory Journal* **39**: 978–984.

Bishayee, A. (2009). Cancer prevention and treatment with resveratrol: from rodent studies to clinical trials. *Cancer Prevention Research* **2**(5): 409–418.

Brown, L. A. S., F. L. Harris et al. (2004). Chronic ethanol ingestion and the risk of acute lung injury: a role for glutathione availability? *Alcohol* **33**(3): 191–197.

Buyken, A. E., V. Flood et al. (2010). Carbohydrate nutrition and inflammatory disease mortality in older adults. *The American Journal of Clinical Nutrition* **92**(3): 634–643.

Chang, S. I., K. El-Bayoumy et al. (2007). 4-(Methylnitrosamino)-I-(3-pyridyl)-1-butanone enhances the expression of apolipoprotein A-I and Clara Cell 17-kDa protein in the lung proteomes of rats fed a corn oil diet but not a fish oil diet. *Cancer Epidemiology Biomarkers & Prevention* **16**(2): 228–235.

Chen, H.-W., J.-Y. Lee et al. (2008). Curcumin inhibits lung cancer cell invasion and metastasis through the tumor suppressor HLJ1. *Cancer Research* **68**(18): 7428–7438.

Chishimba, L., D. R. Thickett et al. (2010). The vitamin D axis in the lung: a key role for vitamin D-binding protein. *Thorax* **65**(5): 456–462.

Choi, S.-W. and S. Friso (2010). Epigenetics: a new bridge between nutrition and health. *Advances in Nutrition: An International Review Journal* **1**(1): 8–16.

Chow, C.-W., M. T. Herrera Abreu et al. (2003). Oxidative stress and acute lung injury. *American Journal of Respiratory Cell and Molecular Biology* **29**(4): 427–431.

Denlinger, C. E., M. D. Keller et al. (2004). Combined proteasome and histone deacetylase inhibition in non-small cell lung cancer. *Journal of Thoracic and Cardiovascular Surgery* **127**(4): 1078–1086.

Egert, S. and G. Rimbach (2011). Which sources of flavonoids: complex diets or dietary supplements? *Advances in Nutrition: An International Review Journal* **2**(1): 8–14.

Finklea, J. D., R. E. Grossmann et al. (2011). Vitamin D and chronic lung disease: a review of molecular mechanisms and clinical studies. *Advances in Nutrition: An International Review Journal* **2**(3): 244–253.

Finley, J. W. (2005). Proposed criteria for assessing the efficacy of cancer reduction by plant foods enriched in carotenoids, glucosinolates, polyphenols and selenocompounds. *Annals of Botany* **95**(7): 1075–1096.

Gebauer, S. K., J.-M. Chardigny et al. (2011). Effects of ruminant trans fatty acids on cardiovascular disease and cancer: a comprehensive review of epidemiological, clinical, and mechanistic studies. *Advances in Nutrition: An International Review Journal* **2**(4): 332–354.

Ginde, A. A., J. M. Mansbach et al. (2009). Association between serum 25-hydroxyvitamin D level and upper respiratory tract infection in the third national health and nutrition examination survey. *Archives of Internal Medicine* **169**(4): 384–390.

Giovannucci, E. (2001). Insulin, insulin-like growth factors and colon cancer: a review of the evidence. *The Journal of Nutrition* **131**(11): 3109S–3120S.

Han, S., X. Sun et al. (2009). Fish oil inhibits human lung carcinoma cell growth by suppressing integrin-linked kinase. *Molecular Cancer Research* **7**(1): 108–117.

Holick, M. F. (2004). Sunlight and vitamin D for bone health and prevention of autoimmune diseases, cancers, and cardiovascular disease. *The American Journal of Clinical Nutrition* **80**(6): 1678S–1688S.

Huang, S., J. M. Rutkowsky et al. (2012). Saturated fatty acids activate TLR-mediated proinflammatory signaling pathways. *Journal of Lipid Research* **53**(9): 2002–2013.

Iyer, S. S., D. P. Jones et al. (2009). Oxidation of plasma cysteine/cystine redox state in endotoxin-induced lung injury. *American Journal of Respiratory Cell and Molecular Biology* **40**(1): 90–98.

Jaudszus, A., M. Krokowski et al. (2008). Cis-9,trans-11-conjugated linoleic acid inhibits allergic sensitization and airway inflammation via a PPARgamma-related mechanism in mice. *The Journal of Nutrition* **138**(7): 1336–1342.

Ju, J., S. C. Picinich et al. (2010). Cancer-preventive activities of tocopherols and tocotrienols. *Carcinogenesis* **31**(4): 533–542.

Jung, A. Y., E. M. Poole et al. (2008). DNA methyltransferase and alcohol dehydrogenase: gene-nutrient interactions in relation to risk of colorectal polyps. *Cancer Epidemiology Biomarkers & Prevention* **17**(2): 330–338.

Kabat, G. C., M. Kim et al. (2008). Body mass index and waist circumference in relation to lung cancer risk in the women's health initiative. *American Journal of Epidemiology* **168**(2): 158–169.

Kim, J.-H., P. Ellwood et al. (2009). Diet and asthma: looking back, moving forward. *Respiratory Research* **10**(1): 49.

Kode, A., S. Rajendrasozhan et al. (2008). Resveratrol induces glutathione synthesis by activation of Nrf2 and protects against cigarette smoke-mediated oxidative stress in human lung epithelial cells. *American Journal of Physiology—Lung Cellular and Molecular Physiology* **294**(3): L478–L488.

Kordonowy, L. L., E. Burg et al. (2012). Obesity is associated with neutrophil dysfunction and attenuation of murine acute lung injury. *American Journal of Respiratory Cell and Molecular Biology* **47**: 120–127.

Kumar, V., A. K. Abbas et al., Eds. (2010). *Robbins and Cotran Pathological Basis of Disease*, 8th edn., Philadelphia, PA: Saunders Elsevier.

Lam, T. K., M. Rotunno et al. (2010). Dietary quercetin, quercetin-gene interaction, metabolic gene expression in lung tissue and lung cancer risk. *Carcinogenesis* **31**(4): 634–642.

Lieber, C. S. (1997). Cytochrome P-4502E1: its physiological and pathological role. *Physiological Reviews* **77**(2): 517–544.

Ling, M. T., S. U. Luk et al. (2012). Tocotrienol as a potential anticancer agent. *Carcinogenesis* **33**(2): 233–239.

Madenspacher, J. H., D. W. Draper et al. (2010). Dyslipidemia induces opposing effects on intrapulmonary and extrapulmonary host defense through divergent TLR response phenotypes. *The Journal of Immunology* **185**(3): 1660–1669.

Maslowski, K. M., A. T. Vieira et al. (2009). Regulation of inflammatory responses by gut microbiota and chemoattractant receptor GPR43. *Nature* **461**(7268): 1282–1286.

McGinley, B. and N. M. M. D. P. Punjabi (2011). Obesity, metabolic abnormalities, and asthma: establishing causal links. *American Journal of Respiratory and Critical Care Medicine* **183**(4): 424–425.

Meja, K. K., S. Rajendrasozhan et al. (2008). Curcumin restores corticosteroid function in monocytes exposed to oxidants by maintaining HDAC2. *American Journal of Respiratory Cell and Molecular Biology* **39**(3): 312–323.

Mir, P. S., T. A. McAllister et al. (2004). Conjugated linoleic acid–enriched beef production. *The American Journal of Clinical Nutrition* **79**(6): 1207S–1211S.

Moiseeva, E. P. and M. M. Manson (2009). Dietary chemopreventive phytochemicals: too little or too much? *Cancer Prevention Research* **2**(7): 611.

Nanua, S., S. M. Zick et al. (2006). Quercetin blocks airway epithelial cell chemokine expression. *American Journal of Respiratory Cell and Molecular Biology* **35**(5): 602–610.

Nithya, R., K. SoHee et al. (2011). Vitamin D and lung cancer. *Expert Review of Respiratory Medicine* **5**(3): 305–309.

O'Shea, M., J. Bassaganya-Riera et al. (2004). Immunomodulatory properties of conjugated linoleic acid. *The American Journal of Clinical Nutrition* **79**(6): 1199S–1206S.

Omenn, G. S., G. E. Goodman et al. (1996). Risk factors for lung cancer and for intervention effects in CARET, the beta-carotene and retinol efficacy trial. *Journal of the National Cancer Institute* **88**(21): 1550–1559.

Pavelić, J., Š. Križanac et al. (2005). The consequences of insulin-like growth factors/receptors dysfunction in lung cancer. *American Journal of Respiratory Cell and Molecular Biology* **32**(1): 65–71.

Pontes-Arruda, A., S. DeMichele et al. (2008). The use of an inflammation-modulating diet in patients with acute lung injury or acute respiratory distress syndrome: a meta-analysis of outcome data. *Journal of Parenteral and Enteral Nutrition* **32**(6): 596–605.

Pöschl, G. and H. K. Seitz (2004). Alcohol and cancer. *Alcohol and Alcoholism* **39**(3): 155–165.

Riboli, E. and T. Norat (2003). Epidemiologic evidence of the protective effect of fruit and vegetables on cancer risk. *The American Journal of Clinical Nutrition* **78**(3): 559S–569S.

Rocksén, D., B. Ekstrand-Hammarström et al. (2003). Vitamin E reduces transendothelial migration of neutrophils and prevents lung injury in endotoxin-induced airway inflammation. *American Journal of Respiratory Cell and Molecular Biology* **28**(2): 199–207.

Salaspuro, V. (2007). Pharmacological treatments and strategies for reducing oral and intestinal acetaldehyde. *Novartis Found Symposium* **285**: 145–153; discussion 153–147, 198–149.

Sanders, Y. Y., T. O. Tollefsbol et al. (2011). Epigenetic regulation of Thy-1 by histone deacetylase inhibitor in rat lung fibroblasts. *American Journal of Respiratory Cell and Molecular Biology* **45**(1): 16–23.

Schabath, M. B., L. M. Hernandez et al. (2005). Dietary phytoestrogens and lung cancer risk. *JAMA* **294**(12): 1493–1504.

Sekhar, R. V., S. G. Patel et al. (2011). Deficient synthesis of glutathione underlies oxidative stress in aging and can be corrected by dietary cysteine and glycine supplementation. *The American Journal of Clinical Nutrition* **94**(3): 847–853.

Shaikh, S. R. and M. Edidin (2006). Polyunsaturated fatty acids, membrane organization, T cells, and antigen presentation. *The American Journal of Clinical Nutrition* **84**(6): 1277–1289.

Simopoulos, A. P. (2008). The importance of the omega-6/omega-3 fatty acid ratio in cardiovascular disease and other chronic diseases. *Experimental Biology and Medicine* **233**(6): 674–688.

Smith-Warner, S. A., J. Ritz et al. (2002). Dietary fat and risk of lung cancer in a pooled analysis of prospective studies. *Cancer Epidemiology Biomarkers & Prevention* **11**(10): 987–992.

Stenius-Aarniala, B., T. Poussa et al. (2000). Immediate and long term effects of weight reduction in obese people with asthma: randomised controlled study. *BMJ* **320**(7238): 827–832.

Traka, M. H. and R. F. Mithen (2011). Plant science and human nutrition: challenges in assessing health-promoting properties of phytochemicals. *The Plant Cell Online* **23**(7): 2483–2497.

Trombetta, A., M. Maggiora et al. (2007). Arachidonic and docosahexaenoic acids reduce the growth of A549 human lung-tumor cells increasing lipid peroxidation and PPARs. *Chemico-Biological Interactions* **165**(3): 239–250.

Vaissière, T., R. J. Hung et al. (2009). Quantitative analysis of DNA methylation profiles in lung cancer identifies aberrant DNA methylation of specific genes and its association with gender and cancer risk factors. *Cancer Research* **69**(1): 243–252.

Wood, L. G., M. L. Garg et al. (2011). A high-fat challenge increases airway inflammation and impairs bronchodilator recovery in asthma. *Journal of Allergy and Clinical Immunology* **127**(5): 1133–1140.

Wright, M. E., Y. Park et al. (2008). Intakes of fruit, vegetables, and specific botanical groups in relation to lung cancer risk in the NIH-AARP diet and health study. *American Journal of Epidemiology* **168**(9): 1024–1034.

Yang, W. S., M. Y. Wong et al. (2012). Meat consumption and risk of lung cancer: evidence from observational studies. *Annals of Oncology* **23**: 3163–3170.

Zoetendal, E. G., J. Raes et al. (2012). The human small intestinal microbiota is driven by rapid uptake and conversion of simple carbohydrates. *International Society for Microbial Ecology Journal* **6**(7): 1415–1426.

# VI

## Gut Microbiome

# 13

# Nutraceutical–Gut Microbiome Interactions in Human Disease

Mukesh Verma

## Contents

## Abbreviations

IBS     Irritable bowel syndrome

LPS    Lipopolysaccharides

## Introduction: The gut microbiome and disease

The largest and earliest source of microbial exposure in human subjects comes from the intestinal tract (gut). Gut microbiota may contain more than 800,000 species of bacteria and other microorganisms. The concentration of these microbes in the gut is estimated to be $10^{12}$ cells per gram of feces (Flint et al., 2007; Laparra and Sanz, 2010). Conditions in the gastrointestinal tract become anaerobic, and the majority of bacteria in the large intestine are anaerobic.

The gut microbiota affects the nutritional and health status of people. This is reflected in processes such as food digestion, energy salvage, micronutrient supply, and xenobiotics and their transformation products. Abnormalities in the gut microbiota result in the development of diseases such as colon cancer, obesity, inflammatory bowel disease, and Crohn's disease (Laparra and Sanz, 2010). Several factors contribute to the composition of the gut microbiota, including diet and consumption of antibiotics, xenobiotics, and other prescription drugs.

Irritable bowel syndrome (IBS) is among the most common intestinal maladies and is one of the most difficult to treat. This disease involves small intestinal bacterial outgrowth (Dahlqvist and Piessevaux, 2011). As a result of this outgrowth, changes such as an increase in mucosal cellularity (enterochromaffin cells, lamina propria T lymphocytes, and mast cells), modified pro-inflammatory/anti-inflammatory cytokine balance, and disordered neurotransmission are observed. Researchers are investigating whether these changes arise from IBS or are the cause of the disease. Intake of probiotics has been shown to decrease IBS symptoms, probably because of the immunomodulatory and analgesic properties of probiotics. Probiotics are microorganisms that supplement the gut's natural bacteria and help to balance intestinal flora. Little improvement in IBS symptoms has been observed with antibiotic treatment.

Obesity and its metabolic complications are major health problems in the United States and worldwide; evidence implicating microbiota in these important health issues is increasing. Experiments with mice have demonstrated that the gut microbiota contributes to obesity because they extract their energy from the diet. Mice raised in the absence of microorganisms were leaner than mice with a normal microbiota (Backhed et al., 2004). Expansion of this study to involve transplantation of the gut from normal mice to germ-free mice demonstrated that mice with a normal microbiota gained weight (body fat), became insulin resistant, and stored triglycerides (Backhed et al., 2007). Musso et al. (2010) demonstrated an association between low-grade inflammation and obesity through activation of innate immunity through the lipopolysaccharide (LPS) Toll-like receptor. In another study, mice fed a high-fat diet for four weeks showed higher LPS levels and ultimately became obese with metabolic endotoxemia (Cani et al., 2007). In obese humans, the concentration of *Firmicutes* is much higher than that of *Bacteroidetes* in the gut (Ley et al., 2006). In the low-calorie diet cohort of an intervention trial, the level of *Bacteroidetes* was reduced significantly. Thus, the microbiota of the gut shows a correlation with obesity phenotype (Cani and Delzenne, 2009). A study conducted by Kalliomaki et al. (2008) in children who were overweight at age 7 and children with normal weight at age 7 showed that the obese children had high concentrations of *Bifidobacterium* and low concentrations of *Staphylococcus*. These examples indicate that the gut microbiota plays a significant role in the pathogenesis of diet-induced obesity and related metabolic disorders. These disorders might be reversible with diet and gut microbiota manipulation.

## Diet, nutraceuticals, and microbiota

The composition of the gut microbiota changes from infancy through childhood and adulthood and again during old age; diet contributes significantly to alterations in the microbiota in different phases of life (Mariat et al., 2009; Claesson et al., 2011). There is a major difference between functional foods and nutraceuticals. Functional foods are supplied in the regular diet, and nutraceuticals are extracts that contain biologically active food components that are taken in addition to the regular diet. Nutraceuticals are food or food products that provide health and medical benefits, including prevention and treatment of disease. Examples of nutraceuticals include dietary supplements, specific diets, genetically engineered foods, herbal products, and processed foods (cereals, soups, beverages). According to a recent estimate, two-thirds of Americans take at least one nutraceutical. The gut microbiome is involved in the metabolism of active food components before they are absorbed in the gut. The colon is the part of the gut in which the most microorganisms are present to metabolize food components. Some food components influence the composition, metabolism, and growth of the gut microbiota (Campbell et al., 1997; Gibson et al., 2005).

Probiotics interact with the mucosal immune system via the same pathways as commensal bacteria to influence both innate and adaptive immune responses. As a result, interventions with immunomodulatory diets, including certain nutrients and probiotics, may be critical in coordinating the adaptive functions necessary for the formation of tolerance and, thus, in the prevention of undesirable metabolic consequences.

## Why nutraceutical and gut microbiome interaction is important

Interaction between microbiota, bioactive food components, and nutraceuticals can affect disease initiation, development, and progression. Therefore, understanding these three major components is important for disease diagnosis and prevention. The gut microbiota performs protective, immune, and metabolomic functions and maintains an individual's nutritional and health status. The initial gut composition can significantly influence immune system development. The colonization of gut microbes starts just after birth and influences the environment and growth of the infant. As expected, the composition of the microbiota is simple at the start; by age 24 months, however, an infant's gut microbiota changes dramatically. In a breast-fed infant, the concentration of *Bifidobacterium* is higher than that of any other bacteria (Gueimonde et al., 2006). *Bifidobacterium* is not

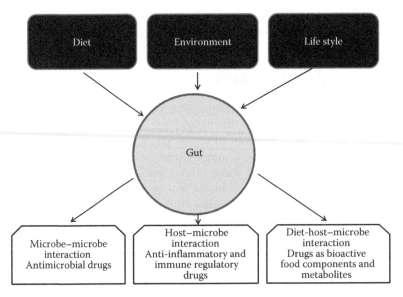

**Figure 13.1** *Therapeutic potential of the gut microbiota.*

the dominant gut bacterium in formula-fed infants, however. The most dramatic changes occur during weaning, when carbohydrate-based substrates promote microbiota expansion. Early life factors such as hygiene, diet, and medication, antioxidant, and nutrient intake associated with different diseases might alter the gut milieu; for example, vitamin D deficiency early in life has been associated with asthma via its influence on inflammation and infection (Zoetendal et al., 2006).

Immune system reactions to the gut microbiota and gut inflammation have been observed. For example, the gut microbiota in Crohn's disease patients differs from that of healthy individuals, and inflammation is common in Crohn's disease patients. A general scheme of the therapeutic implications of the gut microbiota is presented in Figure 13.1.

## Gut microbiome and underlying mechanisms of interaction with nutraceuticals

In adults, the gut microbiota is composed primarily of bacteria from genus *Bacteroides, Firmicutes (Eubacterium, Ruminococcus, and Clostridium)*, and *Bifidobacterium*. Infants predominately have *Bacteroides (Escherichia, Raoultella, and Klebsiella)* in their gut. The composition of microbiota is dynamic and influenced by bioactive food components and nutraceuticals. These microbes interact with the gut epithelium and lymphoid system. The intestinal epithelium protects itself from microbe attack with glycoproteins such as mucin and antimicrobial peptides and via the secretion of biles, acids, and enzymes. The intestinal microbiota influences the postnatal development of the immune system, composition of lamina propria T cells, immunoglobulin-A-producing B cells, intraepithelial T cells, and serum immunoglobulin levels. Probiotic strains are known for their beneficial effects on the treatment of diseases such as diarrhea and eradication of *Helicobacter pylori* and the prevention of atopic eczema (Kalliomaki et al., 2007; Sanz and DePalma, 2009).

## Nutraceuticals and the gut microbiota

The term "nutraceuticals" is used broadly to describe products that are derived from food sources and provide health benefits, in addition to their basic nutrients. Whether proven clinically or not, nutraceuticals are advertised as being able to prevent chronic diseases, improve health, delay the aging process, and promote longer patient survival. Unlike the U.S. Food and Drug Administration (FDA)-approved medicines and prescription drugs, nutraceuticals are loosely regulated in terms of their advertising and benefit claims. The major categories of nutraceuticals are dietary supplements, functional foods, medical foods, and farmaceuticals.

Dietary supplements are nutrients that are derived from food products and concentrated in liquid or capsule form. The dietary ingredients of these products include vitamins, minerals, herbs, botanicals, amino acids, or a mixture of specific metabolites. The marketing of dietary supplements does not require FDA approval, and people take these supplements at their own risk. Functional foods are designed to enrich foods close to their natural state and may be enriched or fortified. The most common example in this category is vitamin D-enriched milk. Medical foods are formulated or administered under the supervision of a doctor during the course of specific dietary management or an abnormal situation that requires a specific nutrient at a specific concentration. Medical foods are regulated by the FDA. "Farmaceuticals" comes from combining the words "farm" and "pharmaceuticals" and refers to medically valuable compounds that are produced from modified agricultural crops or animals. Examples in this category are described as follows, and their medical utility is in the form of antioxidants (resveratrol from red grape products; flavonoids in citrus fruits, tea, wine, and dark chocolate; anthocyanins from berries), cholesterol-reducing agents (psyllium seed husk), cancer-prevention agents (sulforaphane from broccoli), artery health-improving agents (isoflavonoids from soy or clover), and cardiovascular disease-reducing agents (alpha-linoleic acid from flax seeds; omega-3 fatty acids from fish oil). Botanical and herbal extracts, which may have different medical benefits, include ginseng, garlic oil, and turmeric.

Cellulose, xylans, and proteins (some of the polysaccharides in plant cell walls) and storage polysaccharides (inulin and resistant starch) are important

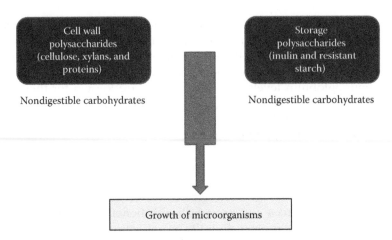

**Figure 13.2** *Nondigestible carbohydrates that support the growth of the microbiota in the gastrointestinal tract (gut).*

substrates for the growth of microorganisms in the gut (Figure 13.2). Specific oligosaccharides, disaccharides, and sugar alcohols do not hydrolyze easily in the gut, and their adsorption is very low. These carbohydrates are not easily digested and serve as prebiotics.

Trace elements such as selenium also influence the growth of the microbiota (Kasaikina et al., 2011). Some trace elements are beneficial for the growth of microorganisms, whereas others have harmful effects on selected microorganisms, even at extremely low concentrations. For example, a diet with decreased iron (Fe) content increases the concentration of anaerobes (*lactobacilli, enterococci*) in the small intestine (Tompkins et al., 2001). Selenium is required for the synthesis of selenoproteins, which affect the host immune system and cellular redox homeostasis, and has anti-inflammatory and anticancer activities. Fortification of foods with selenium has beneficial effects in colon cancer (Irons et al., 2006; Schomburg and Schweizer, 2009).

## Prebiotics and gut microbiota

Prebiotics are defined as selective fermented food ingredients that result in specific changes in the composition of the gut microbiota that are beneficial to the health of the host (Scott et al., 2011). *Bifidobacteria* are considered beneficial to the host, and their growth is supported by inulin-type polysaccharides. However, *Bifidobacteria* are only a minor part of the microbiota that are present in the gut. Two groups of bacteria, *Faecalibacterium prausnitzii* and *Clostridium*, produce butyrate and improve health by inducing an anti-inflammatory response (Sokol et al., 2008). Animal models have demonstrated the growth of *Bacteroidetes* and *Firmicutes* groups of bacteria when fed an inulin-type fructo-oligosaccharide diet (Sokol et al., 2008).

## Polyunsaturated fatty acids and gut microbiota

Dietary fatty acids influence the attachment of microorganisms to the gut, possibly by modifying the fatty acid composition of the intestinal wall. Kankaanpää et al. (2002) conducted a clinical trial in which an infant formula supplemented with *Bifidobacterium* or *Lactobacillus* was fed to infants with atopic eczema. Plasma lipids (neutral lipids and alpha-linolenic acid) were analyzed after the trial, and their levels were correlated with the progression of the disease. The major polyunsaturated fatty acids are linolenic, eicosapentaenoic, and docosahexaenoic acid (omega-3 fatty acids) and linoleic and arachidonic acids (omega-6 fatty acids). These polyunsaturated fatty acids are involved in the synthesis of prostacyclins and thromboxane, pro-inflammatory cytokine production (interleukin-1 and tumor necrosis factor), modulation of hypothalamic–pituitary–adrenal anti-inflammatory response, and acetylcholine release (Laparra and Sanz, 2010).

## Phytochemicals and gut microbiota

This category of compounds, found in fruits, vegetables, and grains, has the property to reduce the risk of chronic diseases. Phytochemicals are classified into carotenoids, phenolics, alkaloids, and organosulfonic compounds. A few of these compounds act as antioxidants, antiestrogens, and anti-inflammatory and immunomodulatory agents (Hertog et al., 1997; Ganry, 2002, 2005; Ruiz et al., 2007; Yuan et al., 2007; Wang et al., 2011). Gut microbes metabolize several phytochemicals by esterase, glucosidase, demethylation, dehydroxylation, and decarboxylation activities. Polyphenols are metabolized by gut microbiota into aglycones (Bialonska et al., 2010; van Duynhoven et al., 2011). Metabolites of polyphenols, generated by microbiota activity, are absorbed in the intestine, as evaluated by urine analysis.

## Technologies to detect microbiota

Advances in molecular biology techniques have facilitated detailed analyses of the composition of the bacterial community resident in the lower gut. High-throughput sequencing is the most common method for quantifying the microbiota. At times, 16S rRNA sequencing is used to identify different groups of microorganisms. Fluorescence in situ hybridization techniques, in combination with culturing bacteria on different substrates, also have been used (Scott et al., 2011). This approach is important specifically for low abundant bacteria. The sensitivity and specificity of assays performed to analyze the gut microbiota can be further improved. NMR mass spectrometry is a common technology that is used routinely in analyzing food components and metabolites (Martin et al., 2010).

## Challenges and opportunities in the field

The intestinal microbiota is essential for gut homeostasis. We are beginning to understand the extent of metabolic interactions between symbiotic intestinal microbes and the human host and their system-wide effects on the physiology of the host. A detailed characterization of the composition and function of the gut microbial ecosystem is required to foster the understanding of its mechanisms and impact.

Probiotics, live bacteria given orally that allow for intestinal colonization, may be beneficial, but longitudinal epidemiological studies in humans have yet to be completed. Previous trials were limited by small sample size, short duration of follow-up, or lack of state-of-the-art analyses of the gut microbiota. Other pathways, such as the vitamin D pathway, are worth exploring to determine their potential therapeutic implications.

Although the attenuated immune response that is associated with chronic inflammation has been studied in animal model systems, longitudinal studies that demonstrate the relationship between the neonatal gut microbiota and the development of allergic diseases and obesity are lacking. Clinical trials involving probiotics have shown some promise, but the results have been far from satisfactory. Dietary factors and vitamin D deficiency, along with lung mechanics and resistance to corticosteroid treatment, are factors that may provide clues to the correlation between asthma, obesity, and underlying chronic inflammation. Initial gut exposure to the microbiota has been identified as the most significant influence on immune system development. Thus, gut exposure in infancy is critical to the development of the immune system in early childhood and later in adult life.

Because the microbiota varies between individuals, the effects of interactions between gut microbiota and nutraceuticals cannot be generalized. However, following the pattern of microbiota in an individual provides information that can be used to analyze the person's health status and plan for improvement. There is a need to develop a reference gut microbiota database from clinically healthy people that can be used to compare with patterns from people with early or advanced disease. Further research also is needed to establish the beneficial role of the microbiota in the gut. Postgenomic approaches that advance discovery of the functionality of intestinal bacteria also should be explored further. In addition, the microbiota is a source of regulatory signals, some of which may be suitable for exploitation for therapeutic purposes. The future of drug discovery in gastroenterology likely will reside in the gut. A better understanding of both nutraceuticals and the microbiome may result in the discovery of new approaches to the prevention and treatment of disease.

## Acknowledgment

I thank Joanne Brodsky of SCG, Inc., for reading the manuscript and providing suggestions.

## References

Backhed, F., Ding, H., Wang, T., Hooper, L.V., and Koh, G.Y. 2004. The gut microbiota as an environmental factor that regulates fat storage. *Proc Natl Acad Sci* 101: 15718–15723.

Backhed, F., Manchester, J.K., Semenkovich, C.F., and Gordon, J.I. 2007. Mechanisms underlying the resistance to diet-induced obesity in germ-free mice. *Proc Natl Acad Sci* 104: 979–984.

Bialonska, D., Ramnani, P., Kasimsetty, S.G., Muntha, K.R., Gibson, G.R., and Ferreira, D. 2010. The influence of pomegranate by-product and punicalagins on selected groups of human intestinal microbiota. *Int J Food Microbiol* 140: 175–182.

Campbell, J.H., Fahey, G.C., and Wolf, B.W. 1997. Selected indigestible oligosaccharides affect the large bowel mass, cecal and fecal short-chain fatty acids, pH, and microflora in rats. *J Nutr* 127: 130–136.

Cani, P.D., Amar, J., Iglesias, M.A., Poggi, M., Knauf, C., Bastelica, D., Neyrinck, A.M. et al. 2007. Metabolic endotoxemia initiates obesity and insulin resistance. *Diabetes* 56: 1761–1772.

Cani, P.D. and Delzenne, N.M. 2009. The role of the gut microbiota in energy metabolism and metabolic disease. *Curr Pharm* 15: 1546–1558.

Claesson, M.J., Cusack, S., O'Sullivan, O., Green-Deniz, R., de Weerd, H., Flannery, E., Marchesi, J.R. et al. 2011. Composition, viability, and temporal stability of the intestinal microbiota of the elderly. *Proc Natl Acad Sci* 108: 4596–4591.

Dahlqvist, G., and Piessevaux, H. 2011. Irritable bowel syndrome: the role of the intestinal microbiota, pathogenesis and therapeutic targets. *Acta Gastroenterol Belg* 74: 375–380.

Flint, H.J., Duncan, S.H., Scott, K.P., and Louis, P. 2007. Interactions and competition within the microbial community of the human colon: links between diet and health. *Environ Microbiol* 9: 1101–1111.

Ganry, O. 2002. Phytoestrogen and breast cancer prevention. *Eur J Cancer Prev* 11: 519–522.

Ganry, O. 2005. Phytoestrogens and prostate cancer risk. *Prev Med* 41: 1–6.

Gibson, G.R., McCartney, A.L., and Rastall, R.A. 2005. Prebiotics and resistance to gastrointestinal infections. *Br J Nutr* 93: S31–S34.

Gueimonde, M., Salminen, S., and Isolauri, E. 2006. Presence of specific antibiotic (tet) resistance genes in infant faecal microbiota. *FEMS Immunol Med Microbiol* 48: 21–25.

Hertog, M.G., Sweetnam, P.M., Fehily, A.M., Elwood, P.C., and Kromhout, D. 1997. Antioxidant flavonols and ischemic heart disease in a Welsh population of men: the Caerphilly Study. *Am J Clin Nutr* 65: 1489–1494.

Irons, R., Carlson, B.A., Hatfield, D.L., and Davis, C.D. 2006. Both selenoproteins and low molecular weight selenocompounds reduce colon cancer risk in mice with genetically impaired selenoprotein expression. *J Nutr* 136: 1311–1317.

Kalliomaki, M., Salminen, S., Poussa, T., and Isolauri, E. 2007. Probiotics during the first 7 years of life: a cumulative risk reduction of eczema in a randomized, placebo controlled trial. *J Allergy Clin Immunol* 119: 1019–1021.

Kalliomaki, M., Collado, M.C., Salminen, S., and Isolauri, E. 2008. Early differences in fecal microbiota composition in children may predict overweight. *Am J Clin Nutr* 87: 534–538.

Kankaanpää, P.E., Yang, B., Kallio, H.P., Isolauri, E., and Salminen, S.J. 2002. Influence of probiotic supplemented infant formula on composition of plasma lipids in atopic infants. *J Nutr Biochem* 13: 364–369.

Kasaikina, M.V., Kravtsova, M.A., Lee, B.C., Servalli, J., Peterson, D.A., Walter, J., Legge, R. et al. 2011. Dietary selenium affects host selenoproteome expression by influencing the gut microbiota. *FASEB J* 25: 2492–2499.

Laparra, J.M. and Sanz, Y. 2010. Interactions of gut microbiota with functional food components and nutraceuticals. *Pharmacol Res* 61: 219–225.

Ley, R.E., Turnbaugh, P.J., Klein, S., and Gordon, J.I. 2006. Microbial ecology: human gut microbes associated with obesity. *Nature* 444: 1022–1023.

Mariat, D., Firmesse, O., Levenez, F., Guimares, V.D., Sokol, H., Dore, J., Corthier, G. et al. 2009. The Firmicutes/Bacteroidetes ratio of the human microbiota changes with age. *BMC Microbiol* 9: 123–127.

Martin, F.P., Sprenger, N., Montoliu, I., Rezzi, S., Kochhar, S., and Nicholson, J.K. 2010. Dietary modulation of gut functional ecology studied by fecal metabonomics. *J Proteome Res* 9: 5284–5295.

Musso, G., Gambino, M., and Cassader, M. 2010. Gut microbiota as a regulator of energy homeostasis and ectopic fat deposition: mechanisms and implications for metabolic disorders. *Curr Opin Lipidol* 21: 76–83.

Ruiz, P.A., Braune, A., Hölzlwimmer, G., Quintanilla-Fend, L., and Haller, D. 2007. Quercetin inhibits TNF-induced NF-kappaB transcription factor recruitment to proinflammatory gene promoters in murine intestinal epithelial cells. *J Nutr* 137: 1208–1215.

Sanz, Y. and DePalma, G. 2009. Gut microbiota and probiotics in modulation of epithelium and gut-associated lymphoid tissue function. *Int Rev Immunol* 28: 397–413.

Schomburg, L. and Schweizer, U. 2009. Hierarchical regulation of selenoprotein expression and sex-specific effects of selenium. *Biochim Biophys Acta* 1790: 1453–1462.

Scott, K.P., Duncan, S.H., Louis, P., and Flint, H.J. 2011. Nutritional influences on gut microbiota and the consequences for gastrointestinal health. *Biochem Soc Trans* 39: 1073–1078.

Sokol, H., Pigneur, B., Watterlot, L., Lakhdari, O., Bermúdez-Humarán, L.G., Gratadoux, J.J., Blugeon, S. et al. 2008. *Faecalibacterium prausnitzii* is an anti-inflammatory commensal bacterium identified by gut microbiota analysis of Crohn disease patients. *Proc Natl Acad Sci USA* 105: 16731–16736.

Tompkins, G.R., O'Dell, N.L., Bryson, I.T., and Pennington, C.B. 2001. The effects of dietary ferric iron deprivation on the bacterial composition of the mouse intestine. *Curr Microbiol* 43: 38–42.

van Duynhoven, J., Vaughan, E.E., Jacobs, D.M., Kemperman, R.A., van Velzen, E.J., Gross, G., Roger, L.C. et al. 2011. Metabolic fate of polyphenols in the human superorganism. *Proc Natl Acad Sci USA* 108: 4531–4538.

Wang, C.Z., Mehendale, S.R., Calway, T., and Yuan, C.S. 2011. Botanical flavonoids on coronary heart disease. *Am J Chin Med* 39: 661–671.

Yuan, J.P., Wang, J.H., and Liu, X. 2007. Metabolism of dietary soy isoflavones to equol by human intestinal microflora: implications for health. *Mol Nutr Food Res* 51: 765–781.

Zoetendal, E.G., Vaughan, E.E., and deVos, W.M. 2006. A microbial world within us. *Mol Microbiol* 59: 1639–1650.

# 14

# Metabolic and Potential Health Benefits of Nutraceuticals on Gut Microbiome

Malay K. Das and Yashwant V. Pathak

## Contents

## Introduction

The Human Genome Project was completed a decade ago, leaving a legacy of process, tools, and infrastructure now leveraged for the study of the microbes that reside in and on the human body as determinants of health and disease. This research focus has been branded "the Human Microbiome Project (HMP)." The human gut houses all forms of life, including bacteria, fungi, and archaea. Although the gut of a newborn is sterile at the time of birth, it gradually acquires hoards of microbial population during postnatal life

(Samanta et al. 2011). The gut microflora is influenced by several factors, including contacts between the child and its environment, mode of delivery, hygiene levels, medication, and type of feeding (Damaskos and Kollios 2008). It has been reported that more than 90% of the total cells of a healthy individual are bacterial cells, largely present in the lower part of the gastrointestinal tract (GIT) (Neish 2002). Approximately 1014 microbes belonging to more than 500 known species live in a complex, thermostable, nutrient-rich environment and play an important role in host physiology and metabolism (Backhed et al. 2005; Samanta et al. 2011). The gut microbiota performs a variety of beneficial functions as shown in Table 14.1.

Given their importance in human health and disease, the HMP was launched by the U.S. National Institutes of Health in 2008 with a total budget of $115 million. The goal of HMP includes:

1. Characterization of the microbial communities of various niches of the human body (e.g., the nasal passages, oral cavity, skin, urogenital system, and GIT)
2. Determination of whether individuals share a core human microbiome
3. Exploration of whether changes in the human microbiome cause or correlate with human disease
4. Development of new technologies and tools for computational analysis
5. Establishment of a resource repository
6. Studies of the ethical, legal, and social implications of human microbiome research (Turnbaugh et al. 2007; Nelson et al. 2010)

The achievements of the HMP were announced by the NIH director Francis Collins on June 13, 2012. The announcement was accompanied with a series of coordinated articles published in Nature and several journals in the Public Library of Science (PLoS) on the same day. By mapping the normal microbial makeup of healthy humans, using genome sequencing techniques, the

**Table 14.1** Important Functions of Gut Microbiome in the Human Body

| Functions | Examples |
| --- | --- |
| Nutrient synthesis | Vitamin biosynthesis including essential chemicals and nutrients such as anti-inflammatory compounds, antibacterial compounds. |
| Digestion and absorption | The final end products of gut digestion and fermentation are SCFAs that are energy source for the human cells. |
| Immune stimulation | Innate and adaptive to protect the human body from opportunistic pathogens. |
| Intestinal epithelium integrity | Intestinal epithelium integrity is very important for the absorption of nutrients as well as other ingredients good for human health; microbiota helps in protecting and maintaining the intestinal epithelium integrity. |
| Xenobiotic metabolism | Various types of chemicals entering orally in the GIT such as drugs, contaminants, food additives, and natural products all are first exposed to the microbiota and get either digested or converted to a form easy for elimination. |

researchers of the HMP have created a reference database and the boundaries of normal microbial variation in humans (Barbara et al. 2012; Curtis et al. 2012). The researchers calculated that more than 10,000 microbial species occupy the human ecosystem and they have identified 81%–99% of the genera. In addition to establishing the human microbiome reference database, the HMP project also discovered several "surprises," which include

1. Microbes contribute more genes responsible for human survival than humans' own genes. It is estimated that bacterial protein-coding genes are 360 times more abundant than human genes.
2. Microbial metabolic activities, for example, digestion of fats, are not always provided by the same bacterial species. The presence of the activities seems to matter more.
3. Components of the human microbiome change over time, affected by a patient disease state and medication.
4. There are huge variations between individuals with a western lifestyle on different microbiomes than individuals living in non-westernized settings. Although populations are essentially identical in terms of the human genome, they are very different in terms of the composition of gut microbes (Barbara et al. 2012; Curtis et al. 2012).

The downside of the HMP was that it only looked at the flora of healthy individuals. Therefore, it provided little information about the microbiome's role in health and disease. After the HMP, the Human Food Project (HFP) is picking up the torch and is launching a project called "American Gut." This project is going to look at the gut flora of humans with different diets, diseases, and activity patterns. Americans can participate and get the chance to see their microbiome compared to people with other diets, health issues, and demographic backgrounds. American Gut will increase knowledge about the specific differences between healthy individuals and people with various health problems (Hunter 2012).

Marked changes in socioeconomic status, cultural traditions, population growth, and agriculture are affecting diets worldwide. Understanding how diet and nutritional status influence the composition and dynamic operations of gut microbial communities, and the innate and adaptive arms of human immune system, represents an area of scientific need, opportunity, and challenge.

## What is microbiome?

The microbiome refers to the totality of microbes, the microbiota, their genetic elements (genomes), and environmental interactions in a defined environment, in this instance, the human gut. The human "metagenome" is a composite of human genes and genes present in the genomes of the microbiome

(Quigley 2011). Unlike our genome, which is inherited, each generation acquires its microbiome from the environment. For example, research demonstrates that the microbiome of babies delivered vaginally initially comes from the mother's vaginal microbiome. But the microbiome of babies born by cesarean section arises from the skin of anyone who has handled the child. The microbiome also changes with age. The microbiomes of infants become increasingly adultlike as infants age, but the microbiomes of the elderly are different from those of infants and adults (Berman and Sudbery 2002; Kellyn and Keegan 2012).

## Composition of gut microbiota

The resident microbiota in the GIT is a heterogeneous microbial ecosystem (Table 14.2) harboring up to $1 \times 10^{14}$ colony-forming units of bacteria (Zboril 2002). It is estimated that the GIT of healthy adults accommodates around 300–400 various cultivable species belonging to more than 190 genera (Holzapfel et al. 1998). Overall, there are between 1000 and 1150 prevalent bacterial species, and each individual person harbors at least 160 such species that are predicted to be largely shared (Qin et al. 2010). The phylogenetic core of a healthy adult fecal microbiota consists of a set of five phyla: Firmicutes, Bacteroidetes, Actinobacteria, Proteobacteria, and Verrucomicrobia (Tap et al. 2009). The predominant genera in the gut microbiota include Faecalibacterium, Ruminococcus, Eubacterium, Dorea, Bacteroides, Alistipes, and Bifidobacterium (Tap et al. 2009). Within the known GIT microbiota, only a few major groups dominate at levels around $10^{11}$ g$^{-1}$ feces, which include the strict anaerobes such as Bacteroides, Eubacterium, Bifidobacterium, and Peptostreptococcus (Holzapfel et al. 1998; Russell et al. 2011).

**Table 14.2** Dominant Types of Microbiota of Human GIT

| | |
|---|---|
| Oral cavity | Gemella (e.g., *Gemella haemolysans*), Granulicatella, Streptococcus (e.g., *Streptococcus mitis*), Veillonella, Prevotella, Porphyromonas, Rothia, Neisseria, Fusobacterium, Lactobacillus |
| | Allochthonous[a] microbes are generally outnumbered by autochthonous[b] microbes. |
| Stomach | *H. pylori* |
| | Allochthonous[a]: Gemella (e.g., *G. haemolysans*), Granulicatella, Streptococcus (e.g., *Streptococcus mitis*), Veillonella, Prevotella, Porphyromonas, Rothia, Neisseria, Fusobacterium, Lactobacillus |
| SI | *E. coli*, Klebsiella, Enterococcus, Bacteroides, Ruminococcus, Dorea, Clostridium, Coprococcus, Weissella, Lactobacillus (some species) |
| | Allochthonous[a]: Granulicatella, Streptococcus (e.g., *S. mitis*), Veillonella, Lactobacillus |
| LI | Five major phyla: Firmicutes, Bacteroidetes, Actinobacteria, Verrucomicrobia, and Proteobacteria, Hundreds of species |
| | Allochthonous[a] microbes are generally outnumbered by autochthonous[b] microbes. |

[a] Allochthonous: derived from outside sources, not found in the majority of hosts, but can be abundant in any one host.
[b] Autochthonous: endogenously derived and common to a majority of hosts.

**Table 14.3** Genome Size of Common Habitant Microbiota of Human GIT

| Microbiota | Genome Size (Mb) | Habitat | Substrate |
|---|---|---|---|
| B. thetaiotaomicron | 6.3 | LI | Polysaccharides |
| E. coli | 4.5 | SI and LI | Simple sugars and amino acids |
| H. pylori | 1.7 | SI and LI | Simple sugars and amino acids |
| L. reuteri | 2.0 | Stomach | Host specific, simple sugars, and 1,2-propanediol |
| Bifidobacterium adolescentis | 2.1 | LI | Dietary carbohydrates |

The organisms that dominate the human large intestine (LI) vary tremendously in genome size and content (Table 14.3), reflecting differences in the niche characteristics and in the ecological and evolutionary strategies used by gut microbes to occupy these niches. Generally, organisms with larger genomes follow more generalist lifestyles. An example is *Bacteroides thetaiotaomicron*, which has evolved to utilize complex and dynamic nutrient sources. Their glycobiomes (microbiome degrading glucose, xylose, and similar carbohydrates) allow them to break down a great variety of polysaccharides. In contrast, organisms with smaller genomes, such as *Helicobacter pylori* and *Lactobacillus reuteri*, appear to be adapted toward a restricted menu of dietary substrates, and these microbes also show specificity toward particular host species (Walter and Ley 2011). The extreme conditions (low pH) in the mammalian stomach and small intestine (SI) select for specialized microbes. *H. pylori* isolates from humans can tolerate the acidic conditions present in the stomach of their hosts (Tomb et al. 1997; Frese et al. 2011). *Enterobacteria* (e.g., *Escherichia coli*) and *Enterococcus* spp., which are commonly found in the human SI (Hayashi et al. 2005), can tolerate high concentrations of bile acids. *H. pylori, lactobacilli, enterobacteria, and enterococci* have limited abilities to utilize complex carbohydrates. Their nutritional requirements reflect the nutrient-rich habitats they occupy, which are proximal to host-nutrient absorption (Hayashi et al. 2005).

The composition of the gut microbiota is influenced by endogenous and environmental factors (diet, antibiotic intake, xenobiotics, etc.). Of these factors, the diet is considered a major driver for changes in gut bacterial diversity that may affect its functional relationships with the host (Ley et al. 2008). In fact, the microbiome of the adult-type and infant-type microbiota has distinct gene contents to accommodate nutrient acquisition strategies to different diets (Kurokawa et al. 2007).

## Gut microbiota and nutraceuticals

The low pH and rapid flow of materials in the stomach and proximal SI contain relatively low numbers of microbes ($10^3$–$10^5$ bacteria/g or mL content). Acid-tolerant lactobacilli and streptococci predominate in the upper SI

(Pathak et al. 2010). The distal SI (ileum) maintains a more diverse microbiota and higher bacterial numbers ($10^8$/g or mL content) than the upper bowel and is considered a transition zone preceding the LI. The LI (colon) is the primary site of microbial colonization because of slow turnover and is characterized by large numbers of bacteria ($10^{10}$–$10^{11}$/g or mL content), low redox potential, and relatively high short-chain fatty acid (SCFA) concentrations. Besides leading to an increasing gradient of indigenous microbes from the stomach to the colon, there are also characteristic spatial distributions of organisms within each gut compartment. Four microhabitats have been described: the intestinal lumen, the unstirred mucus layer or gel that covers the epithelium of the entire tract, the deep mucus layer found in intestinal crypts, and the surface of mucosal epithelial cells (Lee 1984; Berg 1996). The complex processes are involved in the establishment of microbial populations, involving microbial succession as well as microbial and host interactions and eventually resulting in dense, stable populations inhabiting specific regions of the gut.

Babies are born with no bacterial presence and are practically sterile (Samantha et al. 2011). Microbes that originate from the surroundings and from the mother establish themselves with the course of time. *E. coli* from the mother's feces start contaminating the infant in vaginal delivery. The length of delivery process is an important contributing factor (Bettelheim et al. 1974; Brook et al. 1979). Bacteria from the mother's cervix also colonize the alimentary canal of the baby. The nasopharynx of the baby receives bacteria from the mother's vagina (MacGregor and Tunnessen 1973).

There is a difference in exposure and acquisition of bacteria in babies born in developing countries. Enterobacteria and streptococci can be the first group seen in most cases. *E. coli* is seen within 48 h upon contamination (Mata and Urrutia 1971).

The infants in developing countries get exposed to environmental bacteria, regardless of the mode of delivery, as compared to developed countries. Following the exposure to the breast milk, the gut now is in a continuous process of acquiring new microbes (Moughan et al. 1992).

The microbe species like staphylococci, streptococci, corynebacteria, lactobacilli, micrococci, propionibacteria, and bifidobacteria all originate from the mothers nipple, surrounding skin, and milk ducts (Asquith and Harrod 1979; West et al. 1979). In formula-fed infants, the exposure to bacteria comes from dried powder, water, and equipment used in the manufacturing process.

Cooperstock and Zedd (1983) classified the development of intestinal bacteria in infants into four different phases. Phase 1 is the initial acquisition phase lasting over the first 2 weeks. Breast-feeding period is phase 2. Weaning and introduction of nutritional supplements is phase 3. Phase 4 starts after the weaning is complete. The initial colonization of *E. coli*, in large numbers, is later responsible for the establishment of anaerobic genera, Bacteroides, Bifidobacterium, and Clostridium. This happens during the time period of 4–7 days. Bifidobacteria have dominance in breast-fed infants. This changes once the dietary supplementation begins in the breast-fed infants and bifidobacteria are no longer the prominent genera (Pathak et al. 2010).

By the 2nd year of life, with the introduction of solid food almost complete, the fecal microbiota of the baby resembles the adult fecal microbiota. In phases 3 and 4, other bacterial groups including eubacteria, veillonella, staphylococci, propionibacteria, bacilli, and fusobacteria and yeast establish themselves along with the bacteroides and anaerobic gram-positive cocci (Cooperstock and Zedd 1983).

Antimicrobial and antibiotic agents significantly influence in the microflora of infants (Bennet and Nord 1989). A specific component of the microbiota becomes more vulnerable than others, due to this exposure. Mutations in the microbiota can be seen after finishing the drug treatment. The effects of the drug regime can be persistent (Finegold et al. 1983).

Diet is a powerful tool to influence intestinal microbiota. Targeted ingredients in the formula, like oligosaccharides, affect colon fermentation. Breast milk itself contains antimicrobial activity that helps to stimulate development and maturation of intestinal mucosa. This phenomenon now further promotes stability and decreased intestinal disturbances (Palmer et al. 2007).

The gut microbiota develops a number of protective, immune, and metabolic functions, which altogether have an enormous impact on the nutritional and health status of the host. The intestinal tract is inhabited by a complex microbiota that develops strategies to regulate nutrient acquisition and utilization in symbiosis with the host and in response to the diet. The biochemical activity of this complex ecosystem generates healthy as well as potentially harmful compounds from the diet, and their balance is essential to maintain a healthy status. This could be achieved by diverse nutritional strategies, including the administration of probiotic bacteria and other functional food components. Several lines of evidence suggest that dietary factors might profoundly influence microbiotal composition. The lifestyle and ecological roles of microbes affect the genome evolution and content of gut microbes (Roberfroid 2000).

The primary role of diet is providing sufficient nutrients to meet the basic nutritional requirements for maintenance and growth while giving the consumer a feeling of satisfaction and well-being. In addition, fortified food components exert beneficial health effects beyond basic nutrition, leading to the concept of functional foods and nutraceuticals (Roberfroid 2000). Functional foods are those foods that provide benefits beyond basic nutrition when consumed as part of the regular diet. Nutraceuticals are extracts containing the biologically active food components supplied in other than a food form. Dietary components with biological effects are susceptible to be metabolized by intestinal bacteria during the gastrointestinal (GI) passage, prior to being absorbed. The colon has the highest bacterial load and constitutes an active site of metabolism rather than a simple excretion route (Aura 2008). The metabolic activity of the gut microbiota on bioactive food components can modify the host exposure to these components and their potential health effects. Furthermore, some functional food components influence the growth and/or metabolic activity of the gut microbiota and, thereby, its composition and functions (Campbell 1997; Gibson et al. 2005).

## Gut microbiota and probiotics

The word probiotic comes from pro, which means "for," and bios, which means "life." So probiotic literally means "for life." The term probiotic is used to describe nutritional supplements and other products that contain live bacteria. There are literally hundreds of probiotic supplements available to buy. While they all promise to help restore and replenish our gut microflora, many are so deficient in living or viable bacteria that the package they come in has more value than what is inside (Dekker et al. 2007).

In 1905, Dr. Elie Metchnikoff, a Russian scientist working at the famous Institute Pasteur in Paris, was the first to write about the health benefits of probiotics. Dr. Metchnikoff, who later won a Nobel Prize for his research on the immune system, wrote that Bulgarian peasants who consumed large amounts of yogurt lived long, healthy lives. Examination of the yogurt by Dr. Metchnikoff led to his discovery of a unique lactic acid-producing bacterium that helped digestion and improved the immune system (Dekker et al. 2007). The historical association of probiotics with fermented dairy products led to extensive research validating Dr. Metchnikoff's early observations. Investigations during the past several decades have demonstrated numerous health supportive properties of probiotics on human health (Pathak et al. 2010).

Probiotics, as defined by the U.S. Probiotics Organization, are "live microorganisms administered in adequate amounts which confer a beneficial health effect on the host" (Guyton and Hall 2005). Probiotics bacteria are frequently, but not always, chosen from bacteria that normally inhabit the GI system of humans. The genera *Lactobacillus acidophilus* and *Bifidobacterium longum* (normal inhabitants of the healthy intestine) are the most clinically validated of all probiotics strains. Scientific study has repeatedly shown that two-strain probiotics supplements containing *L. acidophilus* and *B. longum* are highly effective in supporting overall human health (Pathak et al. 2010).

When humans are under increased physical, emotional, or intellectual stress, changes often occur within the GI environment (Kailasapathy and Chin 2000). Examples of these changes include slowed secretary responses, increased formation of reactive oxygen species, increased transit times of fecal material, disruption of mucosal cells, and altered epithelium tissues. These changes often result in occasional gas, bloating, and constipation and may interfere with probiotics functionality (Davidson et al. 2007).

The effectiveness of all probiotics is dependent on the ability of the organisms to reach the LI in a viable state and adhere to the intestinal wall. Only then can colonization of the microflora succeed. Researchers have recently discovered that probiotic combinations such as *L. acidophilus, Lactobacillus rhamnosus, Bifidobacterium bifidum, Bifidobacterium breve, B. longum,* and *Bifidobacterium lactis* are able to function well within altered GI environments (Miyake et al. 2006), and these probiotic combinations show promise in disease prevention (Karimi and Pena 2003; Collado et al. 2007).

Probiotic supplements need to be protected from the air, sunshine, artificial light, or moisture The Centers for Disease 2008 and from the digestive juices and enzymes in the stomach (Collado et al. 2007). Research has shown as much as 90% of a supplemented probiotic is destroyed in stomach gastric secretions and/or 50% loss in exposure to environment in storage.

Enteric coating of probiotics is intended to allow the passage of a tablet or capsule through the gastric fluids of the stomach to prevent the release of product contents before it reaches the intestines. However, due to the complexities involved with applying an enteric coating on a tablet or capsule, some enteric coatings do not entirely inhibit stomach acid from entering the encapsulation. As a result, stomach acid can interact with the sensitive bacteria leading to a significant decrease in viability. In addition, the enteric coating manufacturing process frequently utilizes solvents such as methacrylate copolymers. The tablets and capsules are sprayed with these solvents at high temperatures to create the enteric coating. This type of application further exposes microbes to conditions that can dramatically reduce the product shelf life. There is a new technology used in protecting viable ingredients that is a patented, encapsulation technique, known as True Delivery™ Technology, which results in a product that is stable at room temperature for up to 18 months. Additionally, the unique coating protects the bacteria from harsh stomach acid so they can be released live and intact in the intestines, where they need to arrive in live form to perform their beneficial function. Research may show that other methods support viability; for example, combining strains with specific fibers to buffer the probiotic from stomach acidity has shown some results. But as to date, the patented encapsulation has demonstrated the best effects on stability and protection from the environment (Pathak et al. 2010).

## Gut microbiota and prebiotics

Gibson and Roberfroid (1995) define prebiotics as "non-digestible food ingredients that beneficially affect the host by selectively stimulating the growth and/or activity of one or a limited number of bacteria in the colon, and thus improve host health."

Prebiotics modify the balance of the intestinal microbiota by stimulating the activity of beneficial bacteria, such as lactobacilli and bifidobacteria (Gibson and Roberfroid 1995). There is now considerable evidence that manipulation of the gut microbiota by prebiotics can beneficially influence the health of the host (Rastall et al. 2006; Parracho et al. 2007). In particular, some studies have shown to control serum triacylglycerol concentrations through modification of dietary habits with regard to consumption of pre- and probiotics (Parracho et al. 2007).

Prebiotics are not subject to biological viability problems and, thus, can be incorporated into a wide range of alimentary products (such as milk, yogurt, and infant formula), and they target organisms that are natural residents of the gut microbiota (Gibson and Roberfrodi 1995). Oligosaccharides have been suggested to represent the most important prebiotic dietary factor in human milk, promoting the development of a beneficial intestinal microbiota (Bode 2006). When oligosaccharides are consumed, the undigested portion serves as food for the intestinal microflora. Two common supplemental sources are fructo-oligosaccharides (FOSs) and inulin. FOSs, which are found in many vegetables, consist of short chains of fructose molecules. Inulin has a much higher degree of polymerization than FOS and is a polysaccharide. FOS and inulin are found naturally in Jerusalem artichoke, burdock, chicory, leeks, onions, and asparagus. FOS products derived from chicory root contain significant quantities of inulin. Inulin is considered a soluble fiber. As a soluble dietary fiber, inulin also shortens fecal transit time, slightly increases fecal bulk, reduces constipation, has been shown to reduce both serum and hepatic cholesterol and triglycerides, and may provide improved absorption of minerals such as calcium, magnesium, iron, and phosphate. Furthermore, unlike FOS, inulin's longer chain length makes it more easily tolerated by the human intestinal system.

Other benefits noted with FOS or inulin supplementation include increased production of beneficial SCFAs. Regarding SCFAs, about 60 g of carbohydrate is fermented by the bacteria each day to SCFA, which are rapidly absorbed. The SCFAs produced include acetic acid, propionic acid, and butyric acid. These acids have important actions in the colon and in the body as a whole. Acetic acid is an energy source for the body and is a substrate for fat synthesis in the liver. Propionic acid is also an energy source for the liver, is gluconeogenic (i.e., can be used to make glucose), and may reduce cholesterol synthesis. Butyric acid is the major fuel for colonic cells and has been shown to stimulate differentiation and programmed cell death of cancer cells. SCFA enemas have been used effectively in the treatment of ulcerative colitis. SCFAs produced in the colon increase cell proliferation throughout the whole gut. SCFAs are also very important because they promote water absorption and prevent osmotic diarrhea. SCFAs inhibit the growth of pathogenic bacteria (University of Glasgow 2005).

Another supportive factor in modulation of healthy microbiota is lactoferrin. Research using in vitro cell lines shows that supplemental lactoferrin modulates the release of messenger proteins known as essential nutrients, trypsin, and protease inhibitors that protect it from destruction in the GIT (Orsi 2004). It is also rich in antioxidants, and its receptors have been found on most immune cells including lymphocytes, monocytes, macrophages, and platelets (Orsi 2004). Its presence in neutrophils suggests that lactoferrin is also involved in phagocytic immune responses. Owing to its iron-binding properties, lactoferrin has been proposed to play a role in iron uptake by the intestinal mucosa (Legrand et al. 2005).

As it strongly binds iron, lactoferrin supports healthy modulation of gut microflora and assists the attachment of beneficial bacteria to the intestinal wall (Legrand et al. 2005). Lactoferrin may support healthy development and differentiation of T lymphocytes. Preliminary research suggests it supports the healthy production of cytokines and lymphokine, such as tumor necrosis factor (TNF)-alpha and interleukin (IL)-6 (Ward et al. 2005). More recently, lactoferrin receptors have been found in both a variety of immune system cells, including natural killer (NK) cells and intestinal tissue (Legrand et al. 2005; Ward et al. 2005). This discovery demonstrates that supplemental lactoferrin might have a profound impact on immune health.

In order to be effective, however, supplemental lactoferrin must be digested in the SI. Breast milk has a multitude of biological activities benefiting the newborn infant. To reach the SI (where nutrients are digested and released into the infant's bloodstream), the nutrients must be able to withstand exposure to stomach fluids. While the pH of an infant's stomach fluids rarely dips below 4–5 for the first 6 months of life, the pH of an adult's gastric fluids ranges between 1 and 2. This high acidity is lethal to lactoferrin. Therefore, just as probiotics need protection to survive the normal GI fluids of the stomach, so does lactoferrins (Orsi 2004).

Cani et al. (2009) have shown that the gut microbiota fermentation of prebiotics increased the satietogenic and incretin gut peptide resulting in an increase in plasma gut peptide concentrations (glucagon-like peptide 1 and peptide YY), which may contribute in part to changes in appetite sensation and glucose excursion responses after a meal in healthy subjects.

## Clinical applications of probiotics

Probiotics are products containing microorganisms with potential health benefits. They are used to prevent and treat antibiotic-associated diarrhea and acute infectious diarrhea. They may also be effective in relieving symptoms of irritable bowel syndrome and in treating atopic dermatitis in children. These can be used regularly as a preventive measure and help to maintain the gut microbiota. Species commonly used include *Lactobacillus* sp., *Bifidobacterium* sp., *Streptococcus thermophilus*, and *Saccharomyces boulardii*. Typical dosages vary based on the product, but common dosages range from 5 to 10 billion colony-forming units per day for children and from 10 to 20 billion colony-forming units per day for adults. Significant adverse effects are rare, and there are no known interactions with medications.

Several mechanisms are proposed to explain the pharmacological actions of probiotics. It is likely that more than one mechanism is at work simultaneously. In the prevention and treatment of GI infection, it is likely a combination of direct competition between pathogenic bacteria in the gut and immune modulation and enhancement. In children with atopic dermatitis, the mechanism is probably related to the effect of probiotics on the early development of immune tolerance during the first year of life. Probiotics may help downgrade the excessive immune responses to foreign antigens that lead to atopy in some children. They may also contribute to systemic downregulation of inflammatory processes by balancing the generation of pro- and anti-inflammatory cytokines, in addition to reducing the dietary antigen load by degrading and modifying macromolecules in the gut. Probiotics have been shown to reverse the increased intestinal permeability characteristic of children with food allergy, as well as enhance specific serum immunoglobulin A (IgA) responses that are often defective in these children. To be most effective, a probiotic species must be resistant to acid and bile to survive transit through the upper GIT. Most probiotics do not colonize the lower GIT in a durable fashion. Even the most resilient strains generally can be cultured in stool for only 1–2 weeks after ingestion. To maintain colonization, probiotics must be taken regularly (Kligler and Cohrssen 2008).

Several clinical studies have been reported about the applications of probiotics and prebiotics in specific clinical conditions in pediatric and geriatric patients. Table 14.4 outlines the applications reported in various clinical studies.

## Probiotics in infections

Macfarlane and Cummings (2002) summarized the most recent contributions to this rapidly developing area of applications of probiotics in infections. Probiotic bacteria, mainly bifidobacteria and lactobacilli, have shown potential in diseases prevention. Probiotic bacteria are effective in preventing and reducing the severity of acute diarrhea in children. These are also useful in antibiotic-associated diarrhea, but not for elimination of *H. pylori*. In inflammatory bowel disease, especially ulcerative colitis, probiotics offer a safe alternative to current therapy. Probiotics are used to prevent urogenital tract infection with benefit and, perhaps more intriguingly, to reduce atopy in children. However, probiotics do not invariably work and studies of mechanisms are urgently needed.

While a normal microflora is associated with good health, changes in intestinal health are associated with altered immune function. A well-functioning

**Table 14.4** Clinical Indications of Prebiotics and Probiotics

| Clinical Indications | References | Comments |
|---|---|---|
| Prevention and management of GI and non-GI conditions | Saavedra and Tschernia (2002) | Review of preliminary human clinical findings |
| Probiotics, prebiotics, and synbiotics | De Vrese et al. (2008) | Review of German clinical studies |
| Probiotics and prebiotics in diarrhea | De Vrese et al. (2007), Szajewska and Mrukowicz (2001), Guandalini et al. (2011) | Effects in diarrhea |
| Antibiotic-associated diarrhea | Johnston et al. (2011), Nelson et al. (2011) | Probiotics for the prevention of diarrhea |
| Probiotics may reduce the severity of pain and bloating in patients with irritable bowel syndrome. | Guyonnet et al. (2007) | Small clinical trials |
| Probiotics reduce the incidence of atopic dermatitis in at-risk infants. | Kalliomäki (2007) | There is preliminary support for treatment of symptoms. |
| Effects of probiotic bacteria on diarrhea, lipid metabolism, and carcinogenesis | De Roos and Katan (2000) | A review of papers published in 1988–1998 |

GI immune system mediates immune responsiveness at mucosal sites and throughout the entire body via the control of the quality and quantity of foreign substances gaining access to the immune system (Schiffrin et al. 1997). *Lactobacillus acidophilus* and *B. longum* possess immune-protective and immune-modulatory properties. These benefits include modulation of cytokine and various IL production, autoimmunity, NK cell cytotoxicity, lymphocyte proliferation, and antibody production. In an open, randomized, controlled trial, *Lactobacillus acidophilus* and *B. bifidum* were supportive of colon health in older adults. In addition, B cell (important antibody-producing immune cells) levels increased as compared with the untreated group. The probiotics were very well tolerated, with no significant side effects or variations in clinical chemistry or hematologic parameters (Gibson et al. 1997). Bifidobacteria and lactobacilli are gram-positive, lactic acid-producing bacteria constituting a major part of the intestinal microflora in humans and other mammals. Administration of antimicrobial agents may cause disturbances in the ecological balance of the GI microflora with several unwanted effects such as colonization by potential pathogens. To maintain or reestablish the balance in the flora, supplements of intestinal microorganisms, mainly bifidobacteria and lactobacilli, sometimes called probiotics, have been successfully used.

## Probiotics in pediatric urology

The report by Reid (2002) studied the role probiotics therapy in pediatric urology. Many children around the world die of diseases, such as GI infection and HIV, while many have urinary tract infections that subsequently recur frequently in adulthood. Until recently, the role of intestinal and urogenital (vaginal, urethral, and perineal) microflora in health and disease has received little attention. There is mounting evidence that certain strains of lactobacilli and bifidobacteria have a major part in the maintenance and restoration of health in children and adults. Implications for pediatric urology include a decreased risk of infection and stone disease as well as possible positive effects on preventing and managing inflammatory and some carcinogenic diseases.

## Chronic nonspecific diarrhea of infancy

Balli and colleagues (1992) evaluate the effectiveness of oral bacteriotherapy using a combination of anaerobic fecal lactobacilli for chronic, nonspecific diarrhea of infancy. A double-blind study was carried out in a total of 40 children treated with low and high doses of bacteria. The results confirm the importance of fecal flora in this disease and support the hypothesis that oral bacteriotherapy can improve clinical and laboratory presentation, especially when given at high doses.

## Protection of intestinal epithelial cells

The colonic epithelium maintains a lifelong, reciprocally beneficial interaction with the colonic microbiota. Disruption of intestinal mucosa is associated with mucosal injury. Resta-Lenert and Barret (2003) proposed that probiotics may limit epithelial damage induced by enteroinvasive pathogens and promote restitution. Human intestinal epithelial cell lines (HT29/cl.19A and Caco-2) were exposed to enteroinvasive *E. coli* (EIEC 029:NM) and/or probiotics (*Streptococcus thermophilus* (ST), ATCC19258, and *Lactobacillus acidophilus* (LA), ATCC4356). Infected cells and controls were assessed for transepithelial resistance, chloride secretory responses, alterations in cytoskeletal and tight junctional proteins, and responses to epidermal growth factor (EGF) stimulation. Exposure of cell monolayers to live ST/LA, but not to heat inactivated ST/LA, significantly limited adhesion, invasion, and physiological dysfunction induced by EIEC. Antibiotic killed ST/LA reduced adhesion somewhat but were less effective in limiting the consequences of EIEC invasion of cell monolayers. Furthermore, live ST/LA alone increased transepithelial resistance, contrasting markedly with the fall in resistance evoked by EIEC infection, which could also be blocked by live ST/LA. The effect of ST/LA on resistance was accompanied by maintenance (actin, ZO-1) or enhancement (actinin, occludin) of cytoskeletal and tight junctional protein phosphorylation. ST/LA had no effect on chloride secretion by themselves but reversed the increase in basal secretion evoked by EIEC. EIEC also reduced the ability of EGF to activate its receptor, which was reversed by ST/LA. Live ST/LA interacts with intestinal epithelial cells to protect them from the deleterious effect of EIEC via mechanisms that include, but are not limited to, interference with pathogen adhesion and invasion. Probiotics likely also enhance the barrier function of naive epithelial cells not exposed to any pathogen.

## Lactose degradation by probiotic bacteria

Lactose is an important sugar that is converted to lactic acid by lactic acid-producing bacteria, such as *Lactobacillus acidophilus* and *B. longum*. Impaired conversion of lactose to lactic acid can result in symptoms such as occasional gas, bloating, and indigestion, due to accumulated nonabsorbed lactose in the GIT. Lactic acid bacteria can help metabolize the nonabsorbed lactose in the GIT and therefore reduce symptoms of lactose intolerance. In a randomized, controlled clinical trial, *B. longum* was shown to support the breakdown of lactose and reduce flatulence (Lactose 2008).

## Probiotics for occasional constipation

Constipation is one of the most common GI complaints in the United States. More than 4 million Americans have frequent constipation, accounting for 2.5 million physician visits a year. Those reporting constipation most often are women and adults ages 65 and older. Pregnant women may have con-stipation, and it is a common problem following childbirth or surgery. Self-treatment of constipation with over-the-counter (OTC) laxatives is by far the most common aid. Around $725 million is spent on laxative products each year in America (http://digestive.niddk.nih.gov/ddiseases/pubs/constipation/ accessed on November 22, 2012). Constipation is defined as infrequent or dif-ficult defecation that can result from decreased motility of the intestines. It is a common problem, particularly in older adults. When the feces remain in the LI for prolonged periods, there is excessive water absorption, making the feces dry and hard. *Lactobacillus acidophilus* and *B. longum* promote regular bowel movements by contributing to the reestablishment of healthy intestinal flora and stimulation of intestinal peristalsis via lactic acid production (Bennet and Eley 1976).

## Probiotics in putrefactive processes

In unbalanced conditions that are present in the intestines (i.e., unbalanced diet, high acidity, and/or low levels of lactic acid bacteria), organic matter may be putrefied (decomposed or rot) by certain bacteria and produce undesirable compounds (Gibson et al. 1997). Probiotics are recommended in such situations that promote homeostasis (balance) in both the intestine and the vagina (Witsell et al. 1995). These activities are carried out via support of direct production of antibodies, competition with adhesion to intestinal cells, or indirect modulation of the immune system. Probiotics also support a healthy yeast balance (Candida albicans 2000). Normal microflora of the LI helps support and complete digestion via fermentation (Wagner et al. 1997). Oral ingestion of probiotics produces a stabilizing effect on the gut flora. The benefits of probiotics extend beyond digestion support and immune support. *Lactobacillus acidophilus* and *B. longum* also help support the utilization and bioavailability of nutrients, including vitamins, minerals, proteins, fats, and carbohydrates (Witsell et al. 1995).

## Conclusions

Probiotics are products containing living microorganisms that have potential health benefits for infants, adults, and geriatric patients. Probiotics are used to prevent and treat antibiotic-associated diarrhea and acute infectious diarrhea. Probiotics are also effective in relieving symptoms of irritable bowel syndrome and in treating atopic dermatitis in children. In addition, probiotics can be used regularly as a preventive measure to maintain the gut microbiota. Species commonly used include *Lactobacillus* sp., *Bifidobacterium* sp., *Streptococcus thermophilus*, and *Sa. boulardii*. Typical dosages for probiotics vary based on the product, but common dosages range from 5 to 10 billion colony-forming units per day for children and from 10 to 20 billion colony-forming units per day for adults. Significant adverse effects are rare, and there are no known interactions with medications. Several clinical trials have reported efficacies in different clinical scenarios such as from antibiotic-induced diarrhea in children to adult problems of GIT microflora maintenance and enhancement. However, there is a need for more clinical trials to confirm efficacy and interactions with drugs and herbal medicines.

## References

Asquith MT and Harrod JR 1979. Reduction in bacterial contamination in banked human milk. *J Pediatr* 95: 993–994.

Aura AM 2008. Microbial metabolism of dietary phenolic compounds in the colon. *Phytochem Rev* 7: 407–429.

Backhed F, Ley RE, Sonnenburg JL, Peterson DA, and Gordon JI 2005. Host–bacterial mutualism in the human intestine. *Science* 307: 1915–1920.

Balli F, Bertolani P, Giberti G, and Amarri S 1992. High-dose oral bacteria-therapy for chronic non-specific diarrhea of infancy. *Pediatr Med Chir* 14(1): 13–15.

Bennett A and Eley KG 1976. Intestinal pH and propulsion: An explanation of diarrhoea in lactase deficiency and laxation by lactulose. *J Pharm Pharmacol* 28: 192–195.

Bennet R and Nord CE 1989. The intestinal microflora during the first week of life: Normal development and changes induced by caesarean section, preterm birth and antimicrobial treatment. In: Grubb R, Midtvedt T, Norin E, eds. *The Regulatory and Protective Role of the Normal Microflora*. London, U.K.: Macmillan Press, pp. 19–34.

Berg R 1996. The indigenous gastrointestinal micro flora. *Trends Microbiol* 4: 430.

Berman J and Sudbery PE 2002. Candida albicans: a molecular revolution built on lessons from budding yeast. *Nat Rev Genet* 3(12): 918–930.

Bettelheim KA, Breardon A, Faiers MC, and O'Farrell SM 1974. The origin of O serotypes of *Escherichia coli* in babies after normal delivery. *J Hyg* 72: 67–70.

Bode L 2006. Recent advances on structure, metabolism, and function of human milk oligosaccharide. *Int J Probiotics Prebiotics* 1: 19–26, 71.

Brook I, Barett C, Brinkman C, Martin W, and Finegold S 1979. Aerobic and anaerobic bacterial flora of the maternal cervix and newborn gastric fluid and conjunctiva: A prospective study. *Pediatrics* 63: 451–455.

Campbell JH, Fahey GC Jr, and Wolf BW 1997. Selected indigestible oligosaccharides affect large bowel mass, cecal and fecal short-chain fatty acids, pH, and microflora in rats. *J Nutr* 127: 130–136.

Cani PD, Lecourt E, Dewulf EM, Sohet FM, Pachikian BD, Naslain, D, Backer, FD, Neyrinck AM, and Delzenne NM 2009. Gut microbiota fermentation of prebiotics increases satietogenic and incretin gut peptide production with consequences for appetite sensation and glucose response after a meal. *Am J Clin Nutr* 90: 1236–1249.

*Candida albicans* 2002. On-line medical dictionary. Available at: http://www.graylab.ac.uk/cgi-bin/omd?query+candida albicans (accessed on October 10, 2012).

Collado MC, Meriluoto J, and Salminen S 2007. Role of commercial probiotic strains against human pathogen adhesion to intestinal mucus. *Lett Appl Microbiol* 45: 454–460.

Cooperstock MS and Zedd AJ 1983. Intestinal flora of infants. In: Hentges DJ, ed. *Human Intestinal Microflora in Health and Disease*. New York: Academic Press, pp. 79–99.

Damaskos D and Kollios G 2008. Probiotics and prebiotics in inflammatory bowel disease: Microflora on the scope. *Br J Clin Pharmacol* 65: 453–463.

Davidson G, Kritas S, and Butler R 2007. Stressed mucosa. *Nestle Nutr Workshop Ser Pediatr Program* 59: 133–142.

De Roos NM and Katan MB 2000. Effects of probiotic bacteria on diarrhea, lipid metabolism and carcinogenesis: A review of papers published between 1988 to 1998. *Am J Clin Nutr* 71(2): 405–411.

Dekker J, Collett M, Prasad J, and Gopal P 2007. Functionality of probiotics—potential for product development. *Forum Nutr* 60: 196–208.

Finegold SM, Mathisen GE, and George WL 1983. Changes in human intestinal flora related to the administration of antimicrobial agents. In: Hentges DJ, ed. *Human Intestinal Microflora in Health and Disease*. New York: Academic Press, pp. 356–446.

Frese SA, Benson AK, Tannock GW et al. 2011. The evolution of host specialization in the vertebrate gut symbiont *Lactobacillus reuteri*. *PLoS Genet* 7(2): e1001314.

Gibson GR, McCartney AL, and Rastall RA 2005. Prebiotics and resistance to gastrointestinal infections. *Br J Nutr* 93: S31–S34.

Gibson GR and Roberfroid MB 1995. Dietary modulation of the human colonic microbiota: Introducing the concept of probiotics. *J Nutr* 125: 1401–1412.

Gibson GR, Saavedra JM, Macfarlane S, and Macfarlane GT 1997. Probiotics and intestinal infections. In: Fuller R ed. *Probiotics 2 Applications and Practical Aspects*. London, England: Chapman & Hall, pp. 10–39.

Guandalini S 2011. Probiotics for prevention and treatment of diarrhea. *J Clin Gastroenterol* 45: S149–S153.

Guyonnet D, Chassany O, and Ducrotte P 2007. Effect of a fermented milk containing *Bifidobacterium animalis* DN-173 010 on the health-related quality of life and symptoms in irritable bowel syndrome in adults in primary care: A multicentre, randomized, double-blind, controlled trial. *Aliment Pharmacol Ther* 26(3): 475–486.

Guyton AC and Hall JE 2005. Secretory functions of the alimentary canal. In: *Textbook of Medical Physiology*, 11th edn. Philadelphia, PA: W.B. Saunders Company, p. 738.

Hayashi H, Takahashi R, Nishi T, Sakamoto M, and Benno Y 2005. Molecular analysis of jejunal, ileal, caecal and recto-sigmoidal human colonic microbiota using 16S rRNA gene libraries and terminal restriction fragment length polymorphism. *J Med Microbiol* 54: 1093–101.

Holzapfel WH, Haberer P, Snel J, Schillinger U, and Huis in't Veld JHJ 1998. Overview of gut flora and probiotics. *Int J Food Microbiol* 41: 85–101.

Hunter E 2012. We are 90 percent microbes: Human Microbiome Project aims to map gut flora. *http://www.naturalnews.com/036715_Human_Microbiome_Project_gut_flora_bacteria* (accessed on August 7, 2012)

Johnston BC, Goldenberg JZ, Vandvik PO, Sun X, and Guyatt GH 2011. Probiotics for the prevention of pediatric antibiotic associated diarrhea. *Cochrane Database Syst Rev* 9(11): CD 004827.

Kailasapathy K and Chin J 2000. Survival and therapeutic potential of probiotic organisms with reference to *Lactobacillus acidophilus* and Bifidobacterium spp. *Immunol Cell Biol* 78: 80–88.

Kalliomäki M, Salminen S, Poussa T, and Isolauri E 2007. Probiotics during the first 7 years of life: A cumulative risk reduction of eczema in a randomized, placebo-controlled trial. *J Allergy Clin Immunol* 119(4): 1019–1021.

Karimi O and Pena AS 2003. Probiotics: isolated bacteria strain or mixtures of different strains? Two different approaches in the use of probiotics as therapeutics. *Drugs Today* 39: 565–597.

Kellyn B and Keegan S 2012. Emerging Science for environmental health decisions newsletter. Available at: *http://nas-sites.org/emergingscience* (accessed on November 21, 2012)

Kligler B and Cohrssen A 2008. Probiotics. *Am Fam Physician* 78(9): 1073–1078.

Kurokawa K, Itoh T, Kuwahara T et al. 2007. Comparative metagenomics revealed commonly enriched gene sets in human gut microbiomes. *DNA Res* 14: 169–181.

Lactose 2008. On-line medical dictionary. Available at: http://www.graylab.ac.uk/cgi-bin/omd (accessed on November 25, 2012).

Lee A 1984. Neglected niches: The microbial ecology of the gastrointestinal tract. In: Marshall K, ed. *Advances in Microbial Ecology*. New York: Plenum Press, pp. 115–162.

Legrand D, Elass E, Carpentier M, and Mazurier J 2005. Lactoferrin: A modulator of immune and inflammatory responses. *Cell Mol Life Sci* 62: 2549.

Ley RE, Lozupone CA, Hamady M, Knight R, and Gordon JI 2008. Worlds within worlds: Evolution of the vertebrate gut microbiota. *Nat Rev Microbiol* 6: 776–788.

MacFarlane GT and Cummings JH 2002. Probiotics, infections and immunity. *Curr Opin Infect Dis* 15(5): 501–506.

MacGregor RR and Tunnessen WW 1973. The incidence of pathogenic organisms in the normal flora of the neonate's external ear and nasopharynx. *Clin Pediatr* 12: 697–700.

Mata LJ and Urrutia JJ 1971. Intestinal colonization of breast-fed children in a rural area of low socioeconomic level. *Ann N Y Acad Sci* 176: 93–108.

Miyake K, Tanaka, T and McNeil PL 2006. Disruption–induced mucus secretion: repair and protection. *PLoS Biol* 4(9): e276. doi:10.1371/journal.pbio.0040276.

Moughan PJ, Birtles MJ, Cranwell PD et al. 1992. The piglet as a model animal for studying aspects of digestion and absorption in milk-fed human infants. In: Simopoulos AP, ed. *Nutritional Triggers for Health and in Disease*. Basel, Switzerland: Karger, pp. 40–113.

Nelson RL, Kelsey P, Leeman H, Meardon N, Patel H, Paul K, Rees R, Taylor B, Wood E and Malakun R 2011, Antibiotic treatment for *Clostridium difficile* associated diarrhea in adults, *Cochrane Database Syst Rev*, 9(11) CD 004610

Nelson KE, Weinstock GM, Highlander SK et al. 2010. A catalog of reference genomes from the human microbiome. *Science* 328: 994–999.

Neish AS 2002. The gut microflora and intestinal epithelial cells: A continuing dialogue. *Microbes Infect* 4: 309–317.

Orsi N 2004. The antimicrobial activity of lactoferrin: Current status and perspectives. *Biometals* (3): 189–196.

Parracho H, McCartney AL, and Gibson GR 2007. Probiotics and prebiotics in infant nutrition, *Proc Nutr Soc*, 66: 405–411.

Palmer C, Bik EM, DiGiulio DB, Relman DA, and Brown PO 2007. Development of the human infant intestinal microbiota. *PLoS Biology* 5(7): e177. doi:10.1371/journal.pbio.0050177

Pathak SY, Leet C, Simon A, and Pathak YV 2010. Probiotics and prebiotics as nutraceuticals, Chapter 14, In: Pathak YV ed. *Handbook of Nutraceuticals*. Boca Raton, FL: CRC Press, pp. 223–242.

Qin J, Li R, Raes J et al. 2010. A human gut microbial gene catalogue established by metagenomic sequencing. *Nature* 464: 59–65.

Quigley EMM 2011. Gut microbiota and the role of probiotics in therapy. *Curr Opin Pharmacol* 11: 593–603.

Rastall RA, Gibson GR, Gill HS et al. 2006. Modulation of the microbial ecology of the human colon by probiotics, prebiotics and synbiotics to enhance human health: An overview of enabling science and potential applications. *FEMS Microbiol Ecol* 52: 145–152.

Roberfroid M, Gibson GR, Hoyles L et al. 2010. Prebiotic effects: Metabolic and health benefits. *Br J Nutr* 104(Suppl. 2): S1–S63.

Reid G 2002. The potential role of probiotics in pediatric urology. *J Urol* 168: 1512–1517.

Resta-Lenert S and Barrett KE 2003. Live probiotics protect intestinal epithelial cells from the effects of infection with enteroinvasive *Escherichia coli* (EIEC). *Gut* 52(7): 988–997.

Roberfroid MB 2000. Prebiotics and probiotics: Are they functional foods? *Am J Clin Nutr* 71: 1682S–1687S.

Russell DA, Ross RP, Fitzgerald GF, and Stanton C 2011. Metabolic activities and probiotic potential of bifidobacteria. *Int J Food Microbiol* 149: 88–105.

Samanta AK, Kolte AP, Senani S, Sridhar M, and Jayapal N 2011. Prebiotics in ancient Indian diets. *Curr Sci* 101(1): 43–46.

Saavedra JM and Tschernia A 2002. Human studies with probiotics and prebiotics. *Br J Nutr* 87(2): S241–S246.

Szajewska H and Mrukowicz JZ 2001. Probiotics in the treatment and prevention of acute infectious diarrhea in infants and children: A systemic review of published randomized double blind placebo controlled trials. *J Pediatr Gastroenterol Nutr* 33: S17–S25.

Schriffrin EJ, Brassart D, Servin AL, Rochat F, and Donnet-Hughes A 1997. Immune modulation of blood leukocytes in humans by lactic acid bacteria: Criteria for strain selection. *Am J Clin Nutr* 66(Suppl): 515S–520S.

Tap J, Mondot S, Levenez F, Pelletier E, Caron C, Furet J-P, Ugarte E, and Muñoz-Tamayo 2009. Towards the human intestinal microbiota phylogenetic core. *Environ Microbiol* 11, 2574–2584.

Tomb JF, White O, Kerlavage AR et al. 1997. The complete genome sequence of the gastric pathogen *Helicobacter pylori*. *Nature* 388: 539–547.

The Human Microbiome Project Consortium and Barbara A, Nelson Karen E, Pop M et al. 2012. A framework for human microbiome research. *Nature* 486(7402): 215–221. doi:10.1038/nature11209. PMC 3377744. PMID 22699610.

The Human Microbiome Project Consortium and Curtis H, Gevers D, Knight R, et al. 2012. Structure, function and diversity of the healthy human microbiome. *Nature* 486(7402): 207–214. doi:10.1038/nature11234. PMID 22699609.

Turnbaugh PJ, Ley RE, Hamady M, Fraser-Liggett CM, Knight R, and Gordon JI 2007. The human microbiome project. *Nature* 449, 804–810.

University of Glasgow 2005. The normal gut flora. http://www.gla.ac.uk/departments/humannutrition/students/resources/meden/Infection (accessed on November 25, 2012).

Wagner RD, Peirson C, Warner T et al. 1997. Biotherapeutic effects of probiotic bacteria on candidiasis in immunodeficient mice. *Infect Immun* 65: 4165–4172.

Ward PP, Paz E, and Conneely OM 2005. Multifunctional roles of lactoferrin: A critical overview. *Cell Mol Life Sci* 62(22): 2540–2548.

Walter J and Ley R 2011. The human gut microbiome: Ecology and recent evolutionary changes. *Annu Rev Microbiol* 65: 411–429.

West PA, Hewitt JH, and Murphy OM 1979. The influence of methods of collection and storage on the bacteriology of human milk. *J Appl Bacteriol* 46: 269–277.

Witsell DL, Garrett CG, Yarbrough WG et al. 1995. Effect of *Lactobacillus acidophilus* on antibiotic-associated gastrointestinal morbidity: A prospective randomized trial. *J Otolaryngol* 24: 230–233.

Zboril V 2002. Physiology of microflora in the digestive tract. *Vnitřní Lékařství* 48: 17–21.

# VII

## Cognitive Decline and Neurodegeneration

# 15

# Nutraceuticals in the Prevention of Cognitive Decline

Zhenzhen Zhang and L. Joseph Su

## Contents

# Introduction

According to the United Nations' World Population Prospects 2011 report, there were 784 million people aged over 60 in 2011 worldwide and this number will increase to 2 billion by 2050 and 2.8 billion in 2100, suggesting that the older population will triple or quadruple over the next 50–100 years.[1] The oldest-old population (aged 80 or over) are the most rapidly growing portion of the older adult population, numbering 109 million in 2011 in the world, and estimated to number 402 million in 2050 and 792 million in 2100.[1] With the aging of the world population, it is estimated that the number of people living with dementia will increase from 36 million in 2010 to 66 million in 2030 to 115 million in 2050.[2] Mild cognitive impairment (MCI) is a pre-dementia status with objective cognitive impairment but without functional evidence of the impairment.[3,4] The incidence rate of MCI ranges from 51 to 76.8 per 1000 person-years among different studies,[4] and the prevalence is different across countries.[5] Alzheimer's disease (AD) characterized by amyloid deposition and secondary synaptic loss[6] is the most prevalent cause of dementia worldwide.[7] Vascular dementia (VaD) is the second most prevalent cause of dementia characterized by cognitive decline resulting from ischemic or hemorrhagic vascular lesions.[7,8] Except for the commonly diagnosed dementias, many other chronic diseases, such as hypertension,[9] cardiovascular disease,[10] diabetes mellitus,[11] chronic obstructive pulmonary disease,[12] and chronic renal disease,[13,14] are also found to be associated with cognitive decline.

It is widely known that cognitive function declines with age in adulthood. Many behavioral, lifestyle, and pharmaceutical interventions are associated with the progression of cognitive decline.[15-18] Although dietary/nutritional factors associated with the cognitive decline have not been consistently reported, food habits and nutritional factors are among those modifiable lifestyle factors that are considered to play an essential role in cognitive status.[19-21] Physicians normally prescribe medications, such as acetylcholinesterase inhibitors (AChEIs) and N-methyl-d-aspartate (NMDA) antagonist in the current clinical practice. However, these current treatment medications do not appear to halt the increasing trend of cognitive impairment worldwide, and the current treatment strategy hasn't effectively controlled the rising incidence of dementia. Therefore, it is imperative to seek more effective treatment and prevention strategies beyond the conventional treatment products.[22]

The concept of nutraceuticals was first coined by Dr. Stephen DeFelice in 1989.[23] More and more nutraceutical products for maintaining normal cognitive functions are coming into the market nowadays.[24] Nutraceuticals include but are not limited to dietary supplements (vitamins, minerals, herbs, or other botanicals, amino acids, or a concentrate, metabolite, constituent, extract, or combinations of these ingredients) that help to prevent and/or treat disease(s) and/or disorder(s).[23] Products deriving from the complementary and alternative medicine (CAM) are the major nutraceutical products in the market. CAM is normally used for overall health, disease prevention, and medical treatment supplementation by the adults.[25] Many nutraceutical products, including CAM products, can be purchased in the local grocery stores and health food stores or from a neighborhood pharmacy without physician's prescription.[24] It is difficult to quantify the prevalence of nutraceutical products use. Data from the 2007 National Health Interview Survey (NHIS) suggested that the out-of-pocket costs spending on CAM, including visiting CAM practitioners and purchasing CAM products, amounted to $33.9 billion in the United States in 2007.[26] Data from the 2007 NHIS also showed that U.S. adults with functional limitations (FLs) were more likely to use CAM compared to those without FL.[25] This study also suggested more than 50% of U.S. adults with FL discussed the CAM use with their health-care providers.[25] Due to the cost spending on the nutraceutical products and their interaction with conventionally used medications,[27] it is important to advise the general adult population as well as patients with cognitive decline to make wise purchase and use of the nutraceutical products.

This chapter will first introduce the current available FDA-approved drugs for AD; then we will focus on summarizing existing evidence of nutraceutical products including herbal and vitamin/mineral supplements in the prevention of cognitive decline. Functional food including whole, fortified, enriched, or enhanced food is also nutraceutical if it can aid in the prevention and/or treat disease(s) and/or disorder(s),[23] but it is not of this chapter's focus. Instead, we will only briefly review the epidemiological evidence investigating the relationship between certain dietary patterns and cognitive function.

## Current pharmacotherapies

### Acetylcholinesterase inhibitors

AChEIs including tacrine (Cognex®), donepezil (Aricept®), rivastigmine (Exelon®, Exelon Patch®), and galantamine (Razadyne®, Reminyl®) are routinely used in the clinics to treat mild to moderate AD.[28,29] Donepezil is used to treat severe AD.[28] Continued use of donepezil among patients with moderate or severe AD also showed cognitive benefits in a recent clinical trial funded by the U.K. Medical Research Council and the U.K. Alzheimer's Society.[30] These AChEIs are used to inhibit the enzyme that degrades the neurotransmitter acetylcholine, thereby increasing its circulating levels. However, a review of clinical trials conducted on these AChEIs concluded that the recommendations of the AChEIs for AD treatment were questionable because of flawed methodology and small-scale benefits.[31] The long-term effects of these treatments are also questionable and attention is needed on greater rates of adverse effects.[32]

## N-methyl-d-aspartate antagonist

Memantine (Namenda®) is the first drug of this class approved by FDA in 2003.[33] It is thought to treat AD through the mechanisms of antagonizing excessive excitotoxic glutamate activity, increasing long-term potentiating and decreasing tau protein hyperphosphorylation.[22] Memantine is clinically used to treat patients with moderate to severe AD. Recent analyses of nine clinical trials on moderate to severe AD patients have shown a significant effect of delaying clinical worsening in the domains of cognition, functional abilities, and clinical global impression in memantine group compared with placebo group.[34] However, a review on both AChEIs and memantine concluded that the improvement of cognitive function of dementia was clinically marginal, although studies showed statistical significant results.[35]

## Other pharmacotherapies

Different pharmacological drugs other than AChEIs and NMDA antagonist have been or being investigated in clinical trials. A recent review on the pharmacotherapies for AD provided a summary of the major experimental AD therapies and their status in clinical trials.[22] A list of selected drugs in clinical trials conducted in 2011 was summarized by Gravitz.[33] Briefly, other therapeutic approaches under study include other neurotransmitter modulators (latrepirdine/dimebon); amyloid production inhibitors, which are also r-secretase inhibitors (semagacestat/LY450139); amyloid anti-aggregants (tramiprosate/homotaurine, PBT2, scyllo-inositol/ELND005); amyloid removal, which is also immunotherapy (AN1792, CAD106, bapineuzumab, solanezumab, IVIg); and tau-directed therapies (methylthioninium chloride/MTC/TRx0014/Rember, lithium).[22] Although some of these drugs failed to show the anticipated results, many promising drugs are still under investigation.

## Complementary and alternative medicine (herbal medicine)

Herbal medicine has been used in most Asian countries for thousands of years. Older adults in the United States as well as in other countries commonly use prescription medications and herbal supplements together.[27] Dietary supplement is the most common form of herbal medicine and may be beneficial in retaining the cognitive function.[36]

Herbal supplements include single herbs, extracts from single herbs, and herbal formulation (mixture of herbs) developed based on the Chinese medicine theories in a holistic approach.[37] These supplements may have multiple effective ingredients and can be developed into nutraceuticals for cognitive decline prevention. Plants containing components that have acetylcholinesterase (AChE) inhibition effect are potential sources for new drugs against neurodegeneration. Many Asian, European, and South American plants have shown beneficial effect in treating cognitive decline.[38–41] A series of important clinical trials and longitudinal studies on the associations of herbal supplements and cognitive functions have been summarized in this chapter (Table 15.1).

## Single herbs or extracts from single herbs

### Ginseng

Ginseng is the most commonly consumed herbal product in the world and has been suggested to have potentials to improve cognitive performance and mood.[42] The active components in ginseng are a group of saponins called ginsenosides.[43] Out of the 50 already isolated ginsenosides from the *Panax ginseng* root,[43] $Rg_1$, $Rb_1$, and $Rg_3$ in particular may have an antioxidative effect or a NMDA receptor inhibition effect that will attenuate beta-amyloid formation.[44–46] A review of two randomized clinical trials (RCT) using *Panax ginseng* to treat AD patients found ginseng could improve mini-mental state examination (MMSE) and Alzheimer's Disease Assessment Scale (ADAS)-cognitive scores, but both of these two RCTs suffered from serious methodological limitations that prevented the authors to make further conclusions.[47] A systematic review published in 2010 using Cochrane database identified nine randomized, double-blind, placebo-controlled trials on ginseng.[45] Due to the heterogeneity of the outcome definitions in these studies, the authors could not conduct pooled analysis.[45] Overall, their findings suggested there was no convincing evidence to show cognition improvement effect of ginseng due to the small sample size and lower quality evidence.[45]

### Ginkgo biloba

*Ginkgo biloba* has been used for preventing cognitive decline for thousands of years. It has been one of the top eight herbal products sold in the United States.[48] The leaves and extracts of *G. biloba* are widely available in over-the-counter pharmacies or grocery stores.[49] In addition to be used as a memory enhancer, it is used for a wide range of medical conditions such as tinnitus, premenstrual tension, and dizziness.[50] Many clinical and preclinical studies used a standardized *G. biloba* extract, EGb 761, which contains 22%–27% flavonoids and 5%–7% terpene lactones (ginkgolides and bilobalide).[49] A laboratory study supports the neuroprotective properties of EGb 761, especially through suppressing reactive oxygen species (ROS) production in the mitochondria.[51] A meta-analysis evaluating the 6-month treatment effect of standardized *G. biloba* extract among dementia patients showed that it was effective in the improvement of cognitive functions in these patients.[52] Another recent meta-analysis including nine clinical trials concluded that the *G. biloba* extract EGB 761 showed moderate effect in treating Alzheimer's, vascular, or mixed dementia with the treatment duration from 12 to 52 weeks.[53] However, the Ginkgo Evaluation of Memory (GEM) study, which was a large-scale RCT on EGB 761 and included 3069 participants, did not find significant difference in the dementia incidence between the *G. biloba* treatment group and the

**Table 15.1** Epidemiological Studies on the Relationship between Herbal Supplements and Cognition/Dementia

| Study (Author/Year) | Country | Study Design | Population | Total Number of Subjects | Compliance Rate | Product | Duration of Treatment | Outcome/Disease Assessment | Results |
|---|---|---|---|---|---|---|---|---|---|
| Lee et al. (2008)[140] | South Korea | Prospective open-label, randomized study | AD patients (aged 47–83 years, mean = 66.1 Years) | 97 | 84.5% | Panax ginseng | 12 weeks, Panax ginseng powder (4.5 g/day) | MMSE, ADAS | Panax ginseng is effective in treating AD patients. |
| Rafii et al. (2011)[59] | United States | Phase II RCT | Mild to moderate AD patients (50 years or older) | 210 | 84% | Huperzine A | 16 weeks, 200 μg twice daily group and 400 μg twice daily group | ADAS-cog, MMSE, ADCS-ADL, NPI | Huperzine A 200 μg twice daily is ineffective in mild to moderate AD patients; huperzine A 400 μg twice daily improves ADAS-cog. |
| Snitz et al. (2009)[54] | United States | Randomized, double-blind, placebo-controlled clinical trial | Community-dwelling participants (age 72–96 years) | 3069 | 87% | G. biloba | Median follow-up of 6.1 years, twice-daily dose of 120 mg extract | 3MSE, ADCS-Cog, neuropsychological test battery, California Verbal Learning Test, modified Rey–Osterrieth Figure Test, WAIS-R Block Design, | G. biloba administered 120 mg twice daily did not result in less cognitive decline in older adults with normal cognition or with MCI. |

| Reference | Country | Study design | Participants | N | % | Intervention | Dose/Duration | Measures | Findings |
|---|---|---|---|---|---|---|---|---|---|
| Lee et al. (2009)[63] | Korea | Randomized double-blind placebo-controlled study | Healthy participants (mean age 39.9 years for placebo group, 38.04 years for BT-11 group) | 55 | 87% | R. polygalae (BT-11) | 4 weeks, 300 mg/person/day | Boston Naming Test and semantic verbal fluency, WAIS-R Digit Span and the Trail Making Test Part A and Trail Making Test Part B, and Stroop Color/Word Test; K-CVLT, SOPT | BT-11 improves memory in healthy adults. |
| Shin et al. (2009)[64] | Korea | Randomized double-blind placebo-controlled study | Participants aged 60 years and over and have subjective memory complaints | 53 | 87% | R. polygalae (BT-11) | 8 weeks, 300 mg/person/day | CERAD and MMSE | BT-11 improves memory in the elderly humans. |
| Tildesley et al. (2005)[68] | United Kingdom | Placebo-controlled, double-blinded, crossover | Undergraduate volunteers (mean age 23.21 years) | 24 | 100% | S. lavandulaefolia essential oil | 1, 2.5, 4, and 6 h (placebo, 25 μL, 50 μL crossover), 7-day washout | Cognitive Drug Research computerized test battery for cognitive performance | S. lavandulaefolia has acute modulation effect of mood and cognition in healthy young adults. |

*(continued)*

**Table 15.1 (continued)** Epidemiological Studies on the Relationship between Herbal Supplements and Cognition/Dementia

| Study (Author/Year) | Country | Study Design | Population | Total Number of Subjects | Compliance Rate | Product | Duration of Treatment | Outcome/Disease Assessment | Results |
|---|---|---|---|---|---|---|---|---|---|
| Scholey et al. (2008)[69] | United Kingdom | Placebo-controlled, double-blinded, crossover | Older healthy volunteers (mean age 72.95 years) | 20 | 100% | S. officinalis extract capsule | 1, 2.5, 4, and 6 h (placebo, 167, 333, 666, or 1332 mg of standardized sage extract crossover), 7-day washout | Cognitive Drug Research computerized test battery for cognitive performance | S. officinalis can enhance memory in older adults. |
| Akhondzadeh et al. (2003)[70] | Iran | Parallel group placebo controlled | AD patients (ages 65–80 years) | 39 | 77% | S. officinalis | 4 months, 60 drops/day | CDR, ADAScog | S. officinalis can improve cognitive function in AD patients. |
| Nagata et al. (2012)[75] | Japan | Open-label clinical trial | VaD patients (mean age 71.2 years) | 13 | 100% | Yi-Gan San (Yokukansan) | 4 weeks, Yokukansan (7.5 g/day) | NPI, MMSE, Barthel Index, DAD, UPDRS | Yokukansan is beneficial for the treatment of behavioral and psychological symptoms of dementia (BPSD) in VaD patients. |

| Iwasaki et al. (2005)[74] | Japan | Randomized, observer-blind controlled trial | Mild to severe dementia patients (mean age 80.3 years) | 52 | 100% | *Yi-Gan San (Yokukansan)* | 4 weeks, Yi-Gan San powder 2.5 g (1.5 g of extract) 3 times/day | MMSE, NPI, Barthel Index | *Yi-Gan San* improves BPSD and ADL. |
| Iwasaki et al. (2004)[77] | Japan | Randomized, double-blinded, placebo-controlled trial | Dementia patients (mean age 84.4 years) | 33 | 100% | *Ba Wei Di Huang Wan* | 8 weeks, 20 pills (2 g) × 3 times/day | MMSE, Barthel Index | *Ba Wei Di Huang Wan* improves cognitive function and ADL abilities of elderly dementia patients. |

*Source:* See table for multiple data sources.

Abbreviations of the cognitive function tests (outcome/disease assessment): ADAS-cog, Alzheimer disease assessment scale—cognitive subscale; ADAS, Alzheimer disease assessment scale; ADCS-ADL, Alzheimer's Disease Cooperative Study—Clinical Global Impression of Change; ADL, Activities of Daily Living; CERAD, Consortium to Establish a Registry for Alzheimer's Disease Assessment Packet; CDR, Clinical Dementia Rating; DAD, Disability Assessment for Dementia; K-CVLT, Korean version of the California Verbal Learning Test; 3MSE, Modified Mini-Mental State Examination; NPI, Neuropsychiatric Inventory; SOPT, self-ordered pointing test; UPDRS, United Parkinson's Disease Rating Scale; WAIS-R, modified Wechsler Adult Intelligence Scale—Revised.

placebo group.[54] This study also found *G. biloba* could not slow the rate of cognitive decline among all the participants.[54]

## Huperzine A

*Huperzine A* is a plant extract from *Huperzia serrata*. It has been shown to have an effect of AChE inhibition.[55] *Huperzine A* administered at 0.2 mg twice daily is currently used as an approved treatment for AD and VaD in China.[56] Many clinical trials on the effectiveness of *huperzine A* in AD patients have been conducted since 1995. A review of six clinical trials on the efficacy and safety of *huperzine A* in treating AD was conducted by Li et al. using the *Cochrane Collaboration Database* in 2008.[57] This review suggested *huperzine A* had the potential of neuroprotective effect of cognitive function, but the included trials lacked adequate qualities and sample sizes that prohibited the authors to make any valid recommendations.[57] Later, a meta-analysis published in 2009 on four RCTs demonstrated that *huperzine A* (300–500 µg/day) administered for 8–24 weeks significantly improved the cognitive status measured by MMSE and activities of daily living (ADL) scale; no serious adverse effects were found.[58] A phase II clinical trial conducted in the United States among 210 participants with mild to moderate AD reported that *huperzine A* administered at 200 µg twice/day did not influence cognitive status, while *huperzine A* administered at 400 µg twice/day had a modest improvement in cognition.[59] A larger phase III clinical trial in the United States may help to further add evidence for the potential effect of *huperzine A* in treating AD and other dementias.[56]

## Radix polygalae

*Radix polygalae* (Chinese, Yuan Zhi), or BT-11, is the root of *Polygala tenuifolia Willdenow*.[60] It was identified to have neuroprotective effects against neuronal death and cognitive impairments in rats.[61] A triterpenoid saponin (polygalasaponin XXXII, PGS32) isolated from the roots of *P. tenuifolia Willdenow* improves hippocampus-dependent learning and memory in rat and mice models.[62]

Most of the supporting evidence of the effect of *R. polygalae* on maintaining normal cognitive function is from clinical trials completed in Korea. One study conducted among 55 Korean healthy subjects found that BT-11 improved cognitive function in healthy general population and could be used as a functional food for memory improvement.[63] The same research group conducted another study among participants older than 60 years who had subjective memory complaints.[64] They discovered that BT-11 could enhance cognitive function among these elderly population.[64] From those studies, BT-11 has been considered to have an AChEI effect[61] and can be a potential nutraceutical product for cognitive improvement.

## Salvia officinalis and Salvia lavandulaefolia

*Salvia officinalis* and *Salvia lavandulaefolia* are two species of the *Salvia* genus family that originated in European medicine. The extracts and oils from *S. officinalis* and *S. lavandulaefolia* have the effect of AChEI.[65,66] Their major constituent monoterpenoids act synergistically to inhibit AChE.[65,67] A human study conducted in the United Kingdom using the crossover study design found *Salvia* may be able to modulate both the mood and the cognition in healthy young adults.[68] The same study team also found the memory-enhancing effects of *Salvia* in healthy older volunteers.[69] Another study in Iran among mild to moderate AD patients found *S. officinalis'* effect on the improvement of cognitive function.[70] These results indicated a potential beneficial neuroprotective effect of extracts or oils from *S. officinalis* or *S. lavandulaefolia*.

## Combination of herbs (herbal formulations)

In the traditional Chinese medicine (TCM) practice, herbs are frequently prescribed as combinations/formulas under the guidance of TCM theories. The basic theories include yin–yang theory and five-element theory. Under the holistic approach, there is no distinction between different types of dementia or cognitive decline. Instead, the herbal formulas are prescribed based on different patterns or syndromes such as (i) deficiency of vital energy of the marrow, kidney, heart, and spleen or (ii) stagnation of blood and/or phlegm and so on.[71] Therefore, the holistic treatment effects of the herbal formulations are not restricted to just the nervous system.[72] Up to date, a few clinical trials investigating potential herbal formulations and cognitive decline have been published in English literature. As examples, *Yokukansan*, *Yukmijihwang-tang* (*Liu Wei Di Huang Wan*), and *Ba Wei Di Huang Wan* were identified from the literature and included in this chapter.

### Yokukansan

*Yokukansan* originating from the Chinese herbal formula *Yi-Gan San* is also a traditional Japanese medicine formula containing seven prescribed components.[71,73] It has been traditionally used for restlessness and agitation since the sixteenth century in Asia.[38] A randomized, observer-blind controlled trial among 52 subjects (27 mild to severe dementia patients and 25 controls) showed that *Yokukansan* had beneficial effects on behavioral and psychological symptoms and ADL.[74] An open-labeled trial among 13 VaD patients showed *Yokukansan* could improve behavioral and psychological symptoms evaluated by Neuropsychiatric Inventory (NPI), but not cognition measured by MMSE.[75] Another study among 14 patients with AD showed similar results with an improvement of behavioral and psychological symptoms but not MMSE score after 12 weeks of treatment of *Yokukansan*.[73]

## Yukmijihwang-tang/Ba Wei Di Huang Wan

*Yukmijihwang-tang* (Chinese, *Liu Wei Di Huang Tang*) is a basic formula used to supplement yin to treat yin deficiency. The six ingredients in this formula include *Rehmanniae Radix* (Chinese, *shu di huang*), *Corni Fructus* (Chinese, *shan zhu yu*), *Rhizoma Dioscoreae* (Chinese, *shan yao*), *Alismatis Rhizoma* (Chinese, *ze xie*), *Poria* (Chinese, *fu ling*), and *Paeonia suffruticosa Cortex* (Chinese, *mu dan pi*). A study has shown *Yukmijihwang-tang* had the effect of improving the speed of information processing and cognitive ability in a randomized placebo-controlled trial among 35 dementia patients with deficient cognitive ability.[76]

*Ba Wei Di Huang Wan* is an herbal formula based on the formula of *Liu Wei Di Huang Tang*. It includes two more yang-supplementing ingredients *Aconiti Radix Lateralis Preparata* (Chinese, *fu zi*) and *Cinnamomi Cortex* (Chinese, *rou gui*) in addition to the six basic ingredients in *Yukmijihwang-tang*. This modified formula has a variety of yang-supplementing applications. A clinical trial on *Ba Wei Di Huang Wan* conducted in Japan among 33 participants found *Ba Wei Di Huang Wan* could improve the cognitive function measured by MMSE.[77]

The mixture of the herbal formula contains multiple ingredients and these ingredients have multiple important interactions working together to achieve the treatment effects. This makes us very difficult to tease out individual effect of each ingredient.

# Non-herbal dietary supplements/nutrients

## Vitamins and minerals

### Antioxidant vitamin (vitamin C, vitamin E, folic acid, and beta-carotene)

It is gradually becoming clear that aging and cognitive decline are associated with insufficient antioxidant defenses against oxidative stress generated by ROS.[78,79] A majority of ROS are produced during the ATP production in the mitochondria.[80] Antioxidant enzymes and antioxidants such as vitamin C and vitamin E may neutralize ROS and decrease its reactivity. Therefore, nutritional supplements of these antioxidants would provide a low-cost prevention method for cognitive decline, if found effective.

Many epidemiological studies have found the beneficial effect of the nutritional antioxidants in cognitive decline prevention. For example, an Australian cohort study has shown the beneficial effect of vitamin C intake on cognitive function measured by MMSE, although the relationship was weak[81]; a Korean study in elder independent living adults found male subjects with lower vitamin C intake had statistically significant poor cognitive ability[82]; a French cohort study found both vitamin C and E intakes were positively associated with verbal memory scores[83]; a U.S. study among freely living elderly participants

(60–94 years) showed higher vitamin E intake was associated with improved performance on visuospatial recall and/or abstraction tests[84]; another U.S. study with an average follow-up of 3.2 years found that vitamin E, but not vitamin C or carotene, was associated with less age-related cognitive decline[85]; a Spanish study in a sample of institutionalized elderly people found that the subjects with better cognitive testing scores had higher intakes of vitamin C, vitamin E, and folic acid and lower intake of unhealthy food[86]; the Women's Antioxidant and Cardiovascular Study, a trial of vitamin E (402 mg/other day), beta-carotene (50 mg/other day), and vitamin C (500 mg/day), did not find positive effects of these supplemental antioxidants on the cognitive change but suggested a later effect of vitamin C or beta-carotene on cognitive function.[87] A recent review on nutrition and VaD published in 2012 found consistent evidence that micronutrients especially Vitamin C and E intakes were protective against VaD.[19] These studies about the nutrition effects on cognitive decline have been reviewed extensively and suggest that the beneficial effects are not universal, which require further studies in this area.[88–91]

## Vitamin D and calcium

Vitamin D constitutes a family of calciferol hormones. It is synthesized in the skin as vitamin $D_3$ (cholecalciferol) or directly obtained from dietary or supplemental intake of vitamin $D_3$ or $D_2$ (ergocalciferol).[36] Insufficient vitamin $D_3$ has been linked with an increased risk of Parkinson's disease and AD,[92] as well as dementia and cerebrovascular disease.[93] Vitamin $D_3$ insufficiency has also been associated with decline in cognitive function, and particularly executive function, in the elderly.[94] A most recent prospective French EPIDemiology of OSteoporosis (EPIDOS) Toulouse cohort study enrolling 498 community-living women free of supplemental vitamin D intake found that higher dietary vitamin D intake was associated with a lower risk of AD.[95] This was a prospective study that addressed the relationship between dietary Vitamin D intake and cognitive function. The neuroprotective mechanisms of vitamin D have been studied in the labs extensively. Vitamin D receptor has been found in both neurons and glial cells.[96] The neuroprotective effects of vitamin D are thought to be mediated by stimulation of nerve growth factor synthesis in neurons and glial-derived neurotrophic factor in glial cells.[97,98]

Calcium and vitamin D are closely interacted.[99] It is widely known that vitamin D can improve calcium absorption from digestive systems through synergistic effects. On the other hand, calcium intake could influence serum vitamin D status measured by 25(OH)D and 1,25(OH)$_2$D concentration[100] although some literature did not suggest that calcium intake could affect the ability of vitamin D absorption.[99,101]

Calcium has been found to have multiple effects on the neurological systems as well. Calcium is a major component in the multiple signaling pathways of the neurons. Dysregulation of calcium homeostasis may lead to increasing intracellular ROS concentrations progressively and oxidative stress, inducing cell death in the brain.[79] Additionally, calcium homeostasis is thought to be related to cognitive function through regulating the synaptic dysfunction and maintaining synaptic plasticity.[102] The major source of calcium in the human body comes from diet. Therefore, adequate calcium intake helps to maintain calcium homeostasis and normal functions of the brain. A cross-sectional and short-term prospective study has suggested that higher calcium consumption is associated with improved cognitive function.[103] In the future, a large-scale longitudinal study is needed to investigate the temporal relationship between dietary calcium intake and cognitive decline. In addition, whether vitamin D and calcium intake have any interactive effect on cognitive status still needs further studies.

## Fatty acids

Docosahexaenoic acid (DHA) and eicosapentaenoic acid (EPA) belonging to the long-chain omega-3 (n-3) polyunsaturated fatty acids (PUFAs) have been shown to have multiple health benefits.[104,105] PUFAs have been observed to have neuroprotective and cardioprotective effects in several epidemiological studies,[104,106–109] although the protective effect of n-3 fatty acids on cognitive decline was not observed among coronary heart disease patients participating in a large-scale RCT (Alpha Omega Trial) in the Netherlands.[110] The effect of n-3 PUFAs on cognitive decline or AD can be mediated by apolipoprotein E genotype (ApoE epsilon4 allele).[111,112] DHA and EPA are generally derived from marine sources such as fatty fish. It may not be an economical and sustainable long-term approach to have daily intake of fish depending on the geographic location of the population. Therefore, more cost-effective consumption of n-3 fatty acids that can increase tissue concentrations of DHA and EPA is sought.

A lower-cost alternative source for n-3 PUFAs is stearidonic acid (SDA), which is commonly consumed in a variety of products with SDA-enriched soybean oil as an ingredient.[113,114] High levels of SDA is expressed in soybeans and it is an immediate downstream metabolite of n-3 fatty acids alpha-linolenic acid (ALA).[115] SDA has been proved to increase circulating levels of EPA[116] and RCTs have found SDA-enriched soybean oil increased the omega-3 index (red blood cell EPA+DHA).[115,117]

To date, there is no epidemiological study on the SDA intake and cognitive decline yet. Since the plant-based SDA has the potential to improve human health, further studies are warranted on the effect of SDA on maintaining normal cognitive functions.

## Dietary patterns

Mediterranean diet pattern has been proposed to be a healthy eating pattern since 1995.[118] It is characterized by higher intake of plant foods, fresh fruits, and olive oil and lower intake of red meat and saturated fat.[118] Among

1410 participants in a three-city prospective French cohort study, higher Mediterranean dietary pattern score was associated with slower cognitive decline measured by MMSE.[119] The protective effect in this French study was found to be mediated by higher DHA concentration and lower omega-6 (n-6):n-3 PUFA concentration ratio measured in blood.[120] Among 3790 participants in the Chicago Health and Aging Project (CHAP) with an average follow-up time of 7.6 years, a statistically significant inverse association was found between Mediterranean dietary pattern score and cognitive function decline measured by averaging three types of cognitive tests.[121] A study conducted in New York also found significant association between higher adherence to Mediterranean diet and lower risk for developing AD[122,123] and MCI.[124] A nutraceutical effect of Mediterranean diet and cognitive decline has been reviewed in details by Frisardi et al. and Solfrizzi et al.[125,126] Both vascular and nonvascular biological mechanisms, such as oxidative, inflammatory, and metabolic mechanisms, have been suggested.[125]

In addition to Mediterranean dietary pattern, other interpretations of healthy dietary patterns have been investigated. For example, the Supplementation en Vitamines et Mineraux Antioxydant Study conducted in France using principal component analysis identified a healthy dietary pattern.[127] This dietary pattern was positively correlated with intakes of fruit, whole grains, fresh dairy products, vegetables, breakfast cereal, tea, vegetable fat, nuts, and fish and negatively correlated with intakes of meat and poultry, refined grains, animal fat, and processed meat.[127] This study suggested that dietary pattern consumed in middle life was associated with better cognitive performance among subjects with low energy intake.[127]

## Remarks

Current epidemiological evidence has demonstrated potential preventive or treatment effects of various nutraceutical products on cognitive decline and associated dementias. However, definitive recommendations cannot be made without strict multicenter, large-scale, and well-monitored clinical trials to confirm the findings thus far.

In addition to the herbal and vitamin/mineral supplement, many other dietary supplements have been investigated on their relationships with cognitive decline or dementias. For example, other frequently used herbs in treating dementia not discussed earlier in this review include *Rhizoma Chuanxiong* (*Chuanxiong*), *Radix Polygoni Multiflori* (*Heshouwu*), *Radix Astragali* (*Huangqi*), *Rhizoma Acori Tatarinowii* (*Shichangpu*), and *Uncaria rhynchophylla* (*Chinese cat's claw*).[128–131] Some other micronutrients such as selenium (Se) used in preventing cognitive decline have also been suggested recently.[132] Furthermore, lower sodium intake has been found to slow the cognitive decline among lower physical activity older adults.[133] Certain foods have also been reported to relate to cognitive function. For example, tea was found to have beneficial effect on cognitive function.[134] The beneficial effects of alcohol, fruits, and/or vegetable intakes on pre-dementia/dementia have been reviewed by Solfrizzi et al. in details.[126] Additional research is needed to further explore the potential nutraceutical mechanisms of these products.

Proven evidence is still lacking for the prevention and/or treatment effects of the nutraceutical products. Large-scale and rigorous clinical trials and/or prospective studies are needed to move this field forward. There are many issues needed to be considered in the design and analyses of studies on nutraceuticals in the prevention of cognitive decline. First, studies should consider the confounding or effect modification factors such as subjects' disease status, age, gender, and race. A systematic review by Williams et al. showed that the most consistent increased risk factors for AD and cognitive decline were diabetes, epsilon 4 allele of the apolipoprotein E gene (APOE e4), smoking, and depression, while the decreased risk factors for AD and cognitive decline were cognitive engagement and physical activities.[135] Anthropometric risk factors such as body mass index (BMI) and central adiposity may also correlate with cognitive function, which has been suggested by the Women's Health Initiative Memory Study (WHIMS).[136] All these factors should be carefully thought out before the studies initiated. Conceptual framework may help to tease out the complicated confounders and intermediate measurements in the causal pathways. Second, multiple measurements of the exposures and outcomes are warranted. Single time-point measurement of dietary intake cannot make causal inferences concerning the progression of cognitive loss. The measurements of cognitive function are varied. Studies integrating a variety of tests are considered to have more stable measurement than a single test.[137] Third, the assessments of the nutraceutical products face the issue of heterogeneity since different studies use different dosages of supplementation and many dietary factors are highly variable. Regulation of the nutraceutical products is complicated.[138] The effectiveness of the available nutraceutical products can vary for individuals, and consequently, prevention/treatment strategies may need to be made individually, which is a challenge. Fourth, recall bias should be considered since studies among cognitively impaired subjects can be challenging. Recall bias of dietary intake may be more prevalent among those with cognitive impairment. Fifth, the sample size of many of the intervention studies is small. These kinds of intervention studies lack adequate power that is not sufficient to derive convincing conclusions. The attrition bias caused by the loss of participants makes the already small sample size issue even worse, and many studies did not consider this in the analyses.

Nutrient biomarkers including plasma vitamins B, C, D, and E and marine n-3 fatty acids have been reported to be associated with cognitive function as well.[139] Dietary nutrient intake and nutrient blood circulating level need to be compared since dietary nutrient intake may not accurately represent the plasma/serum/red blood cell level nutrient. Therefore, different transporting abilities of nutrients from dietary intake to blood or tissues should be considered in the future studies.

## Conclusions

No definitive conclusion on the nutraceutical products and cognitive decline as a whole can be made. Some studies have provided a glimpse of optimism for the use of nutraceuticals to prevent and/or treat the degeneration of cognitive function. However, many potential confounders have not been considered and/or well managed in many of the published studies. Since cognitive decline is a progressive process, duration of follow-up is crucial to generate valid study results. Nutraceutical products may indeed be able to offer an alternative approach to alleviate suffering of ailments related to cognitive decline, if studies with lengthier follow-up time and well-considered design are available to provide confirming evidence.

## References

1. United Nations. World Population Prospects: The 2010 Revisions. 2011. http://esa.un.org/unpd/wpp/Documentation/pdf//WPP2010_Volume-II_Demographic-Profiles.pdf (accessed on May 18, 2012).
2. Alzheimer's Disease International. *World Alzheimer Report 2009*. London, U.K.: Alzheimer's Disease International, 2009.
3. Luck T, Luppa M, Briel S et al. Mild cognitive impairment: Incidence and risk factors: Results of the leipzig longitudinal study of the aged. *J Am Geriatr Soc* 2010;58:1903–1910.
4. Luck T, Luppa M, Briel S, Riedel-Heller SG. Incidence of mild cognitive impairment: A systematic review. *Dement Geriatr Cogn Disord* 2010;29:164–175.
5. Sosa AL, Albanese E, Stephan BC et al. Prevalence, distribution, and impact of mild cognitive impairment in Latin America, China, and India: A 10/66 population-based study. *PLoS Med* 2012;9:e1001170.
6. Kuller LH, Lopez OL. Dementia and Alzheimer's disease: A new direction. The 2010 Jay L. Foster Memorial lecture. *Alzheimers Dement* 2011;7:540–550.
7. Fratiglioni L, De Ronchi D, Aguero-Torres H. Worldwide prevalence and incidence of dementia. *Drugs Aging* 1999;15:365–375.
8. Roman GC. Vascular dementia revisited: Diagnosis, pathogenesis, treatment, and prevention. *Med Clin North Am* 2002;86:477–499.
9. Manolio TA, Olson J, Longstreth WT. Hypertension and cognitive function: Pathophysiologic effects of hypertension on the brain. *Curr Hypertens Rep* 2003;5:255–261.
10. Okonkwo OC, Cohen RA, Gunstad J, Tremont G, Alosco ML, Poppas A. Longitudinal trajectories of cognitive decline among older adults with cardiovascular disease. *Cerebrovasc Dis* 2010;30:362–373.
11. Szeman B, Nagy G, Varga T et al. Changes in cognitive function in patients with diabetes mellitus. *Orv Hetil* 2012;153:323–329.
12. Hung WW, Wisnivesky JP, Siu AL, Ross JS. Cognitive decline among patients with chronic obstructive pulmonary disease. *Am J Respir Crit Care Med* 2009;180:134–137.
13. Kurella Tamura M, Xie D, Yaffe K et al. Vascular risk factors and cognitive impairment in chronic kidney disease: The Chronic Renal Insufficiency Cohort (CRIC) study. *Clin J Am Soc Nephrol* 2011;6:248–256.
14. Yaffe K, Ackerson L, Kurella Tamura M et al. Chronic kidney disease and cognitive function in older adults: Findings from the chronic renal insufficiency cohort cognitive study. *J Am Geriatr Soc* 2010;58:338–345.

15. Elwood PC, Gallacher JE, Hopkinson CA et al. Smoking, drinking, and other life style factors and cognitive function in men in the Caerphilly cohort. *J Epidemiol Community Health* 1999;53:9–14.
16. Szwast SJ, Hendrie HC, Lane KA et al. Association of statin use with cognitive decline in elderly African Americans. *Neurology* 2007;69:1873–1880.
17. van Gelder BM, Tijhuis MA, Kalmijn S, Giampaoli S, Nissinen A, Kromhout D. Physical activity in relation to cognitive decline in elderly men: The FINE Study. *Neurology* 2004;63:2316–2321.
18. Lautenschlager NT, Cox KL, Flicker L et al. Effect of physical activity on cognitive function in older adults at risk for Alzheimer disease. A randomized trial. *JAMA* 2008;300:1027–1037.
19. Perez L, Heim L, Sherzai A, Jaceldo-Siegl K. Nutrition and vascular dementia. *J Nutr Health Aging* 2012;16:319–324.
20. Morris MC. The role of nutrition in Alzheimer's disease: Epidemiological evidence. *Eur J Neurol* 2009;16(Suppl 1):1–7.
21. Reynish W, Andrieu S, Nourhashemi F, Vellas B. Nutritional factors and Alzheimer's disease. *J Gerontol A Biol Sci Med Sci* 2001;56:M675–M680.
22. Tayeb HO, Yang HD, Price BH, Tarazi FI. Pharmacotherapies for Alzheimer's disease: Beyond cholinesterase inhibitors. *Pharmacol Ther* 2012;134:8–25.
23. Kalra EK. Nutraceutical—definition and introduction. *AAPS Pharm Sci* 2003;5:E25.
24. McDougall GJ, Jr., Austin-Wells V, Zimmerman T. Utility of nutraceutical products marketed for cognitive and memory enhancement. *J Holist Nurs* 2005;23:415–433.
25. Okoro CA, Zhao G, Li C, Balluz LS. Use of complementary and alternative medicine among US adults with and without functional limitations. *Disabil Rehabil* 2012;34:128–135.
26. Nahin RL, Barnes PM, Stussman BJ, Bloom B. Costs of complementary and alternative medicine (CAM) and frequency of visits to CAM practitioners: United States, 2007. *Natl Health Stat Report* 2009:1–14.
27. Gonzalez-Stuart A. Herbal product use by older adults. *Maturitas* 2011;68:52–55.
28. Birks J. Cholinesterase inhibitors for Alzheimer's disease. *Cochrane Database Syst Rev* 2006:CD005593.
29. Howland RH. Drug therapies for cognitive impairment and dementia. *J Psychosoc Nurs Ment Health Serv* 2010;48:11–14.
30. Howard R, McShane R, Lindesay J et al. Donepezil and memantine for moderate-to-severe Alzheimer's disease. *N Engl J Med* 2012;366:893–903.
31. Kaduszkiewicz H, Zimmermann T, Beck-Bornholdt HP, van den Bussche H. Cholinesterase inhibitors for patients with Alzheimer's disease: Systematic review of randomised clinical trials. *BMJ* 2005;331:321–327.
32. Schneider LS. Could cholinesterase inhibitors be harmful over the long term? *Int Psychogeriatr* 2012;24:171–174.
33. Gravitz L. Drugs: A tangled web of targets. *Nature* 2011;475:S9–S11.
34. Hellweg R, Wirth Y, Janetzky W, Hartmann S. Efficacy of memantine in delaying clinical worsening in Alzheimer's disease (AD): Responder analyses of nine clinical trials with patients with moderate to severe AD. *Int J Geriatr Psychiatry* 2012;27:651–656.
35. Raina P, Santaguida P, Ismaila A et al. Effectiveness of cholinesterase inhibitors and memantine for treating dementia: Evidence review for a clinical practice guideline. *Ann Intern Med* 2008;148:379–397.
36. Gestuvo M, Hung W. Common dietary supplements for cognitive health. *Aging Health* 2012;8:89–97.
37. Tian J, Shi J, Zhang X, Wang Y. Herbal therapy: A new pathway for the treatment of Alzheimer's disease. *Alzheimers Res Ther* 2010;2:30.
38. Perry E, Howes MJ. Medicinal plants and dementia therapy: Herbal hopes for brain aging? *CNS Neurosci Ther* 2011;17:683–698.

39. Howes MJ, Houghton PJ. Plants used in Chinese and Indian traditional medicine for improvement of memory and cognitive function. *Pharmacol Biochem Behav* 2003;75:513–527.
40. Feitosa CM, Freitas RM, Luz NN, Bezerra MZ, Trevisan MT. Acetylcholinesterase inhibition by some promising Brazilian medicinal plants. *Braz J Biol* 2011;71:783–789.
41. Howes MJ, Perry NS, Houghton PJ. Plants with traditional uses and activities, relevant to the management of Alzheimer's disease and other cognitive disorders. *Phytother Res* 2003;17:1–18.
42. Kennedy DO, Scholey AB. Ginseng: Potential for the enhancement of cognitive performance and mood. *Pharmacol Biochem Behav* 2003;75:687–700.
43. Wee JJ, Mee Park K, Chung AS. Biological activities of Ginseng and its application to human health. In: Benzie IFF, Wachtel-Galor S, eds. *Herbal Medicine: Biomolecular and Clinical Aspects*. 2012/05/18, 2nd edn., Boca Raton, FL: CRC, 2011.
44. Cheng Y, Shen LH, Zhang JT. Anti-amnestic and anti-aging effects of ginsenoside Rg1 and Rb1 and its mechanism of action. *Acta Pharmacol Sin* 2005;26:143–149.
45. Geng J, Dong J, Ni H et al. Ginseng for cognition. *Cochrane Database Syst Rev* 2010:CD007769.
46. Kim JH, Cho SY, Lee JH et al. Neuroprotective effects of ginsenoside Rg3 against homocysteine-induced excitotoxicity in rat hippocampus. *Brain Res* 2007;1136:190–199.
47. Lee MS, Yang EJ, Kim JI, Ernst E. Ginseng for cognitive function in Alzheimer's disease: A systematic review. *J Alzheimers Dis* 2009;18:339–344.
48. Morris CA, Avorn J. Internet marketing of herbal products. *JAMA* 2003;290:1505–1509.
49. Schneider LS. Ginkgo biloba extract and preventing Alzheimer disease. *JAMA* 2008;300:2306–2308.
50. Aisen PS, Cummings J, Schneider LS. Symptomatic and nonamyloid/tau based pharmacologic treatment for Alzheimer disease. *Cold Spring Harb Perspect Med* 2012;2:a006395.
51. Eckert A, Keil U, Kressmann S et al. Effects of EGb 761 Ginkgo biloba extract on mitochondrial function and oxidative stress. *Pharmacopsychiatry* 2003;36 (Suppl 1):S15–23.
52. Wang BS, Wang H, Song YY et al. Effectiveness of standardized ginkgo biloba extract on cognitive symptoms of dementia with a six-month treatment: A bivariate random effect meta-analysis. *Pharmacopsychiatry* 2010;43:86–91.
53. Weinmann S, Roll S, Schwarzbach C, Vauth C, Willich SN. Effects of Ginkgo biloba in dementia: Systematic review and meta-analysis. *BMC Geriatr* 2010;10:14.
54. Snitz BE, O'Meara ES, Carlson MC et al. Ginkgo biloba for preventing cognitive decline in older adults: A randomized trial. *JAMA* 2009;302:2663–2670.
55. Tang XC, De Sarno P, Sugaya K, Giacobini E. Effect of huperzine A, a new cholinesterase inhibitor, on the central cholinergic system of the rat. *J Neurosci Res* 1989;24:276–285.
56. Ha GT, Wong RK, Zhang Y. Huperzine a as potential treatment of Alzheimer's disease: An assessment on chemistry, pharmacology, and clinical studies. *Chem Biodivers* 2011;8:1189–1204.
57. Li J, Wu HM, Zhou RL, Liu GJ, Dong BR. Huperzine A for Alzheimer's disease. *Cochrane Database Syst Rev* 2008:CD005592.
58. Wang BS, Wang H, Wei ZH, Song YY, Zhang L, Chen HZ. Efficacy and safety of natural acetylcholinesterase inhibitor huperzine A in the treatment of Alzheimer's disease: An updated meta-analysis. *J Neural Transm* 2009;116:457–465.
59. Rafii MS, Walsh S, Little JT et al. A phase II trial of huperzine A in mild to moderate Alzheimer disease. *Neurology* 2011;76:1389–1394.
60. Lin Z, Gu J, Xiu J, Mi T, Dong J, Tiwari JK. Traditional chinese medicine for senile dementia. *Evid Based Complement Alternat Med* 2012;2012:692621.
61. Park CH, Choi SH, Koo JW et al. Novel cognitive improving and neuroprotective activities of Polygala tenuifolia Willdenow extract, BT-11. *J Neurosci Res* 2002;70:484–492.

62. Xue W, Hu JF, Yuan YH et al. Polygalasaponin XXXII from Polygala tenuifolia root improves hippocampal-dependent learning and memory. *Acta Pharmacol Sin* 2009;30:1211–1219.

63. Lee JY, Kim KY, Shin KY, Won BY, Jung HY, Suh YH. Effects of BT-11 on memory in healthy humans. *Neurosci Lett* 2009;454:111–114.

64. Shin KY, Lee JY, Won BY et al. BT-11 is effective for enhancing cognitive functions in the elderly humans. *Neurosci Lett* 2009;465:157–159.

65. Savelev SU, Okello EJ, Perry EK. Butyryl- and acetyl-cholinesterase inhibitory activities in essential oils of Salvia species and their constituents. *Phytother Res* 2004;18:315–324.

66. Perry NS, Houghton PJ, Jenner P, Keith A, Perry EK. *Salvia lavandulaefolia* essential oil inhibits cholinesterase in vivo. *Phytomedicine* 2002;9:48–51.

67. Savelev S, Okello E, Perry NS, Wilkins RM, Perry EK. Synergistic and antagonistic interactions of anticholinesterase terpenoids in *Salvia lavandulaefolia* essential oil. *Pharmacol Biochem Behav* 2003;75:661–668.

68. Tildesley NT, Kennedy DO, Perry EK, Ballard CG, Wesnes KA, Scholey AB. Positive modulation of mood and cognitive performance following administration of acute doses of *Salvia lavandulaefolia* essential oil to healthy young volunteers. *Physiol Behav* 2005;83:699–709.

69. Scholey AB, Tildesley NT, Ballard CG et al. An extract of Salvia (sage) with anticholinesterase properties improves memory and attention in healthy older volunteers. *Psychopharmacology (Berl)* 2008;198:127–139.

70. Akhondzadeh S, Noroozian M, Mohammadi M, Ohadinia S, Jamshidi AH, Khani M. Salvia officinalis extract in the treatment of patients with mild to moderate Alzheimer's disease: A double blind, randomized and placebo-controlled trial. *J Clin Pharm Ther* 2003;28:53–59.

71. Ho YS, So KF, Chang RC. Drug discovery from Chinese medicine against neurodegeneration in Alzheimer's and vascular dementia. *Chin Med* 2011;6:15.

72. Ho YS, So KF, Chang RC. Anti-aging herbal medicine—how and why can they be used in aging-associated neurodegenerative diseases? *Ageing Res Rev* 2010;9:354–362.

73. Monji A, Takita M, Samejima T et al. Effect of yokukansan on the behavioral and psychological symptoms of dementia in elderly patients with Alzheimer's disease. *Prog Neuropsychopharmacol Biol Psychiatry* 2009;33:308–311.

74. Iwasaki K, Satoh-Nakagawa T, Maruyama M et al. A randomized, observer-blind, controlled trial of the traditional Chinese medicine Yi-Gan San for improvement of behavioral and psychological symptoms and activities of daily living in dementia patients. *J Clin Psychiatr* 2005;66:248–252.

75. Nagata K, Yokoyama E, Yamazaki T et al. Effects of yokukansan on behavioral and psychological symptoms of vascular dementia: An open-label trial. *Phytomedicine* 2012;19:524–528.

76. Park E, Kang M, Oh JW et al. Yukmijihwang-tang derivatives enhance cognitive processing in normal young adults: A double-blinded, placebo-controlled trial. *Am J Chin Med* 2005;33:107–115.

77. Iwasaki K, Kobayashi S, Chimura Y et al. A randomized, double-blind, placebo-controlled clinical trial of the Chinese herbal medicine "ba wei di huang wan" in the treatment of dementia. *J Am Geriatr Soc* 2004;52:1518–1521.

78. Berr C, Balansard B, Arnaud J, Roussel AM, Alperovitch A. Cognitive decline is associated with systemic oxidative stress: The EVA study. Etude du Vieillissement Arteriel. *J Am Geriatr Soc* 2000;48:1285–1291.

79. Droge W, Schipper HM. Oxidative stress and aberrant signaling in aging and cognitive decline. *Aging Cell* 2007;6:361–370.

80. Eckert A, Keil U, Marques CA et al. Mitochondrial dysfunction, apoptotic cell death, and Alzheimer's disease. *Biochem Pharmacol* 2003;66:1627–1634.

81. Paleologos M, Cumming RG, Lazarus R. Cohort study of vitamin C intake and cognitive impairment. *Am J Epidemiol* 1998;148:45–50.

82. Lee L, Kang SA, Lee HO et al. Relationships between dietary intake and cognitive function level in Korean elderly people. *Public Health* 2001;115:133–138.

83. Peneau S, Galan P, Jeandel C et al. Fruit and vegetable intake and cognitive function in the SU.VI.MAX 2 prospective study. *Am J Clin Nutr* 2011;94:1295–1303.

84. La Rue A, Koehler KM, Wayne SJ, Chiulli SJ, Haaland KY, Garry PJ. Nutritional status and cognitive functioning in a normally aging sample: A 6-y reassessment. *Am J Clin Nutr* 1997;65:20–29.

85. Morris MC, Evans DA, Bienias JL, Tangney CC, Wilson RS. Vitamin E and cognitive decline in older persons. *Arch Neurol* 2002;59:1125–1132.

86. Aparicio Vizuete A, Robles F, Rodriguez-Rodriguez E, Lopez-Sobaler AM, Ortega RM. Association between food and nutrient intakes and cognitive capacity in a group of institutionalized elderly people. *Eur J Nutr* 2010;49:293–300.

87. Kang JH, Cook NR, Manson JE, Buring JE, Albert CM, Grodstein F. Vitamin E, vitamin C, beta carotene, and cognitive function among women with or at risk of cardiovascular disease: The Women's Antioxidant and Cardiovascular Study. *Circulation.* 2009;119:2772–2780.

88. Harrison FE. A critical review of vitamin C for the prevention of age-related cognitive decline and Alzheimer's disease. *J Alzheimers Dis* 2012;29:711–726.

89. Berr C. Cognitive impairment and oxidative stress in the elderly: Results of epidemiological studies. *Biofactors* 2000;13:205–209.

90. Berr C. Oxidative stress and cognitive impairment in the elderly. *J Nutr Health Aging* 2002;6:261–266.

91. Martin A, Youdim K, Szprengiel A, Shukitt-Hale B, Joseph J. Roles of vitamins E and C on neurodegenerative diseases and cognitive performance. *Nutr Rev* 2002;60:308–326.

92. Evatt ML, Delong MR, Khazai N, Rosen A, Triche S, Tangpricha V. Prevalence of vitamin d insufficiency in patients with Parkinson disease and Alzheimer disease. *Arch Neurol* 2008;65:1348–1352.

93. Buell JS, Dawson-Hughes B, Scott TM et al. 25-Hydroxyvitamin D, dementia, and cerebrovascular pathology in elders receiving home services. *Neurology* 2010;74:18–26.

94. Buell JS, Scott TM, Dawson-Hughes B et al. Vitamin D is associated with cognitive function in elders receiving home health services. *J Gerontol A Biol Sci Med Sci* 2009;64:888–895.

95. Annweiler C, Rolland Y, Schott AM et al. Higher vitamin D dietary intake is associated with lower risk of Alzheimer's disease: A 7-year follow-up. *J Gerontol A Biol Sci Med Sci* 2012;67(11):1205–1211.

96. Prufer K, Veenstra TD, Jirikowski GF, Kumar R. Distribution of 1,25-dihydroxyvitamin D3 receptor immunoreactivity in the rat brain and spinal cord. *J Chem Neuroanat* 1999;16:135–145.

97. Neveu I, Naveilhan P, Menaa C, Wion D, Brachet P, Garabedian M. Synthesis of 1,25-dihydroxyvitamin D3 by rat brain macrophages in vitro. *J Neurosci Res* 1994;38:214–220.

98. Wang Y, Chiang YH, Su TP et al. Vitamin D(3) attenuates cortical infarction induced by middle cerebral arterial ligation in rats. *Neuropharmacology* 2000;39:873–880.

99. Heaney RP. Vitamin D and calcium interactions: Functional outcomes. *Am J Clin Nutr* 2008;88:541S–544S.

100. Berlin T, Bjorkhem I. Effect of calcium intake on serum levels of 25-hydroxyvitamin D3. *Eur J Clin Invest* 1988;18:52–55.

101. Goussous R, Song L, Dallal GE, Dawson-Hughes B. Lack of effect of calcium intake on the 25-hydroxyvitamin d response to oral vitamin D3. *J Clin Endocrinol Metab* 2005;90:707–711.

102. Foster TC. Calcium homeostasis and modulation of synaptic plasticity in the aged brain. *Aging Cell* 2007;6:319–325.

103. Velho S, Marques-Vidal P, Baptista F, Camilo ME. Dietary intake adequacy and cognitive function in free-living active elderly: A cross-sectional and short-term prospective study. *Clin Nutr* 2008;27:77–86.

104. Yokoyama M, Origasa H, Matsuzaki M et al. Effects of eicosapentaenoic acid on major coronary events in hypercholesterolaemic patients (JELIS): A randomised open-label, blinded endpoint analysis. *Lancet* 2007;369:1090–1098.

105. Tavazzi L, Maggioni AP, Marchioli R et al. Effect of n-3 polyunsaturated fatty acids in patients with chronic heart failure (the GISSI-HF trial): A randomised, double-blind, placebo-controlled trial. *Lancet* 2008;372:1223–1230.

106. Burr ML, Fehily AM, Gilbert JF et al. Effects of changes in fat, fish, and fibre intakes on death and myocardial reinfarction: Diet and reinfarction trial (DART). *Lancet* 1989;2:757–761.

107. Marchioli R, Barzi F, Bomba E et al. Early protection against sudden death by n-3 polyunsaturated fatty acids after myocardial infarction: Time-course analysis of the results of the Gruppo Italiano per lo Studio della Sopravvivenza nell'Infarto Miocardico (GISSI)-Prevenzione. *Circulation* 2002;105:1897–1903.

108. Yurko-Mauro K, McCarthy D, Rom D et al. Beneficial effects of docosahexaenoic acid on cognition in age-related cognitive decline. *Alzheimers Dement* 2010;6:456–464.

109. Freund-Levi Y, Eriksdotter-Jonhagen M, Cederholm T et al. Omega-3 fatty acid treatment in 174 patients with mild to moderate Alzheimer disease: OmegAD study: A randomized double-blind trial. *Arch Neurol* 2006;63:1402–1408.

110. Geleijnse JM, Giltay EJ, Kromhout D. Effects of n-3 fatty acids on cognitive decline: a randomized, double-blind, placebo-controlled trial in stable myocardial infarction patients. *Alzheimers Dement* 2012;8:278–287.

111. Samieri C, Feart C, Proust-Lima C et al. Omega-3 fatty acids and cognitive decline: Modulation by ApoEepsilon4 allele and depression. *Neurobiol Aging* 2011;32:2317 e2313–2322.

112. Barberger-Gateau P, Samieri C, Feart C, Plourde M. Dietary omega 3 polyunsaturated fatty acids and Alzheimer's disease: Interaction with apolipoprotein E genotype. *Curr Alzheimer Res* 2011;8:479–491.

113. Kennedy ET, Luo H, Ausman LM. Cost implications of alternative sources of (n-3) fatty acid consumption in the United States. *J Nutr* 2012;142:605S–609S.

114. Deckelbaum RJ, Calder PC, Harris WS et al. Conclusions and recommendations from the symposium, Heart Healthy Omega-3s for Food: Stearidonic Acid (SDA) as a Sustainable Choice. *J Nutr* 2012;142:641S–643S.

115. Harris WS, Lemke SL, Hansen SN et al. Stearidonic acid-enriched soybean oil increased the omega-3 index, an emerging cardiovascular risk marker. *Lipids* 2008;43:805–811.

116. James MJ, Ursin VM, Cleland LG. Metabolism of stearidonic acid in human subjects: Comparison with the metabolism of other n-3 fatty acids. *Am J Clin Nutr* 2003;77:1140–1145.

117. Lemke SL, Vicini JL, Su H et al. Dietary intake of stearidonic acid-enriched soybean oil increases the omega-3 index: Randomized, double-blind clinical study of efficacy and safety. *Am J Clin Nutr* 2010;92:766–775.

118. Willett WC, Sacks F, Trichopoulou A et al. Mediterranean diet pyramid: A cultural model for healthy eating. *Am J Clin Nutr* 1995;61:1402S–1406S.

119. Feart C, Samieri C, Rondeau V et al. Adherence to a Mediterranean diet, cognitive decline, and risk of dementia. *JAMA* 2009;302:638–648.

120. Feart C, Torres MJ, Samieri C et al. Adherence to a Mediterranean diet and plasma fatty acids: Data from the Bordeaux sample of the Three-City study. *Br J Nutr* 2011;106:149–158.

121. Tangney CC, Kwasny MJ, Li H, Wilson RS, Evans DA, Morris MC. Adherence to a Mediterranean-type dietary pattern and cognitive decline in a community population. *Am J Clin Nutr* 2011;93:601–607.

122. Scarmeas N, Stern Y, Tang MX, Mayeux R, Luchsinger JA. Mediterranean diet and risk for Alzheimer's disease. *Ann Neurol* 2006;59:912–921.

123. Scarmeas N, Luchsinger JA, Schupf N et al. Physical activity, diet, and risk of Alzheimer disease. *JAMA* 2009;302:627–637.

124. Scarmeas N, Stern Y, Mayeux R, Manly JJ, Schupf N, Luchsinger JA. Mediterranean diet and mild cognitive impairment. *Arch Neurol* 2009;66:216–225.
125. Frisardi V, Panza F, Seripa D et al. Nutraceutical properties of Mediterranean diet and cognitive decline: Possible underlying mechanisms. *J Alzheimers Dis* 2010;22:715–740.
126. Solfrizzi V, Panza F, Frisardi V et al. Diet and Alzheimer's disease risk factors or prevention: The current evidence. *Expert Rev Neurother* 2011;11:677–708.
127. Kesse-Guyot E, Andreeva VA, Jeandel C, Ferry M, Hercberg S, Galan P. A healthy dietary pattern at midlife is associated with subsequent cognitive performance. *J Nutr* 2012;143:909–915.
128. Man SC, Chan KW, Lu JH, Durairajan SS, Liu LF, Li M. Systematic review on the efficacy and safety of herbal medicines for vascular dementia. *Evid Based Complement Alternat Med* 2012;2012:426215.
129. Xian YF, Lin ZX, Zhao M, Mao QQ, Ip SP, Che CT. *Uncaria rhynchophylla* ameliorates cognitive deficits induced by D-galactose in mice. *Planta Med* 2011;77:1977–1983.
130. Yang ZD, Duan DZ, Du J, Yang MJ, Li S, Yao XJ. Geissoschizine methyl ether, a corynanthean-type indole alkaloid from U*ncaria rhynchophylla* as a potential acetyl-cholinesterase inhibitor. *Nat Prod Res* 2012;26:22–28.
131. Fujiwara H, Iwasaki K, Furukawa K et al. *Uncaria rhynchophylla*, a Chinese medicinal herb, has potent antiaggregation effects on Alzheimer's beta-amyloid proteins. *J Neurosci Res* 2006;84:427–433.
132. Berr C, Arnaud J, Akbaraly TN. Selenium and cognitive impairment: A brief-review based on results from the EVA study. *Biofactors* 2012;38:139–144.
133. Fiocco AJ, Shatenstein B, Ferland G et al. Sodium intake and physical activity impact cognitive maintenance in older adults: The NuAge Study. *Neurobiol Aging* 2012;33:829 e821–828.
134. Song J, Xu H, Liu F, Feng L. Tea and cognitive health in late life: Current evidence and future directions. *J Nutr Health Aging* 2012;16:31–34.
135. Williams JW, Plassman BL, Burke J, Benjamin S. Preventing Alzheimer's disease and cognitive decline. *Evid Rep Technol Assess (Full Rep)* 2010:1–727.
136. Kerwin DR, Gaussoin SA, Chlebowski RT et al. Interaction between body mass index and central adiposity and risk of incident cognitive impairment and dementia: Results from the Women's Health Initiative Memory Study. *J Am Geriatr Soc* 2011;59:107–112.
137. Weuve J, Kang JH, Manson JE, Breteler MM, Ware JH, Grodstein F. Physical activity, including walking, and cognitive function in older women. *JAMA* 2004;292:1454–1461.
138. Fan TP, Deal G, Koo HL et al. Future development of global regulations of Chinese herbal products. *J Ethnopharmacol* 2012;140:568–586.
139. Bowman GL, Silbert LC, Howieson D et al. Nutrient biomarker patterns, cognitive function, and MRI measures of brain aging. *Neurology* 2012;78:241–249.
140. Lee ST, Chu K, Sim JY, Heo JH, Kim M. Panax ginseng enhances cognitive performance in Alzheimer disease. *Alzheimer Dis Assoc Disord* 2008;22:222–226.

# 16

# Nutraceuticals in the Prevention or Treatment of Parkinson's Disease

Elizabeth Mazzio, Fran T. Close, and Karam F.A. Soliman

## Contents

# Definition of relevant or relative evidence

Parkinson's disease (PD) is a progressive neurodegenerative disease associated with loss of dopaminergic (DAergic) neurons in the substantia nigra (SN) resulting in neuromotor, autonomic and balance anomalies. Ultimately, PD is associated with the natural process of aging but can be mimicked by diverse precipitating environmental influences including exposure to/ingestion of natural or synthetic neurotoxins 1-methyl-4-phenyl-1,2,3,6-tetrahydropyridine (MPTP),[1] pesticides, herbicides, fungicides (e.g., ziram, maneb, paraquat),[2] beta-N-methylamino-L-alanine (BMAA), acetogenins, benzylisoquinolines,[3] trichloroethylene,[4] or secondary Parkinson's/mimetic agents (e.g., vascular, viral, drug induced, or manganese).[5] The pathological manifestations involved with loss of SN DAergic neurons are equally complex, involving genetic mutations,[6] mitochondrial abnormalities,[7,8] Lewy bodies/misfolded proteins,[9] impaired autolysosome function, loss of neuromelanin,[10,11] or CNS inflammation.[12] Research providing causal vs. pathological outcome indicators of PD while extensive can elucidate the nature and proclivity of this disease ultimately leading to new areas of therapeutic application.

## Evidence

In reviewing therapeutic *evidence* for practical application of nutraceuticals to treat or prevent PD, it is imperative to define what would be considered sufficient relevant evidence. Scientific evidence should enable growth of knowledge leading to greater understanding as to disease causality, associated risk factors, effective prevention, or treatment regimes. However, evidence can be valid and significant but relative and therefore not necessarily *relevant*. In the case of effective therapeutic nutraceuticals or drugs, which prove to be neuroprotective in experimental animal/mammalian PD models (e.g., MPTP, 1-methyl-4-phenylpyridinium (MPP+), iron overload, 6-OHDA, rotenone, paraquat, or overexpression of disease-related genes), results may or may not be either *reproducible or suitable for extrapolation* to human pathology. We must keep in mind that the ultimate goal of research in the biomedical sciences pertaining to PD is to find an effective means to treat or prevent this disease, for the purpose of ameliorating suffering in *human* PD patients.

Therefore, one of the most important avenues in the search for knowledge, clues, or potential treatments would include a thorough examination of existing evidence on human populations through epidemiological studies, clinical human trials, case studies, or historical documents. Accurate data extrapolation from human studies to derive interpretation for future therapeutics requires elucidation of true relationships between nutraceuticals and PD. This review is specifically focused on human evidence from past to present, which in some instances is further supported by experimental models of PD but not vice versa.

# Historical account of human PD: Nutraceutical therapeutics

Historical documents contain a plethora of wisdom about the use of therapeutic plants as effective treatments for a range of human maladies. The first reported documentation of human PD was described thousands of years ago in the ancient texts of Indian Ayurvedic medicine referred to as *vata dosha* (nervous system out of balance), where a loss of equilibrium precipitated *kampa* (tremors), *vepathu* (shaking, out of alignment), *prevepana* (excessive shaking), and *sirakampa* (head tremor), to where *vata (vyana)* entering *(mamsa dhatu)* was responsible for muscle rigidity and *prana kshaya* (diminished prana) in the *manovaha srota* causing depression.[13] Since then, documents throughout recorded history in diverse parts of the world substantiate a chronological and ever present account of the PD throughout many human generations with described symptomatic manifestations ranging from tremor, slowness of movement, bradykinesia, drooling, reptilian stare, rigidity (no inclination to move), dementia, and salivary drooling to a masked facial expression/fixed position of the eyes.[14] Some of these include various names for the affliction by William Shakespeare in Henry VI (referring to tremors as originating from palsy, not fear), Galen, Leonardo da Vinci, de la Boe Sylvius and van Swieten, Martin Charcot, François Boissier de Sauvages de la Croix ("scelotyrbe festinans"), and ultimately James Parkinson to whom the disease was attributed based on his descriptive review of *paralysis agitans* "Essay on the Shaking Palsy" (1817).[15-17] The chronological recordation of this disease suggests that PD is neither the result of modern civilization nor the result of industrial pollutants, synthetic herbicides/pesticides, or man-made chemicals to which elevated risk is associated. Rather, this is a condition, which has been with us since the dawn of human civilization.

If we look to the knowledge of our ancestors for effective nutraceuticals to treat PD, we also find that natural nervous system tonics were a common practice in Ayurvedic medicine (*rasayanas*),[13] where the first reported treatment for PD was the tropical legume *Mucuna pruriens* (velvet bean) called *Atmagupta*. Despite the passage of several millennia and technological advances, the main treatment for PD is nearly identical today. In the 1960s, the pharmaceutical industry developed what would become a primary palliative treatment for PD, L-3,4-dihydroxyphenylalanine (L-dopa), based on the findings of Carlsson, Birkmayer and Hornykiewicz, Barbeau, and Cotzias.[17,18] In modern pharmaceutics, expansion on that knowledge leads to combining L-dopa with an inhibitor of aromatic amino acid decarboxylation, to enhance its delivery to the brain—sold under the trade names Sinemet® (carbidopa–levodopa) or Madopar® (benserazide–levodopa). Yet, in ancient Ayurvedic medicine, PD disease was referred to as *kampa vata*, *kampa* (tremor) + *vata* (neurological function),[14] and treated effectively with what could be the world's most concentrated natural source of L-dopa, Mucuna pruriens[19,20] (shown in Figure 16.1), which contains up to 8% L-dopa by dry weight (Table 16.1).[21,22]

*Figure 16.1* Mucuna pruriens *(L.) DC. velvet bean. (Image courtesy of USDA-NRCS PLANTS Database, Tracey Slotta, ARS Systematic Botany and Mycology Laboratory.)*

**Table 16.1** Mucuna Germplasm with Geographical Areas of Collection and L-Dopa Content (%) in Seeds

| Accession No. | Species | Area of Collection | % L-DOPA |
|---|---|---|---|
| IC15809A | *M. pruriens* | Palaman/Bihar | 3.62 |
| IC83195 | *M. prurita* | Narmada/Gujarat | 5.36 |
| IC21991 | *M. utilis* | Madhya Pradesh | 3.31 |
| EC4475 | *M. pruriens* | Australia | 2.79 |
| IC202969 | *M. pruriens* | NBPGR, HQ | 2.23 |
| IC21992 | *M. utilis* | Madhya Pradesh | 2.86 |
| IC385926 | *M. prurita* | Jharkhand | 2.82 |
| EC362772 | *M. pruriens* | United States | 2.63 |
| IC43993 | *M. pruriens* | Kerala | 3.45 |
| IC9598 | *M. pruriens* | — | 3.34 |
| IC83298 | *M. pruriens* | NBPGR, HQ | 3.27 |
| IC127363 | *M. pruriens* | NBPGR, HQ | 2.86 |
| IC25333 | *M. pruriens* | Mizoram | 3.13 |
| IC24839 | *M. pruriens* | NBPGR, HQ | 3.53 |
| EC144945 | *M. pruriens* | Italy | 4.04 |
| Mean L-dopa (%) | | | 3.28 |

*Source:* Raina, A.P. and Khatri, R., *Indian J. Pharmaceut. Sci.*, 73, 459, 2011.
L-dopa is expressed as a mean of duplicate analysis.
L-dopa quantified using densitometric high-performance thin-layer chromatographic methods in Mucuna species.

A number of other natural products contain high levels of L-dopa (Table 16.2) such as broad bean (*Vicia faba*)[23] that can also effectively increase plasma levels and improve motor function.[21,22]

While our ancestors accurately described effective nutraceuticals without the technology to understand etiology, today, technology leads extensive understanding of pathology, to which we formulate drugs based on therapeutic targets or mechanism of action. If we compare the efficacy of ancient nutraceuticals (*Mucuna pruriens*) to modern drugs (L-dopa/carbidopa), the effects are strikingly similar, and in some cases, advantages are attributed to nutraceuticals including shorter latency to clinical onset and increased duration of on-time and total latency time from drug ingestion to full on-state.[24] Moreover, in PD animal models (6-OHDA lesions), administration of *Mucuna pruriens* was found to restore SN DA levels.[25]

Other similarities in treatment approach between modern and ancient medicinal practices are the strategic combination of synergistic drugs. Today, medical treatment of PD includes mono- or adjunct therapies that serve to extend therapeutic longevity of L-dopa[26] such as DA receptor agonists (apomorphine, ropinirole, pramipexole), monoamine oxidase inhibitors (selegiline, rasagiline), catechol-O-methyltransferase inhibitors,[27] or anticholinergics[28] (benztropine, biperiden, procyclidine trihexyphenidyl). In Ayurvedic medicine, adjunctive nutraceutical applications used to treat *kampa vata* were referred to as *"medhya rasayana"* or neuronutrients from the plants *ashwagandha*, *Bacopa monnieri*, *Centella asiatica*, and *Sida cordifolia*.[13] These adjunctive natural medicines were believed to be involved with cleansing, but in humans, PD patients are reportedly capable of enhancing the efficacy of L-dopa evidenced by improvement in activities of daily living (ADL) and Unified Parkinson Disease Rating Scale (UPDRS) scores.[29]

Other palliative Ayurvedic regimens for PD-related symptoms have included natural products that aid in constipation, rigidity, or depression such as psyllium, flaxseed, triphala, slippery elm, licorice, jatamansi, gotu kola, shankhpushpi, Saint John's wort, and Gaducci. A number of these Ayurvedic regimens have shown restorative and protective effects in particular on the DAergic system including neuroprotection against PD model toxins,[30,31] reserpine-induced dopamine depletion, orofacial dyskinesia,[32] or inhibition of monoamine oxidase.[33,34]

While historical documents provide some information on palliative treatments to ameliorate symptoms after disease onset, a valuable means to derive information on potential new preventative or therapeutic treatments for PD, examination of epidemiological global disease patterns evidenced by risk assessment must be evaluated.

**Table 16.2** L-Dopa-Bearing Plants

| Plant | Plant Part | L-DOPA (%) |
|-------|-----------|-----------|
| *Alysicarpus rugosus* (Willd.) DC. | Seed | 0.65 |
| *Bauhinia purpurea* Linn. | Seed | 2.2 |
| *Bauhinia racemosa* Lam. | Seed | 0.73 |
| *Canavalia ensiformis* (Linn.) DC. | Seed | 2.46 |
| *Canavalia gladiata* (Jacq.) DC. | Seed | 2.13 |
| *Cassia floribunda* Cav. | Seed | 1.1–1.9 |
| *Cassia hirsute* Linn. | Seed | 2.37–2.82 |
| *Dalbergia retusa* Hemsl. | Seed | 2.2 |
| *Glycine wightii* (W. & A.) Verdc. | Seed | 0.2 |
| *Mucuna andreana* Micheli | Seed (seed coat) | 6.3–8.9 |
| *Mucuna aterrima* (Piper & Tracy) Holland | Seed | 3.31 |
| *Mucuna aterrima* (Piper & Tracy) Holland | Seed (black) | 4.2 |
| *Mucuna birdwoodina* Tutcher | Seed | 9.1 |
| *Mucuna cochinchinensis* (Lour.) A. Chev. | Seed (ash) | 4.2 |
| *Mucuna cochinchinensis* (Lour.) A. Chev. | Pericarp (ash) | 0.14 |
|  | Leaf (ash) | 0.18 |
|  | Stem (ash) | 0.28 |
|  | Root (ash) | 0.14 |
| *Mucuna cochinchinensis* (Lour.) A. Chev. | Seed (gray) | 2.5 |
| *Mucuna cochinchinensis* (Lour.) A. Chev. | Seed | 3–4 |
| *Mucuna deeringiana* (Bort.) Merr. | Seed | 2.7–3.13 |
| *Mucuna gigantea* (Willd.) DC. | Seed | 1.50–3.78 |
| *Mucuna holtonii* (Kuntze) Mold. | Seed | 6.13–7.5 |
| *Mucuna monosperma* DC. ex Wight | Seed | 4.24–4.56 |
| *Mucuna mutisiana* (Kunth.) DC. | Seed | 3.9–6.8 |
| *Mucuna pruriens* (Linn.) DC. | Seed (seed coat) | 5.9–6.4 |
| *Mucuna pruriens* (Linn.) DC. | Seed | 1.25–9.16 |
| *Mucuna pruriens* (Linn.) DC. | Seed (black) | 3.8 |
|  | Pericarp | 0.09–0.22 |
|  | Leaf | 0.35 |
|  | Stem | 0.31 |
|  | Root | 0.16 |
| *Mucuna pruriens* (Linn.) DC. | Whole bean | 4.02 |
|  | Endocarp | 5.28 |
| *Mucuna pruriens f. hirsute* | Seed | 1.4–1.5 |
| *Mucuna pruriens f. utilis* | Seed | 1.8 |

**Table 16.2 (continued)** L-Dopa-Bearing Plants

| Plant | Plant Part | L-DOPA (%) |
|---|---|---|
| Mucuna pruriens var. utilis (Wall. ex Wight) Baker ex Burck | White (whole seed) | 4.96 |
| | Black (whole seed) | 4.1–6.86 |
| | White (dehulled seed) | 5.21 |
| | Black (dehulled seed) | 4.66 |
| Mucuna pruriens var. utilis (Wall. ex Wight) Baker ex Burck | Seed | 8.05 |
| Mucuna pruriens var. utilis (Wall. ex Wight) Baker ex Burck | Seed (white) | 6.08 |
| Mucuna pruriens var. utilis (Wall. ex Wight) Baker ex Burck | Seed (spotted) | 3.6 |
| | Pericarp | 0.16 |
| | Leaf | 0.17 |
| | Stem | 0.19 |
| | Root | 0.12 |
| Mucuna sloanei Fawcett & Rendle | Seed | 3.34–9.0 |
| Mucuna sp. | Seed | 1.96–4.96 |
| Mucuna urens (Linn.) Medik. | Seed | 4.92–7.4 |
| Parkinsonia aculeata Linn. | Seed | 0.64 |
| Phanera vahlii Benth. | Seed | 2.35 |
| Pileostigma malabarica Benth. | Seed | 2.13 |
| Prosopis chilensis Stuntz. | Seed | 1.25 |
| Teramnus labialis (Linn.) Spreng | Seed | Trace |
| Vicia faba Linn. | Green peel of pod | 0.2–0.75 |
| | Flowering green plant | 0.09 |
| | Green seeds | 0.006–0.01 |
| | Dry seeds | Trace |
| Vicia faba var minor | Dry seeds | 0.07 |
| | Green pods (whole unripe fruit) | 0.6 |
| | Green plant with pods | 0.56 |
| | Green flowering plant | 0.40–0.46 |
| | Green vegetative plant | 0.24–0.57 |
| Vicia narbonensis Linn. | Green pods (peel only) | 0.5 |
| | Green plant with pods | 0.6 |
| Vigna aconitifolia (Jacq.) Marechal. | Seed | 0.2 |
| Vigna unguiculata (Linn.) Walp. | Seed | 0.45 |
| Vigna vexillata (Benh.) A. Rich. | Seed | 0.52–0.58 |

*Source:* Modified from Ingle, P. K., *Nat Prod Rad.*, 2, 126, 2003.

## Palliative treatment vs. prevention/reduced risk: Epidemiological research

Epidemiological data sets contain information on patterns of human disease prevalence, which can be analyzed in lieu of diverse correlates including dietary patterns, cultural practices, lifestyle, socioeconomic status, gender, race, or any recorded indicator. The evidence provided can be analyzed across cultural practices in a heterogeneous location or homogeneous demographics with diverse cultural practices. Determination of factors that alter incidence of PD, be it dietary or others, could be useful to advance knowledge leading to future strategies for disease prevention or effective treatment.

### Strengths and weakness of available data

Worldwide estimates reported for prevalence of PD show extreme ranges from ~30 to ~500/100,000 and >2000/100,000 in at-risk populations, which we will discuss in greater detail in this chapter. It is important to understand that the strength of epidemiological evidence is dependent not only on its statistical significance and odds ratios, but also on its consistency of reported effects among diverse designs, corroborated by measure of integrated parallel relationships (such as smoking reduces risk of PD, Amish community does not smoke, Amish show elevated PD prevalence) and substantiated by pharmaceutical or biomedical disease models. Moreover, a major weakness of epidemiological research is its confounding variables, which could lead to incorrect estimates or misinterpretation of data. Some of these include (1) variation in availability and accuracy of documented recordation; (2) variation in population age distributions or life expectancy (in particular when studying diseases most common in the elderly such as PD); (3) incorrect disease diagnosis; (4) variation in sample size; (5) variation in source of data compilation (tracking medications, medical history, door-to-door surveys, medical questionnaires); and (6) subjective biases based on neurological examinations performed or variation in clinical parameter testing such as the use of the UPDRS, Hoehn and Yahr scale, and mini-mental state exams.

In terms of the study of diet, a major confounding variable could be in the manner in which dietary records are collected. A number of interfering variables include (1) variation in surveying techniques such as the use of a food frequency questionnaire, dietary recall, documented dietary recording, diet histories, or 24 h food recalls; (2) biases *on behalf of the study participant* including motivation, literacy, and psychological indices such as memory, cognition, comprehension, illnesses, dementia, or education; and (3) biases *on behalf of the surveyor* including variation and appropriateness of test method; survey procedure (mail, door-to-door, telephone, Internet); proper nutrient accounting for diverse foods, brands, and preparations; combination of foods consumed; cultural cuisine; correct identification of different foods with the same name or same foods with a different name; proper estimation of portion size; choice of nutrient/food database; and/or analysis and measurement to determine nutrient composition of foods.[35] All of these factors make it difficult to substantiate conclusive evidence from global human population studies; however, a number of consistently reported associations with PD risk are reported, despite possibility of error, which are further discussed.

## Global PD patterns

Global patterns of PD are diverse, and at present, little can be concluded from these results due to a lack of worldwide standardized study conditions. Take, for example, one of the lowest PD prevalence rates in the world reported for Tanzania, Hai district. A door-to-door study was conducted in Hai (n = 161,071), using a screening questionnaire—followed by a structured history and examination of positive responders to which 33 cases of PD were detected. The crude prevalence of PD was estimated ~20/100,000, one of the lowest worldwide. Interestingly, of the PD cases identified, 93% were over age of 64, and 78% were previously undiagnosed and untreated, which means that PD patients may have a reduction in life expectancy due to lack of treatment, which could overinflate low prevalence estimates.[36] Moreover, most Hai individuals and health workers within this population never heard of the term "Parkinson's disease" but rather held a belief that symptoms of shaking and tremors were a reflection of being under a spell of witchcraft, poisoning, or being cold. Individuals diagnosed with PD, therefore, were not being treated medically.[37] The Hai district study shows that a number of factors could enter into prevalence estimates to the inclusion of (1) misconceptions about the disease, (2) lack of education or financial means to pursue treatment, and (3) inability to obtain health care—all of which could lead to greater mortality rates in afflicted individuals. On the other hand, low PD rates could be attributable to something in the Hai district environment or diet (e.g., bananas, beans, coffee beans, cassava, maize, and sunflower), but future research will be required to delineate a factorial relationship.

In general, worldwide prevalence rates in PD are often reported as crude or age adjusted. Variation in age frequency distribution of the sample population and life expectancy could contribute to over- or underreporting. The following are reported PD prevalence rates around the world.

| Region | Prevalence | Type | Reference |
|---|---|---|---|
| Antioquia, Colombia | 20/100,000 | Crude prev | [38] |
| Hai district, Tanzania | 39/100,000 | Crude prev | [37,36] |
| Benghazi northeastern Libya | 31/100,000 | Crude prev | [39] |
| Kolkata, India | 53/100,000 | Crude prev | [40] |
| Havana City province, Cuba | 135/100,000 | Age: >15 | [41] |
| Australia | 137/100,000 | DDD | [42] |

| Germany | 144/100,000 | Crude prev | [43] |
|---|---|---|---|
| Antioquia, Colombia | 176/100,000 | Age: >50 | [38] |
| Spain | 186/100,000 | Age: >65–85 | [44] |
| Cordillera Province, Bolivia | 286/100,000 | Age: >40 | [45] |
| Singapore | 300/100,000 | Age: >50 | [46] |
| Italy | 326/100,000 | Age: >65–85 | [47] |
| Swedish Twin Registry | 496/100,000 | Crude prev | [48] |
| United States | 563/100,000 | Age: >40 | [49] |
| United States | 1588/100,000 | Age: >65 | [50] |
| Kin-Hu, Republic of China | 620/100,000 | Age: >50 | [51] |
| Assiut Governorate, Egypt | 659/100,000 | Crude prev | [52] |
| Argentina | 656/100,000 | Age: >40 | [53] |
| Rotterdam, the Netherlands | 979/100,000 | Age: >60 | [54] |
| *Chamorro Guam 1975–1979 parkinsonism–dementia complex (PDC) | 3384/100,000 | Age: >50 | [55] |
| *Amish community | 5703/100,000 | Age: >60 | [56] |

*At-risk populations.

## Correlations and patterns within population subsets

In the process of examining human evidence, we look at consistent patterns within and between populations across the world, with large-scale studies and high population sample numbers first. One of the largest epidemiological studies ever preformed on PD occurred in the United States, using the most inclusive, population-based U.S. health-care database, which is the U.S. Medicare system. The information contains data of the U.S. Medicare-eligible individuals, comprising 98% of Americans over the age of 65 from the years 1995 and 2000–2005 (Centers for Medicare and Medicaid Services, 2008).[50]

Most epidemiological studies consist of a population number from the 100s to 1000s, however, in the Medicare study (n = 58 million), to which over 450,000 PD cases per year were documented. These data provide a platform for in-depth analysis by a serial cross-sectional analysis using geographic information to combine spatial and temporal relationships across the country. The Medicare study most appropriately uses a sample population aged 65 and older. The mean prevalence of PD was 1,588.43 per 100,000 among Medicare beneficiaries over age 65 from 1995 and 2000–2005, where prevalence was a function of increasing age over 65.

### Age

Epidemiological studies across the world consistently show that age is the single most important correlate to development of PD. This is not surprising

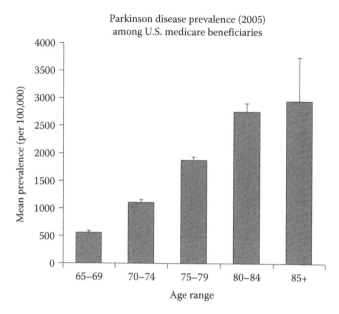

Parkinson disease prevalence (2005)
among U.S. medicare beneficiaries

**Figure 16.2** *PD prevalence and annual incidence (per 100,000) by age among U.S. Medicare beneficiaries years 1995 and 2000–2005: The data represent mean prevalence + 95% CI. (Modified from Wright, W.A. et al.,* Neuroepidemiology, *34, 143, 2010.)*

given that normal aging of the brain involves a gradual loss of postmitotic cells, which would also include DAergic neurons of the basal ganglia. Age-specific incidence rates increase sharply at age 60 years, peak at 85–89 years, and decline beginning at 90 years. In the Physicians' Health Study, *a crude annual* incidence rate of PD was 120.5 cases/100,000 person-years. The age-specific incidence rate increased sharply beginning at age 60 years, peaked at 85–89 years, and declined in those aged 90–99 years.[49] Age-related PD occurrence in the U.S. Medicare study is shown in Figure 16.2.

## Gender and ethnicity

According to the U.S. Medicare study, age-standardized PD prevalence was 2168/100,000 (white men), 1,036.41/100,000 (African Americans), and 1138/100,000 for the Asian population. The prevalence was greater in men than in women for all races, with a mean prevalence sex ratio of 155 males per 100 females (Figure 16.3).[50]

Greater incidence of PD in males has been reported in many studies. Similar findings were reported by the Kaiser Permanente Medical Care Program of Northern California, where incidence rapidly increased over the age of 60 years and, again, the rate for men (19.0 per 100,000; 95% CI: 16.1, 21.8) was 91% higher than that for women (9.9 per 100,000; 95% CI: 7.6, 12.2).

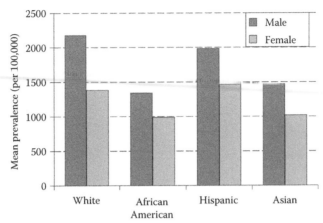

Parkinson disease prevalence (2005)
among U.S. medicare beneficiaries
<65 years age by gender and race

**Figure 16.3** *PD prevalence and annual incidence (per 100,000) by age and race among U.S. Medicare beneficiaries year 2005. (Modified from Wright, W.A. et al.,* Neuroepidemiology, 34, 143, 2010.)

The age- and gender-adjusted rate is often highest among Hispanics, followed by non-Hispanic Whites, Asians, and Blacks[57] and higher incidence for men than women at all ages.[58]

### United States: PD spatial distribution

The incidence and prevalence of PD vary significantly across the United States. Disease rates are highest in the Midwest and Northeast regions vs. Western and Southern U.S. counties. PD incidence for urban counties (476.81/100,000) was significantly greater than rural counties (413.2/100,000) (Figure 16.4).

## Disease correlates

### Cigarette smoking and tobacco products

One of the most protective effects consistently reported is that of cigarette smoking (CS), which is associated with reduced risk for idiopathic PD and essential tremor.[59] This inverse relationship is consistently reported in some of the largest and most well carried out studies in the world. Most studies show an ~50% reduction in PD occurrence with heavy use of cigarettes, including reports from the Honolulu Heart study,[60] the National Institutes of Health (NIH)-AARP (organization for people aged 50 and over) Diet and Health study, and ACS Cancer Prevention Study II, where longevity/extent of habit correlates to gradually lower risks of developing PD.[59,61–63] Similar findings are also

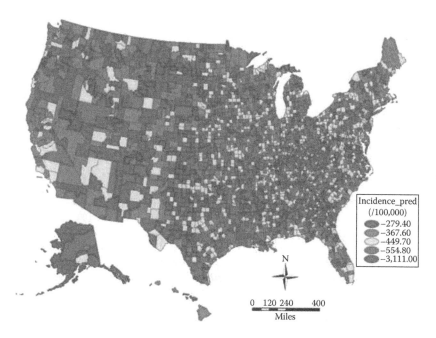

**Figure 16.4** *County level age- and race-standardized prevalence (per 100,000) of PD among Medicare beneficiaries in the United States (year=2003). (From Wright, W.A. et al.,* Neuroepidemiology, *34, 143, 2010.)*

reported for the use of other tobacco products such as cigars and/or pipes and chewing tobacco in male subjects.[64,65] In an extensive review of human evidence for environmental and lifestyle factors associated with risk of PD by Weirdefelt et al. (2011), the authors concluded that while there is an inadequate or inconsistent evidence to define a protective role for most dietary factors, there is a sufficient evidence to suggest protective effects of cigarette smoking against risk of PD, where 30/44 case-control studies showed an odds ratio (0.32–0.77) and nearly all of the major prospective studies reported reduced risk with pooled meta-analysis risk ratio of ~0.5.[66] These findings were also substantiated in the Physicians' Health Study (Figure 16.5) demonstrating that smoking was associated with a reduction in lifetime risk of PD[49] where age-associated risk of PD heightens after 70 years but is substantially reduced by smoking.

## What is in the smoke?

Cigarette smoking is a protective measure against risk for PD, but counterintuitively, tobacco smoke contains carbon monoxide, arsenic, benzene, formaldehyde, toluene, heavy metals, and a diverse range of chemicals that in theory should act as neurotoxins. Cigarette smoking is dangerous and estimated to be a leading cause of preventable deaths and in the United States alone accounts

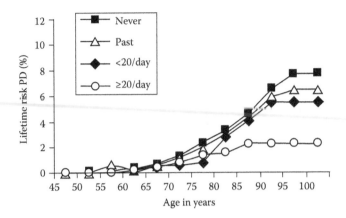

Cigarette smoking reduces lifetime risk of PD

*Figure 16.5* Cumulative incidence and lifetime risk curves: lifetime risk of PD by four categories of baseline smoking status (never, past, current <20 cigarettes/day, current ≥20 cigarettes/day). (From Driver, J.A. et al., Neurology, 3, 72(5), 432, 2009.)

for approximately 443,000 deaths each year.[67] How cigarette smoking can be neuroprotective remains a mystery. Several studies suggest protective facets of smoking may be attributable to the effects of nicotine,[68] on nicotinic acetylcholine receptors that could enhance motor function and/or promote survival of dopamine neurons.[69] However, it is also known that cigarette smoke (CS) contains hundreds of *nutraceuticals* that warrant further investigations.[70] Natural constituents of cigarette smoke are listed in Table 16.3.

## Nutraceuticals in cigarettes

Future research would be required to evaluate individual or combined effects of the possible hundreds of nutraceutical substances in CS to elucidate the true mechanism of action for preventative long-term habitual smoking on risk for PD, if unrelated to nicotine.

## Fruits and vegetables vs. pesticides/herbicides

Epidemiological studies investigating fruit and vegetable consumption with risk of PD are inconsistent, likely attributable to a multiple of confounding variables. First, agricultural and crop yields are dependent on proper control of insects, and the practice and usage of controlling agents vary, as does type of pesticide (e.g., organochlorines (DDT, dieldrin, and heptachlor)), some of which could elevate risk of PD.[71-74] This is also evidenced by some studies reporting higher PD prevalence in plantation workers,[75] field-crop workers, and landscapers[71-74] although not always consistent.[66] It is also possible that insecticides or other environmental contaminants could leach into

**Table 16.3** Additives Approved by the U.S. Government for Use in the Manufacture of Cigarettes

## Natural herbal and plant additives in cigarette tobacco

| | | | |
|---|---|---|---|
| • Alfalfa extract | • Cinnamon leaf oil, bark oil | • Lavender oil | • Piperonal |
| • Allspice extract, oleoresin, and oil | • Citronella oil | • Lemongrass oil | • Pipsissewa leaf extract |
| • Amyris oil | • Clover tops, red solid extract | • Licorice root | • Plum juice |
| • Angelica root extract, oil, seed oil | • Cocoa | • Lime oil | • Prune juice and concentrate |
| • Anise | • Coconut oil | • Linden flowers | • Raisin juice concentrate |
| • Anise star, extract, and oils | • Coffee | • Lovage oil and extract | • Rose absolute and oil |
| • Apple juice concentrate, extract | • Cognac white and green oil | • Mace powder, extract, and oil | • Rosemary oil |
| • Apricot extract, juice concentrate | • Copaiba oil | • Malt and malt extract | • Rum |
| • Asafetida fluid extract and oil | • Coriander extract and oil | • Maltodextrin | • Rum ether |
| • Ascorbic acid | • Corn oil | • Mandarin oil | • Rye extract |
| • Balsam Peru and oil | • Corn silk | • Maple syrup and concentrate | • Sage, sage oil |
| • Basil oil | • Costus root oil | • Mimosa absolute and extract | • Sandalwood oil, yellow |
| • Bay leaf, oil, and sweet oil | • Cubeb oil | • Molasses extract and tincture | • Spearmint oil |
| • Beeswax white | • Dill seed oil and extract | • Mountain maple solid extract | • Tannic acid |
| • Beet juice concentrate | • Eucalyptol | • Mullein flowers | • Tartaric acid |
| • Bergamot oil | • Fennel sweet oil | • Myrrh oil | • Tea leaf and absolute |
| • Bisabolene | • Fenugreek, extract, resin | • Nutmeg powder and oil | • Thyme oil, white and red |
| • Black currant buds absolute | • Fig juice concentrate | • Oak chips extract and oil | • Thymol |
| • Buchu leaf oil | • Food starch modified | • Oak moss absolute | • Tobacco extracts |
| • Capsicum oleoresin | • Galbanum oil | • Orange blossoms | • Tocopherols (mixed) |
| • Caramel color | • Gentian root extract | • Orange oil and extract | • Tolu balsam gum and extract |
| • Caraway oil | • Geranium rose oil | • Origanum oil | • Valerian root extract |
| • Carrot oil | • Ginger oil and oleoresin | • Orris concrete oil and root | • Vanilla extract and oleoresin |
| • Cascarilla oil and bark extract | • Grape juice concentrate | • Palmarosa oil | • Vanillin |

*(continued)*

**Table 16.3 (continued)** Additives Approved by the U.S. Government for Use in the Manufacture of Cigarettes

### Natural herbal and plant additives in cigarette tobacco

| | | | |
|---|---|---|---|
| • Cassia bark oil | • Guaiac wood oil | • Palmitic acid | • Violet leaf absolute |
| • Cassie absolute and oil | • Guar gum | • Parsley seed oil | • Walnut hull extract |
| • Castoreum extract, tincture, absolute | • Hops oil | • Patchouli oil | • Wild cherry bark extract |
| • Cedar leaf oil | • Hyssop oil | • Pepper oil, black and white  • | • Wine and wine sherry |
| • Celery seed extract, solid, oil | • Immortelle absolute, extract | • Peppermint oil | • Xanthan gum |
| • Chamomile flower oil, extract | • Jasmine absolute, oil | • Pimenta leaf oil | |
| • Chicory extract | • Kola nut extract | • Pine needle oil | |
| • Chocolate | • Lavandin oil | • Pineapple juice concentrate | |

Submitted by the major American cigarette companies to the Department of Health and Human Services in April of 1994: American Tobacco Company, Brown and Williamson, Liggett Group, Inc., Philip Morris Inc., R.J. Reynolds Tobacco Company.

surrounding water systems or soil spreading toward residential communities, contributing to high prevalence rates of PD in rural communities or associated with well-water drinking.[71,72] Contaminants in the environment can also include heavy metals, solvents, petroleum waste, or man-made chemicals that could contribute to groundwater pollution. This could ultimately alter the quality of drinking water and edible food supplies, in particular livestock, meats, fruits, and vegetables. As a whole, while there is some evidence supporting that greater intake of fruits and vegetables is a protective correlate to PD[76] with flavonoid-rich foods (tea, berry fruits, apples, red wine, and orange/orange juice) specific to male populations,[77-79] these effects are not consistently reported throughout the world, leaving a considerable question as to causality. A number of studies show no protective effects of vegetarian diets[80] and consumption of fruits and vegetables[81] or foods containing vitamins A, C, and E[82] or antioxidant activity.[61,83]

## Vitamins

There is also considerably weak evidence to support the hypothesis that vitamins such as niacin; riboflavin; vitamins A, B, C, and $B_{12}$; pantothenic acid;[84] fruits/vegetables high in vitamin content;[85] folic acid; or multivitamins can lower the risk of PD except for vitamin E (foods only) to which a confounding variable could include the carrier oil.[85-87] A meta-analysis of published studies between 1966 and March 2005 shows some evidence to suggest protective effects of vitamin E,[88] but most reports are inconclusive and inconsistent,[89,90] including the Deprenyl and Tocopherol Antioxidant Therapy of Parkinsonism (DATATOP) trial showing that 2 g/day in early stage PD did not delay disease progression.[91,92] Similarly, studies in PD patients suggest that deficiency in vitamins A, C, and E are not likely to contribute to the onset or progress of PD.[93]

## Vitamin B$_6$ and vitamin D

Interestingly, two well-known studies including the Rotterdam prospective cohort study report that low levels of vitamin B$_6$ could be a risk factor in PD.[94,95] In the Rotterdam study, higher dietary intake of vitamin B$_6$ was associated with a significantly reduced risk of PD (0.69 0.46 [95% CI: 0.22–0.96]).[95] Moreover, administration of vitamin B$_6$ reportedly improves neurological motor function associated with severity of L-dopa-induced dyskinesia in patients with PD.[96] Vitamin B$_6$ is in relative higher abundance in fish (cod, salmon, tuna, halibut), bananas, nuts (peanuts, sunflower seeds, cashews), meats, whole grains, and legumes and plays a major role in transamination reactions,[97] anaerobic metabolism,[98,99] and neurotransmission, all required for health, viability, and function of DAergic neurons.[100] The positive effects of vitamin B$_6$ have been observed for a number of neurological-related conditions such as carpel tunnel syndrome,[101] seizures,[102] and attention deficit disorder.[103] In addition, it is believed that PD patients could benefit from taking vitamin B$_6$ (along with folate and B$_{12}$) not necessarily in prevention but to counteract side effects of L-dopa that include mediated rise in neurotoxic homocysteine[104–106] and associated neuropathy.[107] In total, it appears that vitamin B$_6$ may be protective, but further research is required to confirm this effect.

Another vitamin that could elevate risk for PD, *if deficient*, is vitamin D.[108,109] Several cohort studies report that individuals with higher serum levels of vitamin D have reduced risk of[108,110] and severity of PD assessed by Hoehn and Yahr stage and UPDRS by low serum vitamin D.[111] The effects of vitamin D deficiency and the greater prevalence/severity of PD are fairly consistent,[112] with some exceptions,[113] also indirectly corroborated by evidence showing that outdoor workers/sunlight exposure is a protective measure against PD.[114] Not only is vitamin D deficiency associated with greater PD prevalence, but a loss of bone density could also pose a significant problem for patients due to greater likelihood of falling events and hip fractures.[115,116] While there is a relationship between physiological vitamin D levels and PD, the mechanism of action for deficiency is uncertain but is suggested to involve vitamin D receptor gene polymorphisms[117] leading to lower levels of protective substances such as neurotrophic factors[118] or calretinin.[119] Interestingly, this relationship is not reflected by dietary epidemiological data where studies suggest that consumption of dairy products in particular milk products (high in vitamin D) is associated with elevated risk of PD in men.[66] In contrast, diets high in fish/fish oils (high in vitamin D) but also polyunsaturated fatty acids (PUFA) could lower the risk for PD, that latter of which alone is believed to be a protective factor.

# Fish, lipids, and legumes

There is considerable evidence to suggest that diets high in fish (PUFA, omega 3 fatty acids) could lower risk for PD.[76,78] In a prospective population-based Rotterdam cohort study, an association between the intake of unsaturated fatty acids and the risk of incidence for PD among 5289 subjects over 6 years showed a reduced hazard ratio for dietary monounsaturated fatty acids (MUFAs) and PUFAs [120] Diets low in saturated fat but high in MUFA (e.g., virgin olive oil) are also typically described in the Mediterranean diet, which also correlates to lower prevalence of PD[121–125] and longevity.[126] The practices of a Mediterranean diet include consumption of fruits and vegetables and olive oil as the primary source of fat, where fish, poultry, and red wine are consumed at moderate amounts along with lofty consumption of legumes (with some salt-cured foods). In the Mediterranean diet, displacement of dietary meats or animal fat with elevated fish could be a protective measure against risk of PD. Epidemiological studies tend to show a pattern of elevated PD risk with consumption of saturated animal fats, raw meat[82] (OR 5.3; 95% CI: 1.8–15.5; p trend=0.001),[81,83,120] cholesterol,[127] whale meat/blubber,[128] and arachidonic acid (animal-based foods), although this too is not always observed.[84,129] Whether or not saturated fats or the subsequent elevated risk of cardiovascular disease is associated with PD risk remains uncertain. However, the Nurses' Health Study and the Health Professionals Follow-Up Study suggest no causal link, where incidence of PD was not associated with hypertension or high cholesterol[130] but showed a risk elevation with higher intake of saturated fats.[76] Why there is higher PD risk with consumption of saturated fats also remains uncertain; however, massive lipid alterations are reported in the brain of PD patients, which could be affected by the diet. Analysis of postmortem frontal cortical tissue of human idiopathic PD patients vs. healthy controls shows a higher saturated fatty acid composition/omega-3 ratio in PD patients (47.01) vs. controls (7.75).[131] A diet high in saturated fats could contribute to the deposition of fats in part through elevated triglycerides/LDLs (also a PD risk factor), effects that can be attenuated by greater intake of fish oils/omega-3 fatty acids.[76,78,132–134] These effects have also been corroborated in experimental animal studies where administration of omega-3 fatty acids such as docosahexaenoic acid or eicosapentaenoic acid can reduce DAergic degeneration within the brain in experimental models of PD.[135–137]

The Mediterranean diet also encourages consumption of legumes, also associated with lower risk of PD.[138] Both omega-3 PUFA and legumes can lower plasma lipids in addition to stabilizing blood glucose/insulin levels over time.[139,140] In turn, insulin contributes to normal brain function by regulating uptake and use of glucose to produce energy, which is very important in the process of neuroprotection. There are reported associations between systemic glucose homeostasis and elevated risk of PD in correlation with type II adult onset diabetes, further evidenced by the risk reduction observed by adhering to a Mediterranean diet.[141] The connection between diabetes, glucose homeostasis, and PD is still unclear; however, a study using 30 year Danish Hospital Register Pharmacy records (1986–2006) found a 36% elevated risk of developing PD in individuals using antidiabetic drugs such as insulin, sulfonylureas, or biguanides[142]; this is consistent with a greater risk of PD reported among diabetic patients[143] and reproducible findings in experimental models.[144] Ultimately, habitual consumption of legumes could alter glycemic index, stabilize blood glucose levels, and reduce risk of PD.[145,146] In total, it appears that consumption of legumes, fish oil, or other sources rich in omega-3 PUFA may be a protective component against PD risk, but further research is warranted.

## Caffeine

An inverse association between caffeine intake and PD risk in both men and women is consistently reported.[61,147,148] The NIH-AARP Diet and Health Study reported a higher caffeine intake associated with lower PD risk OR 0.75 (95% CI: 0.60, 0.94; p trend=0.005) for men and 0.60 (95% CI: 0.39, 0.91; p trend=0.005) for women.[147] Increasing intensity of coffee drinking and dosage was also inversely associated with high dosage presenting a significant inverse correlation (OR=0.58; 95% CI: 0.34–0.99).[149] Likewise, a 30 year follow-up of 8004 Japanese American men (aged 45–68 years) in the prospective longitudinal Honolulu Heart Program (1965 and 1968) showed that an incidence of PD declined consistently with increased amounts of coffee intake, from 10.4 per 10,000 person-years in men who drank no coffee to 1.9 per 10,000 person-years in men who drank at least 28 oz/d (p < 0.001 for trend). These results are also reported for total caffeine intake and caffeine from noncoffee sources.[150] In total, it appears consumption of caffeine may be protective, but further research is required to corroborate, validate, and elucidate the mechanism of action.

## At-risk populations

Populations with high reported PD prevalence can provide clues about the disease pathology, with the most well-known examples including high frequency in the Amish community and the Chamorro natives of the Mariana Islands in the Western Pacific Ocean and Guadeloupe, a French Caribbean island.

The Amish dwell in communities culturally clustered from society, with practices that include occupational farming, drinking of well water, rural living, and abstaining from tobacco; all factors that epidemiological studies suggest would elevate occurrence of PD. This seems to be the case, where a population-based screening for PD in subjects in an Amish community over age 60 established a very high PD prevalence, 5703/100,000 (95% CI: 5095–6225), one of the highest in the world reported for any subset population.[56] Likewise, Chamorro natives of the Mariana Islands in the Western Pacific Ocean also have a high rate of amyotrophic lateral sclerosis/Parkinsonism–dementia[55] that is not observed in surrounding areas such as islands of Rota, Tinian, and Saipan and the four remote islands within the Mariana's chain.[151] Studies on lifestyle and environment of this population show that major and unique risk factors may arise due to the traditional Chamorro diet, found to contain a potent neurotoxin, BMAA, derived from indigenous cycad.[152] BMAA is believed to be magnified up through the food chain from soil (cyanobacteria) to plant (cycad seeds), plant to human (cycad seed flour), plant to animal, (flying foxes, Pteropus bats that feed on cycad seeds), and animal to human where flying foxes are served at traditional Chamorro feasts as part of the culture.[152] Similarly, abnormally high frequency of atypical Parkinsonism has been reported in Guadeloupe, a French Caribbean island with 422, 000 inhabitants that regularly consume *Annona muricata* (soursop) containing another neurotoxic class of compounds called acetogenins and alkaloids, one of which is annonacin, an extremely potent complex I activity inhibitor. In rats, annonacin decreases brain ATP levels and induces neurodegeneration in the basal ganglia area in a manner similar to patients with atypical Parkinsonism of Guadeloupe.[153] Further investigation as to the deleterious mechanism evoked by BMAA or annonacin may lead to pathological models and effective treatments in establishing PD therapeutic strategies.

## Conclusion

In summary, there are a large number of specific, known, and unknown factors that precipitate PD pathology and initiate destruction of DAergic neurons in the SN, which are complicated by the natural process of aging. Further influence on the disease process by dietary practices or inherent environmental factors could exacerbate or prevent the neurodegenerative process. Isolation and identification of positive or negative influences on PD risk will require large-scale and highly controlled studies of consistent methodology. As of today, the evolution of drug and pharmaceutical industry has diverged from which it was originated—use of natural plants, diet, and botanical products to treat disease—now classified as complementary and alternative medicine practiced by two-thirds of PD patients.[154] From the data in their current form, we conclude that there is extensive research to be conducted in both epidemiological and clinical trials in order to extrapolate, validate, corroborate, and justify effective preventative or post-onset therapeutics with the use of nutraceuticals specific to PD.

## Acknowledgments

This research was supported by a grant from NIH NCRR RCMI program (G12RR 03020) and the National Institute of Minority Health and Health Disparities, NIH (8G12MD007582–28 and 1P20 MD006738–01).

# References

1. Williams A. MPTP toxicity: Clinical features. *J Neural Transm Suppl* 1986; 20:5–9.
2. Wang A, Costello S, Cockburn M, Zhang X, Bronstein J, Ritz B. Parkinson's disease risk from ambient exposure to pesticides. *Eur J Epidemiol* 2011; 26:547–555.
3. Kotake Y, Ohta S. MPP+ analogs acting on mitochondria and inducing neuro-degeneration. *Curr Med Chem* 2003; 10:2507–2516.
4. Goldman SM, Quinlan PJ, Ross GW, Marras C, Meng C, Bhudhikanok GS, et al. Solvent exposures and parkinson disease risk in twins. *Ann Neurol* 2012; 71:776–784.
5. Racette BA, Aschner M, Guilarte TR, Dydak U, Criswell SR, Zheng W. Pathophysiology of manganese-associated neurotoxicity. *Neurotoxicology* 2011; 32(4):383–391.
6. Ali SF, Binienda ZK, Imam SZ. Molecular aspects of dopaminergic neurodegenera-tion: Gene-environment interaction in parkin dysfunction. *Int J Environ Res Public Health*2011; 8:4702–4713.
7. Wang X, Petrie TG, Liu Y, Liu J, Fujioka H, Zhu X. Parkinson's disease-associated DJ-1 mutations impair mitochondrial dynamics and cause mitochondrial dysfunction. *J Neurochem* 2012; 121:830–839.
8. McCoy MK, Cookson MR. Mitochondrial quality control and dynamics in Parkinson's disease. *Antioxid Redox Signal* 2012; 16:869–882.
9. Ulusoy A, Decressac M, Kirik D, Bjorklund A. Viral vector-mediated overexpression of alpha-synuclein as a progressive model of Parkinson's disease. *Prog Brain Res* 2010; 184:89–111.
10. Tanaka M, Aihara Y, Ikeda S, Aihara Y. Neuromelanin-related contrast in the substantia nigra semiquantitatively evaluated by magnetic resonance imaging at 3T: Comparison between normal aging and Parkinson disease. *Rinsho Shinkeigaku* 2011; 51.14–20.
11. Fahn S, Sulzer D. Neurodegeneration and neuroprotection in Parkinson disease. *NeuroRx J Am SocExp NeuroTherapeut* 2004; 1:139–154.
12. Rohn TT, Catlin LW. Immunolocalization of influenza A virus and markers of inflamma-tion in the human Parkinson's disease brain. *PloS One* 2011; 6:e20495.
13. Mishra LC. *Scientific Basis for Ayurvedic Therapies*, Boca Raton, FL: CRC Press LLC, 2004.
14. M.D. RBR. *Understanding Parkinson's Disease: A Personal And Professional View.* Westport, CT: Praeger Publishers, 2006.
15. Lieberman AN. *Parkinson's Disease: The Complete Guide for Patients and Caregivers,* 1st edn., Austin, TX: Touchstone, February 1, 1993.
16. Finger S. *Origins of Neuroscience: A History of Explorations into Brain Function,* New York: Oxford University Press, October 11, 2001.
17. Lanska DJ. Chapter 33: The history of movement disorders. In: PJ Vinken and GW Bruyn, *Handbook of Clinical Neurology,* vol. 95, Amsterdam, the Netherlands: North-Holland Pub. Co., 2010, pp. 501–546.
18. Tolosa E, Marti MJ, Valldeoriola F, Molinuevo JL. History of levodopa and dopamine agonists in Parkinson's disease treatment. *Neurology* 1998; 50:S2–S10; discussion S44–S48.
19. Pras N, Woerdenbag HJ, Batterman S, Visser JF, Van Uden W. Mucuna pruriens: Improvement of the biotechnological production of the anti-Parkinson drug L-dopa by plant cell selection. *Pharm World Sci* 1993; 15:263–268.
20. Raina AP, Khatri R. Quantitative determination of L-DOPA in seeds of *Mucuna pruriens* germplasm by high performance thin layer chromatography. *Indian J Pharmaceut Sci* 2011; 73:459–462.
21. Ramya KB, Thaakur S. Herbs containing ʟ-Dopa: An update. *Ancient Sci Life* 2007; 27:50–55.
22. Ingle PK. ʟ-Dopa bearing plants. *Nat Prod Rad* 2003; 2:126–133.
23. Rabey JM, Vered Y, Shabtai H, Graff E, Harsat A, Korczyn AD. Broad bean (Vicia faba) consumption and Parkinson's disease. *Adv Neurol* 1993; 60:681–684.

24. Katzenschlager R, Evans A, Manson A, Patsalos PN, Ratnaraj N, Watt H, et al. Mucuna pruriens in Parkinson's disease: A double blind clinical and pharmacological study. *J Neurol Neurosurg Psychiatry* 2004; 75: 1672–1677.

25. Manyam BV, Dhanasekaran M, Hare TA. Neuroprotective effects of the antiparkinson drug Mucuna pruriens. *Phytother Res* 2004; 18:706–712.

26. Ahlskog JE. Treatment of Parkinson's disease. From theory to practice. *Postgrad Med* 1994; 95:52–54, 7–8, 61–64 passim.

27. Hauser RA, Panisset M, Abbruzzese G, Mancione L, Dronamraju N, Kakarieka A, et al. Double-blind trial of levodopa/carbidopa/entacapone versus levodopa/carbidopa in early Parkinson's disease. *Mov Disord* 2009; 24:541–550.

28. Bene R, Antic S, Budisic M, Lisak M, Trkanjec Z, Demarin V, et al. Parkinson's disease. *Acta Clin Croat* 2009; 48:377–380.

29. Nagashayana N, Sankarankutty P, Nampoothiri MR, Mohan PK, Mohanakumar KP. Association of L-DOPA with recovery following Ayurveda medication in Parkinson's disease. *J Neurol Sci* 2000; 176:124–127.

30. Xu J, Li Y, Guo Y, Guo P, Yamakuni T, Ohizumi Y. Isolation, structural elucidation, and neuroprotective effects of iridoids from *Valeriana jatamansi. Biosci Biotechnol Biochem* 2012; 76:1401.

31. Xu J, Guo Y, Xie C, Jin DQ, Gao J, Gui L. Isolation and neuroprotective activities of acylated iridoids from *Valeriana jatamansi. Chem Biodiv* 2012; 9:1382–1388.

32. Patil RA, Hiray YA, Kasture SB. Reversal of reserpine-induced orofacial dyskinesia and catalepsy by *Nardostachys jatamansi. Indian J Pharmacol* 2012; 44:340–344.

33. Dhingra D, Goyal PK. Evidences for the involvement of monoaminergic and GABAergic systems in antidepressant-like activity of *Tinospora cordifolia* in mice. *Indian J Pharmaceut Sci* 2008; 70:761–767.

34. Hatano T, Fukuda T, Miyase T, Noro T, Okuda T. Phenolic constituents of licorice. III. Structures of glicoricone and licofuranone, and inhibitory effects of licorice constituents on monoamine oxidase. *Chem Pharm Bull* (Tokyo) 1991; 39:1238–1243.

35. Thompson FE SA. *Nutrition in the Prevention and Treatment of Disease*, vol. 2. San Diego, CA: Academic Press, 2008.

36. Dotchin C, Msuya O, Kissima J, Massawe J, Mhina A, Moshy A, et al. The prevalence of Parkinson's disease in rural Tanzania. *Mov Disord* 2008; 23:1567–672.

37. Mshana G, Dotchin CL, Walker RW. 'We call it the shaking illness': Perceptions and experiences of Parkinson's disease in rural northern Tanzania. *BMC Public Health* 2011; 11:219.

38. Sanchez JL, Buritica O, Pineda D, Uribe CS, Palacio LG. Prevalence of Parkinson's disease and parkinsonism in a Colombian population using the capture-recapture method. *Int J Neurosci* 2004; 114:175–182.

39. Ashok PP, Radhakrishnan K, Sridharan R, Mousa ME. Epidemiology of Parkinson's disease in Benghazi, North-East Libya. *Clin Neurol Neurosurg* 1986; 88:109–113.

40. Das SK, Misra AK, Ray BK, Hazra A, Ghosal MK, Chaudhuri A, et al. Epidemiology of Parkinson disease in the city of Kolkata, India: A community-based study. *Neurology* 2010; 75:1362–1369.

41. Giroud Benitez JL, Collado-Mesa F, Esteban EM. Prevalence of Parkinson disease in an urban area of the Ciudad de La Habana province, Cuba. Door-to-door population study. *Neurologia* 2000; 15:269–273.

42. Hollingworth SA, Rush A, Hall WD, Eadie MJ. Utilization of anti-Parkinson drugs in Australia: 1995–2009. *Pharmacoepidemiol Drug Saf* 2011; 20:450–456.

43. Kleinhenz J, Vieregge P, Fassl H, Jorg J. The prevalence of Parkinson disease in West Germany—are general practice data a suitable survey instrument?. *Offentl Gesundheitswes* 1990; 52:181–190.

44. Benito-Leon J, Bermejo-Pareja F, Morales-Gonzalez JM, Porta-Etessam J, Trincado R, Vega S, et al. Incidence of Parkinson disease and parkinsonism in three elderly populations of central Spain. *Neurology* 2004; 62:734–741.

45. Nicoletti A, Sofia V, Bartoloni A, Bartalesi F, Gamboa Barahon H, Giuffrida S, et al. Prevalence of Parkinson's disease: A door-to-door survey in rural Bolivia. *Parkinsonism Related Disorders* 2003; 10:19–21.

46. Tan LC, Venketasubramanian N, Hong CY, Sahadevan S, Chin JJ, Krishnamoorthy ES, et al. Prevalence of Parkinson disease in Singapore: Chinese vs Malays vs Indians. *Neurology* 2004; 62:1999–2004.

47. Baldereschi M, Di Carlo A, Rocca WA, Vanni P, Maggi S, Perissinotto E, et al. Parkinson's disease and parkinsonism in a longitudinal study: Two-fold higher incidence in men. ILSA Working Group. Italian Longitudinal Study on Aging. *Neurology* 2000; 55:1358–1363.

48. Wirdefeldt K, Gatz M, Bakaysa SL, Fiske A, Flensburg M, Petzinger GM, et al. Complete ascertainment of Parkinson disease in the Swedish Twin Registry. *Neurobiol Aging* 2008; 29:1765–1773.

49. Driver JA, Logroscino G, Gaziano JM, Kurth T. Incidence and remaining lifetime risk of Parkinson disease in advanced age. *Neurology* 2009; 72:432–438.

50. Wright Willis A, Evanoff BA, Lian M, Criswell SR, Racette BA. Geographic and ethnic variation in Parkinson disease: A population-based study of US Medicare beneficiaries. *Neuroepidemiology* 2010; 34:143–151.

51. Wang SJ, Fuh JL, Liu CY, Lin KP, Chang R, Yih JS, et al. Parkinson's disease in Kin-Hu, Kinmen: A community survey by neurologists. *Neuroepidemiology* 1994; 13:69–74.

52. Khedr EM, Al Attar GS, Kandil MR, Kamel NF, Abo Elfetoh N, Ahmed MA. Epidemiological study and clinical profile of Parkinson's disease in the Assiut Governorate, Egypt: A community-based study. *Neuroepidemiology* 2012; 38:154–163.

53. Melcon MO, Anderson DW, Vergara RH, Rocca WA. Prevalence of Parkinson's disease in Junin, Buenos Aires Province, Argentina. *Mov Disord* 1997, 12:197–205.

54. de Lau LM, Giesbergen PC, de Rijk MC, Hofman A, Koudstaal PJ, Breteler MM. Incidence of parkinsonism and Parkinson disease in a general population: The Rotterdam Study. *Neurology* 2004; 63:1240–1244.

55. Plato CC, Garruto RM, Galasko D, Craig UK, Plato M, Gamst A, et al. Amyotrophic lateral sclerosis and parkinsonism-dementia complex of Guam: Changing incidence rates during the past 60 years. *Am J Epidemiol* 2003; 157:149–157.

56. Racette BA, Good LM, Kissel AM, Criswell SR, Perlmutter JS. A population-based study of parkinsonism in an Amish community. *Neuroepidemiology* 2009; 33:225–230.

57. Van Den Eeden SK, Tanner CM, Bernstein AL, Fross RD, Leimpeter A, Bloch DA, et al. Incidence of Parkinson's disease: Variation by age, gender, and race/ethnicity. *Am J Epidemiol* 2003; 157:1015–1022.

58. Bower JH, Maraganore DM, McDonnell SK, Rocca WA. Incidence and distribution of parkinsonism in Olmsted County, Minnesota, 1976–1990. *Neurology* 1999; 52:1214–1220.

59. Louis ED, Benito-Leon J, Bermejo-Pareja F, Neurological Disorders in Central Spain Study G. Population-based prospective study of cigarette smoking and risk of incident essential tremor. *Neurology* 2008; 70:1682–1687.

60. Morens DM, Grandinetti A, Davis JW, Ross GW, White LR, Reed D. Evidence against the operation of selective mortality in explaining the association between cigarette smoking and reduced occurrence of idiopathic Parkinson disease. *Am J Epidemiol* 1996; 144:400–404.

61. Campdelacreu J. Parkinson disease and Alzheimer disease: Environmental risk factors. *Neurologia* (in press).

62. Scott WK, Zhang F, Stajich JM, Scott BL, Stacy MA, Vance JM. Family-based case-control study of cigarette smoking and Parkinson disease. *Neurology* 2005; 64:442–447.

63. Thacker EL, O'Reilly EJ, Weisskopf MG, Chen H, Schwarzschild MA, McCullough ML, et al. Temporal relationship between cigarette smoking and risk of Parkinson disease. *Neurology* 2007; 68:764–768.

64. Ritz B, Ascherio A, Checkoway H, Marder KS, Nelson LM, Rocca WA, et al. Pooled analysis of tobacco use and risk of Parkinson disease. *Arch Neurol* 2007; 64:990–997.

65. Chen H, Huang X, Guo X, Mailman RB, Park Y, Kamel F, et al. Smoking duration, intensity, and risk of Parkinson disease. *Neurology* 2010; 74:878–884.

66. Wirdefeldt K, Adami HO, Cole P, Trichopoulos D, Mandel J. Epidemiology and etiology of Parkinson's disease: A review of the evidence. *Eur J Epidemiol* 2011; 26 Suppl 1:S1–S58.

67. Moeller DW, Sun LS. Chemical and radioactive carcinogens in cigarettes: Associated health impacts and responses of the tobacco industry, U.S. Congress, and federal regulatory agencies. *Health Phys* 2010; 99:674–679.

68. Quik M, Wonnacott S. alpha6beta2* and alpha4beta2* nicotinic acetylcholine receptors as drug targets for Parkinson's disease. *Pharmacol Rev* 2011; 63:938–966.

69. Toulorge D, Guerreiro S, Hild A, Maskos U, Hirsch EC, Michel PP. Neuroprotection of midbrain dopamine neurons by nicotine is gated by cytoplasmic Ca2+. *FASEB J* 2011; 25:2563–2573.

70. Gaworski CL, Heck JD, Bennett MB, Wenk ML. Toxicologic evaluation of flavor ingredients added to cigarette tobacco: Skin painting bioassay of cigarette smoke condensate in SENCAR mice. *Toxicology* 1999; 139:1–17.

71. Zuber M, Alperovitch A. Parkinson's disease and environmental factors]. *Rev Epidemiol Sante Publique* 1991; 39:373–387.

72. Garcia Ruiz PJ. Prehistory of Parkinson's disease. *Neurologia* 2004; 19:735–737.

73. Weisskopf MG, Knekt P, O'Reilly EJ, Lyytinen J, Reunanen A, Laden F, et al. Persistent organochlorine pesticides in serum and risk of Parkinson disease. *Neurology* 2010; 74:1055–1061.

74. van der Mark M, Brouwer M, Kromhout H, Nijssen P, Huss A, Vermeulen R. Is pesticide use related to Parkinson disease? Some clues to heterogeneity in study results. *Environ Health Perspect* 2012; 120:340–347.

75. Petrovitch H, Ross GW, Abbott RD, Sanderson WT, Sharp DS, Tanner CM, et al. Plantation work and risk of Parkinson disease in a population-based longitudinal study. *Arch Neurol* 2002; 59:1787–1792.

76. Gao X, Chen H, Fung TT, Logroscino G, Schwarzschild MA, Hu FB, et al. Prospective study of dietary pattern and risk of Parkinson disease. *Am J Clin Nutr* 2007; 86:1486–1494.

77. Gao X, Cassidy A, Schwarzschild MA, Rimm EB, Ascherio A. Habitual intake of dietary flavonoids and risk of Parkinson disease. *Neurology* 2012; 78:1138–1145.

78. Okubo H, Miyake Y, Sasaki S, Murakami K, Tanaka K, Fukushima W, et al. Dietary patterns and risk of Parkinson's disease: A case-control study in Japan. *Eur J Neurol* 2012; 19:681–688.

79. Saaksjarvi K, Knekt P, Lundqvist A, Mannisto S, Heliovaara M, Rissanen H, et al. A cohort study on diet and the risk of Parkinson's disease: The role of food groups and diet quality. *Br J Nutr* 2012:1–9.

80. Behari M, Srivastava AK, Das RR, Pandey RM. Risk factors of Parkinson's disease in Indian patients. *J Neurol Sci* 2001; 190:49–55.

81. Hellenbrand W, Seidler A, Boeing H, Robra BP, Vieregge P, Nischan P, et al. Diet and Parkinson's disease. I: A possible role for the past intake of specific foods and food groups. Results from a self-administered food-frequency questionnaire in a case-control study. *Neurology* 1996; 47:636–643.

82. Anderson C, Checkoway H, Franklin GM, Beresford S, Smith-Weller T, Swanson PD. Dietary factors in Parkinson's disease: The role of food groups and specific foods. *Mov Disord* 1999; 14:21–27.

83. Logroscino G, Marder K, Cote L, Tang MX, Shea S, Mayeux R. Dietary lipids and antioxidants in Parkinson's disease: A population-based, case-control study. *Ann Neurol* 1996; 39:89–94.

84. Abbott RD, Ross GW, White LR, Sanderson WT, Burchfiel CM, Kashon M, et al. Environmental, life-style, and physical precursors of clinical Parkinson's disease: Recent findings from the Honolulu-Asia Aging Study. *J Neurol* 2003; 250(Suppl 3): III30–III39.

85. Miyake Y, Fukushima W, Tanaka K, Sasaki S, Kiyohara C, Tsuboi Y, et al. Dietary intake of antioxidant vitamins and risk of Parkinson's disease: A case-control study in Japan. *Eur J Neurol* 2011; 18:106–113.

86. Chen H, Zhang SM, Schwarzschild MA, Hernan MA, Logroscino G, Willett WC, et al. Folate intake and risk of Parkinson's disease. *Am J Epidemiol* 2004; 160:368–375.

87. Zhang SM, Hernan MA, Chen H, Spiegelman D, Willett WC, Ascherio A. Intakes of vitamins E and C, carotenoids, vitamin supplements, and PD risk. *Neurology* 2002; 59:1161–1169.

88. Etminan M, Gill SS, Samii A. Intake of vitamin E, vitamin C, and carotenoids and the risk of Parkinson's disease: A meta-analysis. *Lancet Neurol* 2005; 4:362–365.

89. Ayuso-Peralta L, Jimenez-Jimenez FJ, Cabrera-Valdivia F, Molina J, Javier MR, Almazan, et al. Premorbid dietetic habits and risk for Parkinson's disease. *Parkinson Rel Disord* 1997; 3:55–61.

90. Jimenez-Jimenez FJ, Ayuso-Peralta L, Molina JA, Cabrera-Valdivia F. Do antioxidants in the diet affect the risk of developing Parkinson disease?. *Rev Neurol* 1999; 29:741–744.

91. Poewe WH, Wenning GK. The natural history of Parkinson's disease. *Ann Neurol* 1998; 44:S1–S9.

92. Impact of deprenyl and tocopherol treatment on Parkinson's disease in DATATOP patients requiring levodopa. Parkinson Study Group. *Ann Neurol* 1996; 39:37–45.

93. King D, Playfer JR, Roberts NB. Concentrations of vitamins A, C and E in elderly patients with Parkinson's disease. *Postgrad Med J* 1992; 68:634–637.

94. Murakami K, Miyake Y, Sasaki S, Tanaka K, Fukushima W, Kiyohara C, et al. Dietary intake of folate, vitamin B6, vitamin B12 and riboflavin and risk of Parkinson's disease: A case-control study in Japan. *Br J Nutr* 2010; 104:757–764.

95. de Lau LM, Koudstaal PJ, Witteman JC, Hofman A, Breteler MM. Dietary folate, vitamin B12, and vitamin B6 and the risk of Parkinson disease. *Neurology* 2006; 67:315–318.

96. Sandyk R, Pardeshi R. Pyridoxine improves drug-induced parkinsonism and psychosis in a schizophrenic patient. *Int J Neurosci* 1990; 52:225–232.

97. Bolander FF. Vitamins: Not just for enzymes. *Curr Opin Invest Drugs* 2006; 7:912–915.

98. Sato S, Ohi T, Nishino I, Sugie H. Confirmation of the efficacy of vitamin B6 supplementation for McArdle disease by follow-up muscle biopsy. *Muscle a d Nerve* 2012; 45:436–440.

99. Butler PE, Cookson EJ, Beynon RJ. Accelerated degradation of glycogen phosphorylase in denervated and dystrophic mouse skeletal muscle. *Biosci Rep* 1985; 5:567–572.

100. Gokcan H, Konuklar FA. Theoretical study on HF elimination and aromatization mechanisms: A case of pyridoxal 5'¢ phosphate-dependent enzyme. *J Org Chem* 2012; 77:5533–5543.

101. Aufiero E, Stitik TP, Foye PM, Chen B. Pyridoxine hydrochloride treatment of carpal tunnel syndrome: A review. *Nutr Rev* 2004; 62:96–104.

102. Kluger G, Blank R, Paul K, Paschke E, Jansen E, Jakobs C, et al. Pyridoxine-dependent epilepsy: Normal outcome in a patient with late diagnosis after prolonged status epilepticus causing cortical blindness. *Neuropediatrics* 2008; 39:276–279.

103. Mousain-Bosc M, Roche M, Polge A, Pradal-Prat D, Rapin J, Bali JP. Improvement of neurobehavioral disorders in children supplemented with magnesium-vitamin B6. I. Attention deficit hyperactivity disorders. *Magnes Res* 2006; 19:46–52.

104. Lamberti P, Zoccolella S, Armenise E, Lamberti SV, Fraddosio A, de Mari M, et al. Hyperhomocysteinemia in L-dopa treated Parkinson's disease patients: Effect of cobalamin and folate administration. *Eur J Neurol* 2005; 12:365–368.

105. Bostantjopoulou S, Katsarou Z, Frangia T, Hatzizisi O, Papazisis K, Kyriazis G, et al. Endothelial function markers in parkinsonian patients with hyperhomocysteinemia. *J Clin Neurosci Official J Neurosurg Soc Aust* 2005; 12:669–672.

106. Lokk J. Treatment with levodopa can affect latent vitamin B 12 and folic acid deficiency. Patients with Parkinson disease runt the risk of elevated homocysteine levels. *Lakartidningen* 2003; 100:2674–2677.

107. Rajabally YA, Martey J. Neuropathy in Parkinson disease: Prevalence and determinants. *Neurology* 2011; 77:1947–1950.

108. Evatt ML, Delong MR, Khazai N, Rosen A, Triche S, Tangpricha V. Prevalence of vitamin d insufficiency in patients with Parkinson disease and Alzheimer disease. *Arch Neurol* 2008; 65:1348–1352.

109. Newmark HL, Newmark J. Vitamin D and Parkinson's disease—A hypothesis. *Mov Disord* 2007; 22:461–468.

110. Knekt P, Kilkkinen A, Rissanen H, Marniemi J, Saaksjarvi K, Heliovaara M. Serum vitamin D and the risk of Parkinson disease. *Arch Neurol* 2010; 67:808–811.

111. Suzuki M, Yoshioka M, Hashimoto M, Murakami M, Kawasaki K, Noya M, et al. 25-hydroxyvitamin D, vitamin D receptor gene polymorphisms, and severity of Parkinson's disease. *Mov Disord* 2012; 27:264–271.

112. Evatt ML, DeLong MR, Kumari M, Auinger P, McDermott MP, Tangpricha V, et al. High prevalence of hypovitaminosis D status in patients with early Parkinson disease. *Arch Neurol* 2011; 68:314–319.

113. Chen H, Zhang SM, Hernan MA, Willett WC, Ascherio A. Diet and Parkinson's disease: A potential role of dairy products in men. *Ann Neurol* 2002; 52:793–801.

114. Kenborg L, Lassen CF, Ritz B, Schernhammer ES, Hansen J, Gatto NM, et al. Outdoor work and risk for Parkinson's disease: A population-based case-control study. *Occup Environ Med* 2011; 68:273–278.

115. Sato Y, Kikuyama M, Oizumi K. High prevalence of vitamin D deficiency and reduced bone mass in Parkinson's disease. *Neurology* 1997; 49:1273–1278.

116. Iwamoto J, Takeda T, Matsumoto H. Sunlight exposure is important for preventing hip fractures in patients with Alzheimer's disease, Parkinson's disease, or stroke. *Acta Neurol Scand* 2012; 125:279–284.

117. Kim JS, Kim YI, Song C, Yoon I, Park JW, Choi YB, et al. Association of vitamin D receptor gene polymorphism and Parkinson's disease in Koreans. *J Korean Med Sci* 2005; 20:495–498.

118. Sanchez B, Lopez-Martin E, Segura C, Labandeira-Garcia JL, Perez-Fernandez R. 1,25-Dihydroxyvitamin D(3) increases striatal GDNF mRNA and protein expression in adult rats. *Brain Res Mol Brain Res* 2002; 108:143–146.

119. Mouatt-Prigent A, Agid Y, Hirsch EC. Does the calcium binding protein calretinin protect dopaminergic neurons against degeneration in Parkinson's disease? *Brain Res* 1994; 668:62–70.

120. de Lau LM, Bornebroek M, Witteman JC, Hofman A, Koudstaal PJ, Breteler MM. Dietary fatty acids and the risk of Parkinson disease: The Rotterdam study. *Neurology* 2005; 64:2040–2045.

121. Demarin V, Lisak M, Morovic S. Mediterranean diet in healthy lifestyle and prevention of stroke. *Acta Clin Croat* 2011; 50:67–77.

122. Perez-Lopez FR, Chedraui P, Haya J, Cuadros JL. Effects of the Mediterranean diet on longevity and age-related morbid conditions. *Maturitas* 2009; 64:67–79.

123. Sofi F, Cesari F, Abbate R, Gensini GF, Casini A. Adherence to Mediterranean diet and health status: Meta-analysis. *BMJ* 2008; 337:a1344.

124. Di Giovanni G. A diet for dopaminergic neurons? *J Neural Transm Suppl* 2009:317–331.

125. Alcalay RN, Gu Y, Mejia-Santana H, Cote L, Marder KS, Scarmeas N. The association between Mediterranean diet adherence and Parkinson's disease. *Mov Disord* 2012; 27:771–774.

126. Perez-Jimenez F, Alvarez de Cienfuegos G, Badimon L, Barja G, Battino M, Blanco A, et al. International conference on the healthy effect of virgin olive oil. *Eur J Clin Invest* 2005; 35:421–424.

127. Johnson CC, Gorell JM, Rybicki BA, Sanders K, Peterson EL. Adult nutrient intake as a risk factor for Parkinson's disease. *Int J Epidemiol* 1999; 28:1102–1109.

128. Petersen MS, Halling J, Bech S, Wermuth L, Weihe P, Nielsen F, et al. Impact of dietary exposure to food contaminants on the risk of Parkinson's disease. *Neurotoxicology* 2008; 29:584–590.

129. Miyake Y, Sasaki S, Tanaka K, Fukushima W, Kiyohara C, Tsuboi Y, et al. Dietary fat intake and risk of Parkinson's disease: A case-control study in Japan. *J Neurol Sci* 2010; 288:117–122.

130. Simon KC, Chen H, Schwarzschild M, Ascherio A. Hypertension, hypercholesterolemia, diabetes, and risk of Parkinson disease. *Neurology* 2007; 69:1688–1695.

131. Fabelo N, Martin V, Santpere G, Marin R, Torrent L, Ferrer I, et al. Severe alterations in lipid composition of frontal cortex lipid rafts from Parkinson's disease and incidental Parkinson's disease. *Mol Med* 2011; 17:1107–1118.

132. Vanschoonbeek K, Wouters K, van der Meijden PE, van Gorp PJ, Feijge MA, Herfs M, et al. Anticoagulant effect of dietary fish oil in hyperlipidemia: A study of hepatic gene expression in APOE2 knock-in mice. *Arterioscler Thromb Vasc Biol* 2008; 28:2023–2029.

133. Duda MK, O'Shea KM, Stanley WC. Omega-3 polyunsaturated fatty acid supplementation for the treatment of heart failure: Mechanisms and clinical potential. *Cardiovasc Res* 2009; 84:33–41.

134. Varga Z. Omega-3 polyunsaturated fatty acids in the prevention of atherosclerosis. *Orv Hetil* 2008; 149:627–637.

135. Bousquet M, Gue K, Emond V, Julien P, Kang JX, Cicchetti F, et al. Transgenic conversion of omega-6 into omega-3 fatty acids in a mouse model of Parkinson's disease. *J Lipid Res* 2011; 52:263–271.

136. Tanriover G, Seval-Celik Y, Ozsoy O, Akkoyunlu G, Savcioglu F, Hacioglu G, et al. The effects of docosahexaenoic acid on glial derived neurotrophic factor and neurturin in bilateral rat model of Parkinson's disease. *Folia Histochem Cytobiol* 2010; 48:434–441.

137. Meng Q, Luchtman DW, El Bahh B, Zidichouski JA, Yang J, Song C. Ethyl-eicosapentaenoate modulates changes in neurochemistry and brain lipids induced by parkinsonian neurotoxin 1-methyl-4-phenylpyridinium in mouse brain slices. *Eur J Pharmacol* 2010; 649:127–134.

138. Morens DM, Grandinetti A, Waslien CI, Park CB, Ross GW, White LR. Case-control study of idiopathic Parkinson's disease and dietary vitamin E intake. *Neurology* 1996; 46:1270–1274.

139. De Natale C, Annuzzi G, Bozzetto L, Mazzarella R, Costabile G, Ciano O, et al. Effects of a plant-based high-carbohydrate/high-fiber diet versus high-monounsaturated fat/low-carbohydrate diet on postprandial lipids in type 2 diabetic patients. *Diabetes Care* 2009; 32:2168–2173.

140. Crochemore IC, Souza AF, de Souza AC, Rosado EL. Omega-3 polyunsaturated fatty acid supplementation does not influence body composition, insulin resistance, and Lipemia in women with type 2 diabetes and obesity. *Nutr Clin Pract* 2012; 27:553–560.

141. Mello VD, Laaksonen DE. Dietary fibers: Current trends and health benefits in the metabolic syndrome and type 2 diabetes. *Arq Bras Endocrinol Metabol* 2009; 53:509–518.

142. Schernhammer E, Hansen J, Rugbjerg K, Wermuth L, Ritz B. Diabetes and the risk of developing Parkinson's disease in Denmark. *Diabetes Care* 2011; 34:1102–1108.

143. Xu Q, Park Y, Huang X, Hollenbeck A, Blair A, Schatzkin A, et al. Diabetes and risk of Parkinson's disease. *Diabetes Care* 2011; 34:910–915.

144. Morris JK, Bomhoff GL, Stanford JA, Geiger PC. Neurodegeneration in an animal model of Parkinson's disease is exacerbated by a high-fat diet. *Am J Physiol Regul Integr Comp Physiol* 2010; 299:R1082–R1090.

145. Murakami K, Miyake Y, Sasaki S, Tanaka K, Fukushima W, Kiyohara C, et al. Dietary glycemic index is inversely associated with the risk of Parkinson's disease: A case-control study in Japan. *Nutrition* 2010; 26:515–521.

146. Trinidad TP, Mallillin AC, Loyola AS, Sagum RS, Encabo RR. The potential health benefits of legumes as a good source of dietary fibre. *Br J Nutr* 2010; 103:569–574.

147. Liu R, Guo X, Park Y, Huang X, Sinha R, Freedman ND, et al. Caffeine intake, smoking, and risk of Parkinson disease in men and women. *Am J Epidemiol* 2012; 175:1200–1207.

148. Hellenbrand W, Boeing H, Robra BP, Seidler A, Vieregge P, Nischan P, et al. Diet and Parkinson's disease. II: A possible role for the past intake of specific nutrients. Results from a self-administered food-frequency questionnaire in a case-control study. *Neurology* 1996; 47:644–650.

149. Hancock DB, Martin ER, Stajich JM, Jewett R, Stacy MA, Scott BL, et al. Smoking, caffeine, and nonsteroidal anti-inflammatory drugs in families with Parkinson disease. *Arch Neurol* 2007; 64:576–580.

150. Ross GW, Abbott RD, Petrovitch H, Morens DM, Grandinetti A, Tung KH, et al. Association of coffee and caffeine intake with the risk of Parkinson disease. *JAMA* 2000; 283:2674–2679.

151. Yanagihara RT, Garruto RM, Gajdusek DC. Epidemiological surveillance of amyotrophic lateral sclerosis and parkinsonism-dementia in the Commonwealth of the Northern Mariana Islands. *Ann Neurol* 1983; 13:79–86.

152. Steele JC, Guzman T. Observations about amyotrophic lateral sclerosis and the parkinsonism-dementia complex of Guam with regard to epidemiology and etiology. *Can J Neurol Sci* 1987; 14:358–362.

153. Shaw CA, Hoglinger GU. Neurodegenerative diseases: Neurotoxins as sufficient etiologic agents? *Neuromol Med* 2008; 10:1–9.

154. Zesiewicz TA, Evatt ML. Potential influences of complementary therapy on motor and non-motor complications in Parkinson's disease. *CNS Drugs* 2009; 23:817–835.

# Dietary Supplements and Other Alternative Substances for the Treatment of Multiple Sclerosis

Angela Senders, Rebecca I. Spain, and Vijayshree Yadav

## Contents

# Introduction

Multiple sclerosis (MS) is a chronic disease affecting the central nervous system (CNS) that is thought to be caused by autoimmune dysfunction [1]. MS affects about 400,000 people in United States and usually has its onset in young adulthood. Clinically, MS is divided into subtypes, namely, relapsing–remitting MS (RRMS), secondary progressive MS (SPMS), and primary progressive MS (PPMS) [2]. RRMS is characterized by periodic neurologic dysfunction (relapses) that can last for days to weeks but eventually improve (remit). Most people with RRMS transition into SPMS after 10–15 years whereupon they incur progressive neurologic decline with or without relapses [3,4]. People with PPMS have progressive neurologic decline from disease onset and usually do not have relapses. Despite the development of several disease-modifying therapies (DMTs) for relapsing MS in the last two decades, MS treatment remains suboptimal [5]. Limitations of DMTs include cost, side effects, and potentially fatal complications [6,7]. Additionally, DMTs may not provide symptomatic relief, and hence, people with MS often take other medications for symptoms such as pain, fatigue, depression, cognitive and balance impairments, and urinary and bowel dysfunction. Because both conventional symptomatic therapies and DMTs are often inadequate to control MS-related symptoms, patients and physicians alike often explore alternative options for MS treatment [8]. This chapter reviews evidence-based CAM approaches that are commonly explored by people with MS.

# Diet

Diet modification and nutritional interventions are the most commonly used CAM approaches for MS management. In 2008, the World Health Organization and the Multiple Sclerosis International Federation examined MS occurrence and treatment in 112 countries, representing 88% of the world's population. This report indicated that diet modification and nutrition are used in 88.3% of respondents [9]. Despite its widespread use, few studies have systematically addressed the role of diet in MS.

Neurologist Roy Swank believed that diets rich in saturated animal fats were deleterious in MS. He devised the "Swank diet," which limited saturated fat intake (no more than 10–15 g a day) and promoted omega-3-rich cod liver oil supplementation. In a 50-year long prospective but uncontrolled study, Dr. Swank evaluated his diet's effect on the survival and disability in 144 people with MS [10]. The 70 "good dieters" who consumed less than 20 g of fat per day were compared to the 74 "bad dieters" [11,12]. Thirty-four years later, Swank observed that 58 of 74 of the bad dieters had died—more than double the death rate of the good dieters [13]. After 50 years, only 15 of the original participants survived, all of them belonging to the good dieters group and the majority (13 of 15) still ambulatory [10].

Dietary and lifestyle-related illness may increase the risk of MS disease severity and progression. A 2010 study found a high prevalence of such diseases, including high blood pressure (21%), diabetes (~8%), and hyperlipidemia (14%) in a survey of 8900 people with MS [14]. This study demonstrated that the presence of one or more vascular comorbidities increased the risk of walking disability in MS and the risk increased with the number of vascular conditions. The heightened risk was not trivial—those people with MS and vascular comorbidities needed walking assistance up to 6 years prior to those without vascular comorbidities [14]. In a 2011, 2-year study, Weinstock-Guttman and colleagues demonstrated that higher low-density lipoprotein (LDL, or "bad" cholesterol) and higher total cholesterol levels were associated with greater MS disability, while higher high-density lipoprotein level (HDL, or "good" cholesterol) was associated with lower inflammatory disease activity on brain magnetic resonance imaging (MRI) [15].

The MS Center at Oregon Health & Science University is currently conducting a 1-year clinical trial of the effects of a very low-fat vegan diet (total fat < 20% of daily intake) on RRMS disease activity, including brain MRI, relapse rate, disability, and changes in blood biomarkers. In this single-blind, randomized controlled trial (RCT), 60 people are randomly assigned to either an active group that undergoes a 10-day residential training program of vegan diet and exercise or a wait-listed group. The study is expected to be completed in March 2013.

While diet is the most commonly used CAM approach for MS, there is currently insufficient evidence about efficacy of any particular diet until rigorous studies are completed. However, the recent associations between vascular comorbidities and MS disability suggest that dietary choices that can decrease blood pressure, lower cholesterol, and balance blood sugars may also positively impact MS.

## Essential fatty acids

Omega-3 and omega-6 fatty acids (FAs) are important nutritional building blocks that play a critical role in immune and inflammatory reactions. These polyunsaturated fatty acids (PUFAs) are major structural components of the phospholipid cell membrane, thereby influencing membrane permeability, flexibility, and activity of membrane-bound enzymes. During an inflammatory response, omega-6 FAs can be converted into either anti-inflammatory prostaglandin E1 series molecules or pro-inflammatory agents, including prostaglandin 2-series, thromboxane A2, and leukotriene B4 [16]. Omega-3 eicosapentanoic acid (EPA) can be converted into anti-inflammatory 3-series prostaglandins, thromboxane A3, and leukotriene B5 [16]. Figure 17.1 shows the conversion pathways of essential fatty acids to pro- and anti-inflammatory agents [17].

Because humans cannot synthesize omega-3 or omega-6 FAs, these nutrients are considered essential and must be supplied in the diet. Dietary sources of

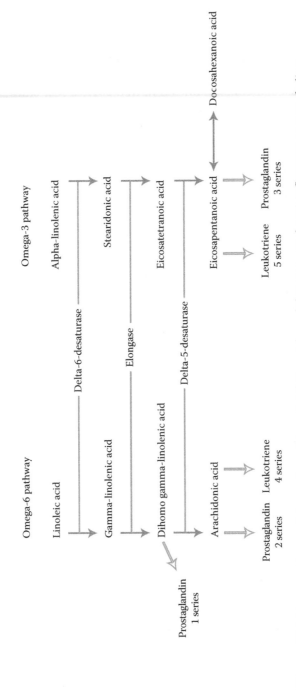

**Figure 17.1** *Desaturation and elongation of essential fatty acids and their pro- and anti-inflammatory metabolites.*

essential fatty acids are provided in Table 17.1. The current typical Western diet supplies about 15 times more omega-6 FA than omega-3 FA [16], and some believe this overabundance will promote pro-inflammatory reactions and increase the risk for chronic conditions known to have an underlying inflammatory component, such as cardiovascular disease and cancer [16]. Others argue, however, that evidence of linoleic acid's pro-inflammatory effects in the current dietary range is lacking [18] and, indeed, studies of omega-6 supplementation have shown benefit in cardiovascular trials [19]. Despite the debate, experts generally agree that a reduction in dietary omega-6 FAs along with increased omega-3 intake will promote good health, although optimal dosing and ratios have yet to be determined.

## Omega-3 fatty acids

Epidemiologic and case-control studies suggest an inverse relationship between fish consumption and the risk of MS [20,21], and laboratory studies have demonstrated anti-inflammatory and immunomodulatory effects of omega-3 supplementation in humans with MS. Gallai et al. found that 6 g of fish oil per day (containing 3.0 g EPA, 1.8 g DHA) for 3 months significantly decreased levels of pro-inflammatory cytokines secreted from peripheral blood mononuclear cells for both MS subjects and healthy controls [22]. An open-label pilot study of ten RMMS patients who received 8 g of fish oil per day (containing 2.9 g EPA, 1.9 g DHA) for 3 months demonstrated that omega-3 supplementation caused a significant decrease in matrix metalloproteinase-9 levels, a molecule that appears to be important for T-cell migration into the CNS in MS [23].

Despite these intriguing findings, there have been only two studies evaluating the effects of omega-3 FA on MS disease activity. Bates et al. conducted a double-blind, placebo-controlled RCT in which 312 MS patients received either fish oil (containing 1.71 g EPA and 1.41 g DHA per day) or olive oil placebo for 2 years [24]. The authors reported a trend toward improvement in disease severity in omega-3-treated subjects compared to controls (measured by Expanded Disability Status Score, p=0.07). In a 2012 RCT, Torkildsen et al. administered 5 g of fish oil per day (1.3 g of EPA, .850 g of DHA; n=46) or placebo (corn oil, n=45) to people with MS who were not taking DMT [25]. After 6 months, all subjects began interferon beta-1a for additional 18 months. There was no difference between groups in the cumulative number of gadolinium-enhancing MRI lesions during the first 6 months of fish oil (median difference, 1; 95% CI, 0–3; p =.09). The addition of interferon reduced disease activity for both groups equally. No differences were seen in relapse rate or disability progression between groups after 6 and 24 months.

## Omega-6 fatty acids

Several animal studies have demonstrated that supplementation with linoleic acid can reduce the severity of experimental autoimmune encephalomyelitis

**Table 17.1** Dietary and Supplement Sources of Nutraceuticals for MS

| Neutraceutical | Food Source | Dietary Supplement |
|---|---|---|
| Alpha-linolenic acid | Flaxseeds, walnuts, chia seeds, hemp, canola oil, butternuts | Flaxseed oil |
| Eicosapentanoic acid | Cold-water fish, including herring, lake trout, anchovies, albacore tuna, mackerel, salmon. Certain algae | Fish, krill, and cod liver oils |
| Docosahexanoic acid | Cold-water fish, including herring, lake trout, anchovies, albacore tuna, mackerel, salmon. Certain algae | Fish, krill, and cod liver oils |
| Linoleic acid | Safflower oil, sunflower seeds and oil, pine nuts, corn oil, soybean oil, pecans, brazil nuts, sesame oil | Supplements of conjugated linoleic acid are marketed for weight control |
| Gamma-linolenic acid | Borage seed oil, evening primrose oil, black currant seed oil | Borage, evening primrose, and black currant seed oils |
| Arachidonic acid | Red meat, duck, poultry, fish, eggs, dairy | Typically not supplemented, sometimes found in essential fatty acid complex formulas |
| Vitamin B12 | Salmon, trout, tuna, haddock, clams, red meat, chicken, egg yolk, dairy | Individual supplement or included in B complex or multivitamin formula. Intramuscular injections or IV preparations |
| Vitamin C | Oranges, grapefruit, kiwi, strawberries, bell peppers, broccoli, cauliflower, Brussels sprouts, cantaloupe, tomato, spinach, green peas | Individual supplement or part of a multivitamin formula. IV preparations |
| Vitamin D | Cod liver oil, swordfish, salmon, tuna, sardines, liver, egg yolk, mushrooms, fortified dairy products | Cod liver oil, fortified fish oils, individual D2 or D3 supplements, part of multivitamin formula. Intramuscular or IV preparations |
| Vitamin E | Wheat germ, sunflower seeds, almonds, hazelnuts, peanuts, olive oil, spinach, broccoli, kiwi, mango, tomato | Alpha-tocopherol is often isolated and sold as individual supplement or in multivitamin formula. Vitamin E complexes of mixed tocopherols are available. IV preparations |
| Magnesium | Wheat bran and wheat germ, almonds, spinach, cashews, soybeans, peanuts, potato, legumes, brown rice, avocado, banana, halibut, dairy products | Individual supplement or part of multivitamin or multimineral formula. IV preparations |
| Alpha-LA | Red meat, organ meats, brewer's yeast, broccoli, spinach, chard, and collard greens | Individual supplement or IV preparation |
| Cannabis | | Prescription only for medically necessary circumstances |
| Ginkgo biloba | | Available in tea, tincture, capsule, or tablet form |
| Ginseng | | Available in tea, tincture, capsule, or tablet form |
| Echinacea | | Available in tea, tincture, capsule, or tablet form |
| Green tea | | Available in tea, tincture, capsule, or tablet form. Most preparations are decaffeinated |
| Probiotics and helminthes | Probiotics are found in fermented foods like yogurt, kefir, sauerkraut, kimchi, tempeh, and kombucha tea | Helminthes are not commercially available. Probiotics are sold over the counter in capsule, tablet, powder, and liquid forms |

(EAE) [26,27], but the small number of human RCTs has shown little difference in MS relapse rate or disability [28–31]. The majority of these studies used olive oil as a placebo that was thought to be immunologically inactive, although evidence exists that components of olive oil are capable of reducing EAE disease severity at the clinical and molecular level [32,33], suggesting that olive oil may have confounded the data interpretation.

In summary, while the anti-inflammatory and immune effects of PUFAs are compelling in the laboratory, the few clinical trials conducted to date have not shown a beneficial effect of PUFAs on disease progression or clinical relapse of MS. Uncertainty also exists regarding optimal dosing and the desired ratio of omega-6 and omega-3 FAs in the diet. However, PUFAs do appear safe in trials up to 2 years in length and can improve cardiovascular health that may secondarily benefit MS. Therefore, supplementation of PUFAs is a safe and reasonable option while awaiting further evidence to determine their ultimate efficacy in MS outcomes.

## Vitamins

### Vitamin B12

Vitamin B12, or cobalamin, plays a structural and functional role in the CNS as it is essential for normal myelin formation. Vitamin B12 deficiency can lead to CNS demyelination and may mimic both MS symptoms and MRI appearance. Vitamin B12 cannot be synthesized by the body and must be supplied by diet, mainly from meat and dairy products (Table 17.1), and deficiency results from inadequate intake (e.g., vegetarianism) or malabsorption. Malabsorption can occur due to inadequate production or function of intrinsic factor (IF) necessary for B12 uptake in the ileum, disorders of the ileum such as Crohn's, or competition from intestinal parasites.

While some studies have found an association between B12 deficiency and MS [34,35], others have not [36,37]. Hypotheses regarding the etiology of B12 deficiency in MS include increased consumption due to proliferation of inflammatory cells and during myelin repair, or even as an etiological cause of MS [38]. Additional evidence points to an immunoregulatory role of vitamin B12 via modulation of TNFα and other cytokine activity [39], providing further argument for supplementation in MS. While combination therapy of vitamin B12 and beta-interferon was shown to be superior to interferon alone in the EAE model [40], to date few trials of B12 supplementation have been conducted in MS [41].

The role of vitamin B12 in the pathogenesis and treatment of MS is unclear; however, given the ease of both testing and supplementation, testing of all people with MS at least once is recommended. If low or low normal, the cause of vitamin B12 deficiency should be considered before supplementation as it may affect successful therapy.

## Vitamin C

The antioxidant vitamin C is the active (L-) form of ascorbic acid. Ascorbic acid is obtained exclusively from dietary sources and crosses the blood–brain barrier. In a study of serum metabolites in MS, ascorbic levels were discovered to be 35% lower in people with MS than controls; the authors concluded the deficit might be due to a combination of impaired energy metabolism and increased production of toxic radical species [42]. In regard to MS therapy, ascorbic acid is part of an experimental compound GEMSP currently being studied in EAE. GEMSP, a mixture of fatty acids, ascorbic acid, α-tocopherol succinate (vitamin E), and amino acids, has effectively prevented leukocyte trafficking and crossing of the blood–brain barrier in mice with EAE [43]. In a 1990 study, 10 people with MS were treated with a combination of vitamin C (2000 mg), selenium (6 mg), and vitamin E (480 mg) for 5 weeks. The combination was found to be safe and improved glutathione peroxidase activity; however, no clinical outcomes were collected [44].

In summary, there is insufficient evidence to recommend routine supplementation of vitamin C for the prevention or treatment of MS; however, it appears safe and well tolerated.

## Vitamin D

Vitamin D, a fat-soluble steroid hormone, has long been known to play a crucial role in bone health and the prevention of osteoporosis, rickets, hypocalcemia, and hypophosphatemia. As vitamin D levels fall, intestinal absorption of calcium also falls, leading to a decrease in serum calcium. Low calcium causes serum parathyroid hormone (PTH) concentrations to rise, which in turn causes calcium resorption from the bone and can result in a weakening of the bone structure. The role of vitamin D in extra-skeletal health is being explored in many diseases, including MS, and excellent reviews of vitamin D in MS are available [45,46].

Vitamin D is obtained from the conversion of a precursor form in the skin after exposure to ultraviolet B rays from sunlight. Additional sources are obtained from vitamin D-containing foods (fish, eggs, liver, mushrooms, and fortified foods such as dairy products) and supplements (Table 17.1). Both sun-catalyzed and dietary sources are converted into an inactive vitamin D (calcidiol, or 25-hydroxyvitamin D) in the liver, levels of which are routinely measured in the blood. Inactive calcidiol is then converted to a short-lived active form (calcitriol) primarily in the kidney. Receptors for calcitriol exist on many cell types including immune cells and in many organs (bone, muscle, kidney, breast, brain, and intestine) as well as on vitamin D-regulated genes, which account for its widespread effects.

Growing data support an association between low vitamin D intake and risk of MS in adults [47], children [48], and offspring of mothers with low vitamin D intake [49]. Low sunlight exposure may also contribute to risk of MS; in fact, studies using geospatial mapping of sunlight exposure better predict the risk of MS than the long-known association between place of origin, or latitude, and MS incidence [50,51]. Support for the causative role of low vitamin D in MS is shown by a higher risk of MS in families with a genetic mutation that prevents conversion of calcidiol to the active calcitriol [52]. Low vitamin D levels are associated with more relapses [53] and worse disability levels in both adult [54] and pediatric MS [48].

The role of vitamin D in EAE is less clear with some studies reporting a reduction of EAE symptoms with supplementation [55], while others suggest deficiency actually lowers the risk of EAE development [56]. The few longitudinal studies of vitamin D supplementation in people with MS have also had mixed outcomes with both beneficial [57] and non-beneficial results [58]. Longitudinal studies to determine the therapeutic potential of vitamin D are ongoing.

Because of the widespread prevalence of vitamin D insufficiency and deficiency based on blood levels, determination of optimal blood levels of vitamin D in MS and the general population is also under investigation. Normal serum levels of the inactive calcidiol form of vitamin D are generally ≥30 ng/mL depending on the specific laboratory. Levels of 20–29 ng/mL are considered insufficient, and levels <20 ng/mL are deemed deficient. Standards have recently been called into question and many research studies involving vitamin D are ongoing to determine ideal blood levels in both general and specific (e.g., MS) populations.

Low vitamin D levels can be treated with scheduled sun exposure. However, as unfiltered UVB rays also increase the risk of skin cancers, many health providers recommend oral supplementation. The most concentrated dietary sources include oily fish (sardines, mackerel, herring, tuna, and salmon). Supplemental vitamin D is supplied as D2 (ergocalciferol) generally derived from fungal and plant sources and vitamin D3 (cholecalciferol) derived from animal sources. There is some evidence that D3 raises vitamin D levels more efficiently than D2 [59]. The 2011 updated recommendations from the Institute of Medicine are 600 IU/day vitamin D increasing to 800 IU/day after age 70, along with 1000 mg/day of calcium, increasing to 1200 mg/day after age 70 [60]. However, based on an individual's geographic location, sun exposure, and diet, these daily allowances may still be inadequate to maintain normal levels.

The safety of vitamin D and calcium supplementation has been studied. In a 52-week dose-escalation study, patients were treated with up to 40,000 IU of vitamin D3 along with 1200 mg calcium daily without any serious adverse events [61]. Even so, toxic effects of over supplementation of vitamin D have been reported, including tremors, confusion, hypercalcemia, and elevated vitamin D levels after taking daily doses of 5500 IU vitamin D and 2020 mg calcium [62]. Therefore, supplementation should be periodically monitored.

Because of increasing evidence associating vitamin D deficiency with risk of MS and MS disability, it is reasonable to check vitamin D levels in people with MS and their family members. Patients should have periodic monitoring of vitamin D and calcium levels and a review of all individual and combination supplements to ensure appropriate dosing.

## Vitamin E

The antioxidant vitamin E is a well-known reactive oxygen species (ROS) scavenger that protects cells from apoptosis and cell death. Vitamin E is an essential nutrient found primarily in plant oils such as sunflower, safflower, wheat germ, and peanut oil, although small amounts can also be found in cereals, fruits, vegetables, and nuts. It is stored in fat and transported by specific receptors.

Neurodegeneration in MS is thought due in part to oxidative damage; thus, neuroprotective effects of vitamin E supplementation are theoretically possible. In fact, hippocampal damage was reduced and remyelination promoted in rats who received a combination of vitamin E and D3 supplementation using an ethidium bromide model of MS [63]. However, the few epidemiological studies of dietary vitamin E intake and MS have not demonstrated a benefit [64,65], and no prospective studies of vitamin E supplementation in MS have been undertaken to date. Therefore, there is insufficient evidence to recommend routine checking of vitamin E levels or for supplementation with vitamin E for the prevention or treatment of MS.

## Magnesium

Magnesium (Mg) is a trace electrolyte with widespread effects on many metabolic functions. Hypomagnesemia can lead to impaired electrolyte transport and recycling, abnormal vasoactive responses to hormones, and poorly functioning thiamine (vitamin B1). Abnormalities of these processes may result in hypertension, mitochondrial dysfunction, Wernicke's encephalopathy, pseudogout, and other ill effects [66]. Magnesium also plays roles in nitric oxide function, inhibition of osteoclasts, and muscle contraction and may have a role in migraines [67]. Studies finding decreased levels of Mg in the CNS of people with MS have sparked interest in the role of low Mg in the pathogenesis and treatment of MS [68].

Mg requires both PTH and vitamin D for absorption in the gut. Therefore, people at risk for vitamin D deficiency or who have low PTH are also at risk for Mg deficiency [67]. An epidemiological study from Denmark reviewed 14-day food diaries and found that the MS population had lower folate, magnesium, and copper intake than the overall Dutch population. Furthermore, people with SPMS (n=32) had lower levels of magnesium, calcium, and iron intake than other forms of MS (n=48). All together, the data suggested a role of micronutrients in MS, although the study did not specifically implicate Mg [69].

A 1986 open-label trial of combined Mg, vitamin D, and calcium found a decreased relapse rate among 16 people with MS using their prior relapse history as a control [70]. A more recent (2000) case report of the symptomatic benefit of Mg found that Mg successfully reduced leg spasticity when given at a dose of 100 mg daily [71]. No current trials of Mg in MS are under way.

There is currently insufficient data to determine a role of Mg in the pathogenesis or treatment of MS. However, because of the widespread effects of Mg and the many ways to become Mg deficient, screening for Mg deficiency is advised. Treatment of Mg deficiency should take into account coexisting vitamin D, calcium, PTH, and nutritional statuses.

## Antioxidants

### Lipoic acid

Lipoic acid (LA) is an antioxidant and dietary supplement, also known by the names alpha-LA or α-LA. LA is made naturally in nearly every cell of the body during the normal metabolism of fatty acids by the action of the enzyme LA synthase. LA is also found in small quantities in foods such as kidney, liver, broccoli, spinach, and yeast. Endogenously produced LA is primarily enzyme bound, while oral and intravenous forms are free and detectable in the blood along with its reduced form dihydrolipoic acid (DHLA). LA occurs in two isoforms: the naturally occurring R form and the synthetically produced S enantiomers. While the R form is reduced many times faster than the S form [72], racemic (50:50) LA was found to suppress EAE equally well to purified R and S forms alone [73]; thus, the racemic form continues to be used in MS research.

LA may be beneficial in MS by its antioxidant activity and potential for immune modulation. LA and DHLA together form a redox couple that functions as a cofactor for several mitochondrial enzymes necessary for energy production [72]. In vitro studies credit LA with the ability to scavenge free radicals, chelate metals, regenerate glutathione, and repair oxidative damage [74]. LA inhibits a number of inflammatory cytokines in the blood as well as suppresses activation and cytotoxicity of natural killer cells, thereby suppressing inflammation [75,76]. By stimulating cAMP production, LA may inhibit microglial activation, thereby serving to slow neurodegeneration in MS as well as other neurodegenerative disorders such as Alzheimer's and Parkinson's disease [77,78].

Several laboratories have shown LA to be effective in the animal EAE model, both to reduce disability and confer neuroprotection [73,79]. In 37 people with MS, a double-blind, placebo-controlled, dose-finding trial of orally administered racemic LA was conducted to relate serum LA concentrations to changes in serum markers of inflammation [80]. The 2-week trial found that the 1200 mg daily dose was significantly better than 600 mg in producing reliable serum LA concentrations, although there was still considerable intersubject variability in peak concentrations. The 1200 mg dose was also well tolerated and produced lower levels of serum-soluble intercellular adhesion molecule-1 and matrix metalloproteinase-9, markers of inflammation, in a dose–response fashion. The most common adverse reactions to LA in this trial were gastro-intestinal intolerance and malodorous urine, similar to previous diabetic neu-ropathy trials [81,82]. Because LA may improve insulin resistance, diabetics should monitor their blood sugars when starting LA supplementation.

Ongoing RCTs at Oregon Health & Science University and the Portland Veterans Affairs Medical Center in Portland, OR, are examining the anti-inflammatory effects of LA in acute optic neuritis and the neuroprotective effects of LA in SPMS, respectively (www.clinicaltrials.gov). Because of the proven long-term safety and potential for widespread health benefits, LA is a safe but unproven option for use in MS while awaiting definitive results from clinical trials.

# Herbs
## Cannabis

Cannabis has been used medicinally for thousands of years to relieve symptoms of pain, nausea, anorexia, muscle stiffness, bladder dysfunction, and others; however, only more recently has cannabis been rigorously studied to determine its objective benefits. The *Cannabis sativa* plant produces over 421 chemical compounds, many of which are unique to the plant. The compound delta-9 tetrahydrocannabinol ($\Delta^9$–THC) produces much of the well-known psychotropic effects. Other cannabis compounds such as cannabidiol (CBD) and cannabigerol (CBG) produce weak or no psychotropic effects and thus are under investigation for medicinal purposes [83].

A number of case reports as well as small, single-center trials have examined the effects of various cannabis preparations on MS spasticity, cognitive, sleep, pain, and pathophysiological outcomes. A trial of the effects of smoked cannabis (containing 4% $\Delta^9$–THC) for 3 days followed by an 11-day washout and then crossover period demonstrated a significant improvement in muscle spasticity using the Ashworth scale among the 30 of 37 completers. Pain also decreased but cognitive performance worsened [84]. A larger multicenter trial in the United Kingdom and Romania demonstrated a significant subjective benefit in spasticity among people with MS (n = 124) using the oromucosal spray Sativex, a roughly equal mixture of $\Delta^9$–THC and CBD, versus placebo (n=64), although secondary objective measures were not different [85]. An open-label trial of long-term Sativex use for an average of 434 days (range, 21–814) demonstrated a high dropout rate due to perceived lack of efficacy (18%) and adverse events (12%); however, for those that received an initial benefit, the efficacy was maintained [86].

A 2003 large, multicenter, RCT of cannabis in MS treated 630 people with MS at 33 centers in the United Kingdom [87]. Subjects received either oral cannabis extract (n=211), Δ⁹–THC (n=206), or placebo (n=213) over a 15-week period. While there was no change in the primary outcome of objective improvement in spasticity using the Ashworth scale (p=0.40), there was patient-reported improvement in both pain and spasticity of about 15% over that of the improvement reported by the placebo group for both cannabis preparations. The 12-month extension of the study found improvement in both objective (Ashworth scale) and patient-reported spasticities in the cannabis groups [88]. The same U.K. team conducted a second (2012) double-blind, placebo-controlled RCT but this time used a patient-reported measure of muscle stiffness with a validated 11-point category rating scale as the primary outcome [89]. Subjects with SPMS (67%), PPMS (24%), and RRMS (9%) received either Δ⁹–THC (n=144) or placebo (n=135). After a 2-week titration phase from 5 to 25 mg Δ⁹–THC daily doses, subjects were maintained on their dose for an additional 10 weeks. Overall, more people in the Δ⁹–THC group reported improvement in stiffness than the placebo group (29.4% vs. 15.7%, p=0.004) regardless of baseline severity of stiffness, ability to walk, geographic area, and use of additional spasticity and analgesic medications. Notably, subjects on Δ⁹–THC were more likely to drop out (21%) than those on placebo (6.7%).

Serious adverse events in the 2012 U.K. trial were higher in the Δ⁹–THC group (4.9%) than placebo (2.2%) with urinary tract infection as the most common event [89]. Reporting of dizziness, disturbance in attention, balance disorder, somnolence, dry mouth, nausea, diarrhea, fatigue, asthenia, feeling abnormal, disorientation, confused state, and falls was also more common in the Δ⁹–THC than placebo group. The high rate of adverse effects (93% vs. 75%) in the Δ⁹–THC group was attributed by the authors to the rapid dose escalation [89].

Safety concerns regarding abuse and intoxication as a result of medicinal cannabis use were explored in a 2011 review of Sativex [90]. The author found that euphoria rates were low, that abrupt cessation did not produce a withdrawal syndrome, and that no reports of abuses or diversions of Sativex existed. While this author noted that the potential for abuse was related to the amount of cannabis consumed, other reports more clearly describe a negative impact of street cannabis use on cognitive function in people with MS [91].

Interestingly, while basic science research is exploring the symptomatic and immunomodulatory effects of cannabinoid subtypes, clinical trials to date have all utilized the psychotropic Δ⁹–THC cannabinoid alone, in combination products, or as part of the *Cannabis sativa* plant. The subjective benefits of cannabis in the trials are difficult to interpret in the context of the psychotropic effects of Δ⁹–THC that may also influence safety analyses. Because of the limited data, use of cannabis for symptom management in MS should be used with caution, particularly in those with cognitive deficits and balance problems. However, cannabis preparations could be considered for those failing conventional treatments for subjective spasticity. Based on the basic science literature, there exists the potential for improved efficacy of cannabis with reduced side effects once clinical trials using targeted cannabinoids without psychotropic properties are conducted.

## Ginkgo biloba

*Ginkgo biloba* is a Chinese herb that has traditionally been used to improve vascular insufficiency and memory impairment. Three clinical studies have explored *Ginkgo biloba* and cognitive function in MS. In a 2006 double-blind, placebo-controlled, parallel group study, Johnson et al. found significant improvements in self-reported fatigue and a quality of life scale for the gingko group (n=11, daily dose of 240 mg) compared with placebo (n=11) [92]. A 2007 pilot study by Lovera et al. found significant improvement in the objective Stroop color word memory test (p=0.015) and the self-rated Retrospective Memory Scale of the Perceived Deficits Questionnaire (p=0.016) for the gingko group (n=20, daily dose of 240 mg) compared with placebo (n=18) [93]. However, a larger trial (n=120, daily dose 240 mg) by this same group found ginkgo did not produce significant improvements in cognitive performance [94]. Ginkgo was generally well tolerated in the aforementioned studies. At present, there is insufficient evidence to determine the effect of *Ginkgo biloba* on MS-related symptoms.

## Ginseng

Ginseng is an ancient Chinese herb that is thought to have antioxidant and corticosteroidal activity. Common forms of ginseng include Asian ginseng (*Panax ginseng*), also known as Chinese or Korean ginseng, and American ginseng (*Panax quinquefolius*). Ginsenosides are the primary active constituents in ginseng, the quantity and composition of which varies depending on the species of ginseng, age of the plant, cultivation methods, season of harvest, and extraction and preservation methods [95].

Ginseng's effect on disease activity has been tested in two animal studies in MS [96,97]. While one study used aqueous American ginseng and the other an acidic polysaccharide of Asian ginseng, both studies showed decrease in EAE disease activity. No human studies have assessed ginseng's effect on disease activity in MS. However, one study using ginseng for MS fatigue has been conducted. This double-blind, placebo-controlled, crossover trial of American ginseng (escalating daily doses of 20–80 mg active component over 6 weeks, n=56) failed to produce significant changes in self-reported fatigue on the Fatigue Severity Scale, the Modified Fatigue Impact Scale, or the Real-Time Digital Fatigue Score. There were no significant side effects with American ginseng, though some subjects reported nausea, insomnia, headache, rash, and a flu-like syndrome. Other safety reports indicate that ginseng may reduce the effects of warfarin [98] and that high doses can cause a "ginseng abuse syndrome" including hypertension, nervousness, irritability, insomnia, rash, and diarrhea [99]. At present, there is no clinical evidence that ginseng can improve MS symptoms or disease activity.

## Echinacea

Echinacea is an herb that is thought to stimulate the immune system in ways that can treat the common cold. There are several different species of Echinacea, of which the leaves, stems, roots, and flowers are used medicinally. *Echinacea purpurea* extract has been shown to affect cytokine and chemokine levels in healthy people [100], and it is suspected that modulation of these mediators, and of leukocyte activity, may influence immune response to upper respiratory infections. However, there is concern that stimulation of the immune system, which may be beneficial for the common cold, could be harmful for people with MS. Therefore, while there are no specific studies of Echinacea's effect on MS, MS CAM experts generally recommend against use of Echinacea [101].

## Green tea

Green tea has been consumed by humans for millennia and is thought to have medicinal value based on its high content of catechins, the most abundant and biologically active of which is epigallocatechin-3-gallate (EGCG). Both in vitro and in vivo studies suggest a therapeutic role of EGCG in treating immune disorders based on its ability to inhibit B- and T-cell proliferation and to modulate immune responses to stimulation [102]. In EAE, green tea extracts have demonstrated a downregulation of matrix metalloproteinases that are believed to be involved in MS pathogenesis [103]. In addition, the combination of glatiramer acetate and EGCG induced regeneration of hippocampal axons in an EAE model [104]. Both pretreatment [105] and posttreatment [106] with EGCG ameliorate clinical symptoms in mice induced with EAE.

A pilot trial of the safety and neuroprotective effects of green tea for people with MS is currently under way [107], but at this time, there is insufficient evidence to determine the efficacy of green tea extracts for MS.

# Miscellaneous

## Low-dose naltrexone

Naltrexone is an opiate antagonist that is approved by the U.S. Food and Drug Administration for the treatment of alcohol and opiate dependence. A typical daily dose of 50–100 mg will counteract the effect of opiates by blocking opioid receptors. However, a lower dose (3–5 mg daily) of naltrexone (LDN) can trigger a prolonged release of endogenous opioids such as β-endorphin [108,109] and [met]enkephalin [110]. Endogenous opioids help regulate pain, inflammation, learning, memory, and mood [111], and opioid imbalances may occur in MS [112]. For this reason, LDN is used off-label to treat a variety of MS symptoms.

Despite anecdotal reports of the symptomatic benefit of LDN, evidence supporting its use in MS is limited. An uncontrolled, open-label pilot study (n=40) in PPMS found significant reduction in objective spasticity (using the modified

Ashworth scale) after 6 months of daily treatment with 4.5 mg of LDN compared to baseline (p<0.01) [109]. No differences were seen in fatigue, pain, quality of life, or EDSS scores. A double-blind, placebo-controlled, crossover trial by Cree et al. (n=60; RRMS, SPMS, and PPMS) found that 8 weeks of treatment with LDN was associated with improvement in self-reported mental health measures (p values from <0.01 to 0.05), but not in self-rated physical outcome measures on the global Multiple Sclerosis Quality of Life Inventory [113]. In contrast, a similarly designed 8-week crossover trial by Sharafaddinzadeh et al. (n=86; RRMS and SPMS) found that LDN (4.5 mg/day) had no effect on mental health outcomes [114]. Notably, the Cree study claims to be the first patient-funded trial in MS, which may have introduced selection bias and also included subjects on DMTs that make comparison of the two studies more difficult.

No significant adverse events were observed in any of the aforementioned described clinical trials, and LDN appeared to be well tolerated. The most frequent side effects reported were nausea, epigastric pain, mild irritability, headache, and joint pain, all of which were minor and did not interfere with daily function. LDN appears safe and may provide improvement in mental and physical symptoms in people with MS. There is potential for opiate withdrawal symptoms for people who are taking concomitant opiate medication, and practitioners should be alert to this.

## Prokarin

Prokarin is a proprietary blend of histamine and caffeine developed by Elaine DeLack, a nurse who herself has MS, for the treatment of MS symptoms. The product, available at compounding pharmacies, is delivered transdermally by applying a medicated disk to the trunk of the body during waking hours. A case series reported by DeLack et al. found Prokarin was safe and well tolerated, but only 55% of 167 patients experienced at least some improvement in their MS symptoms (strength, balance, bladder control, fatigue, heat tolerance, cognitive function, peripheral edema, and activities of daily living), and 45% either discontinued treatment or found no benefit [115,116]. An underpowered 2002 double-blind, RCT of Prokarin (n=21) versus placebo (n=5) found significant improvement in fatigue for those taking Prokarin (p<0.02) [117].

Prokarin appears to be safe and no serious adverse events have been reported. Side effects reported in the case series included rash at the site of application, transient headache, diarrhea, abdominal discomfort, and sleep disturbance. Other than anecdotal accounts, there is little evidence to support its use in MS at this time.

## Probiotics and helminthes

Probiotics are live microorganisms that confer a health benefit on the host. Traditional probiotic supplements contain healthy human gastrointestinal organisms; however, recent research indicates that administration of organisms typically thought of as pathogenic may have benefit in MS. Rationale for this emerging therapy stems from the hygiene hypothesis, which states that excessively clean living conditions and a lack of early childhood exposure to infectious agents (e.g., bacteria, parasites) can alter the normal development of the immune system, thereby increasing susceptibility to asthma, allergies, and some autoimmune diseases. Indeed, epidemiologic evidence suggests improved hygiene and higher socioeconomic conditions are associated with a higher prevalence of MS [9,118].

In a 2007 observational study in Argentina, Correale et al. identified a group of 12 RRMS patients with mild, asymptomatic intestinal parasitosis who were not treated for the infection per standard medical practice. The infected MS patients were matched to 12 uninfected RRMS patients and followed prospectively for 7.5 years. During the first 5 years, parasite-infected MS patients had significantly less clinical and radiological disease activity compared with uninfected MS individuals [119]. After 5.25 years, four of the infected patients required antiparasitic treatment, after which both their clinical and radiological disease activities increased to levels observed in uninfected MS patients [120].

In 2011, Fleming et al. conducted an open-label pilot study at the University of Wisconsin in which sterilized *Trichuris suis* (whipworm) ova (TSO) were administered to five newly diagnosed, treatment-naive RRMS patients every 2 weeks [121]. At the end of 3 months, the mean number of new gadolinium-enhancing MRI lesions fell from 6.6 at baseline to 2.0. Similarly to the Argentinean study, the beneficial effect of the parasite exposure ended after the parasites were removed; the mean number of new gadolinium-enhancing lesions rose to 5.8 two months after TSO was discontinued. Fleming and his group are currently conducting an extension of the pilot study with 18 MS patients; results are expected at the end of 2014 [122].

No significant adverse effects were observed in the Fleming study and TSO was well tolerated. Transient mild gastrointestinal symptoms (e.g., 2–3 loose stools per day) occurred in three of five subjects [121]. Safety of oral TSO administration has also been demonstrated in controlled clinical trials of ulcerative colitis [123], Crohn's disease [124], and allergic rhinitis [125], the longest study lasting 6 months. At this time, further research is required to fully explore the safety and effects of helminthes in MS, and administration of TSO or other helminth preparations outside of controlled trials with continuous safety monitoring is discouraged.

## Bee venom therapy

Bee venom contains many pharmacologically active substances and is purported to reduce MS disease activity and cause clinical improvement. Bees are placed on specific body parts with tweezers until they sting, and stingers are left in place for up to 15 min. Treatments are typically given two to three times a week, often administered by private beekeepers. Bee venom, although not an FDA-approved therapy, is also provided by some CAM health-care providers by subcutaneous injection of venom, rather than live bees. Two clinical trials of bee venom therapy have been conducted in MS. In an open-label study, Castro et al. divided nine people with progressive forms of MS into four groups. Each group received a total of 0.025 mg/week, 0.1 mg/week, 0.5 mg/week, or 2 mg/week of intradermal bee venom extract for 4–72 weeks, depending on tolerability. The authors described a worsening of neurological symptoms in four people with MS, subjective improvement in three, and objective improvement in two. There was no correlation between the degree of worsening and the dose of bee venom [126]. A randomized crossover trial of 26 people with RRMS and SPMS did not demonstrate any beneficial effects of bee venom on MRI activity, relapse frequency, disability, fatigue, or quality of life [127].

No anaphylactic reactions or other serious adverse events were recorded in either of the trials [126,127]. While the therapy appears to be generally well tolerated, reactions to bee venom can be varied depending on the individual and the location of the sting. Optic neuritis has been reported after bee stings near the eye [128], hepatotoxicity has been described [129], and anaphylaxis is always a concern due to the rate of allergies to bee venom in the general population. Based on the limited evidence of efficacy and potential for adverse effects, bee venom therapy is not recommended for use in MS at this time.

# References

1. Noseworthy JH, Lucchinetti C, Rodriguez M, Weinshenker BG. Multiple sclerosis. *N. Engl. J. Med.* 2000 September 28;343(13):938–952.
2. Lublin FD, Reingold SC. Defining the clinical course of multiple sclerosis: Results of an international survey. National Multiple Sclerosis Society (USA) Advisory Committee on Clinical Trials of New Agents in Multiple Sclerosis. *Neurology* 1996 April;46(4):907–911.
3. Weinshenker BG, Bass B, Rice GP, Noseworthy J, Carriere W, Baskerville J et al. The natural history of multiple sclerosis: A geographically based study. I. Clinical course and disability. *Brain* 1989 February;112 (Pt 1):133–146.
4. Vukusic S, Confavreux C. Prognostic factors for progression of disability in the secondary progressive phase of multiple sclerosis. *J. Neurol. Sci.* 2003 February 15;206(2):135–137.
5. Derwenskus J. Current disease-modifying treatment of multiple sclerosis. *Mt. Sinai J. Med.* 2011 April;78(2):161–175.
6. Center for Drug Evaluation and Research. FDA Drug Safety Communication: Safety update on progressive multifocal leukoencephalopathy (PML) associated with Tysabri (natalizumab). Available from http://www.fda.gov/drugs/drugsafety/ucm252045.htm (cited March 25, 2013).
7. Center for Drug Evaluation and Research. FDA Drug Safety Communication: Safety review of a reported death after the first dose of multiple Sclerosis drug Gilenya (fingolimod). Available from http://www.fda.gov/Drugs/DrugSafety/ucm284240.htm (cited August 31, 2012).

8. Bowling AC. Complementary and alternative medicine and multiple sclerosis. *Neurol. Clin.* 2011 May;29(2):465–480.

9. Organisation mondiale de la santé, Multiple Sclerosis International Federation. *Atlas Multiple Sclerosis Resources in the World 2008.* Geneva, Switzerland: World Health Organization; 2008.

10. Swank RL, Goodwin J. Review of MS patient survival on a Swank low saturated fat diet. *Nutrition* 2003 February;19(2):161–162.

11. SWANK RL. Treatment of multiple sclerosis with low-fat diet. *AMA Arch. Neurol. Psychiatr.* 1953 January;69(1):91–103.

12. Swank RL. Multiple sclerosis: Twenty years on low fat diet. *Arch. Neurol.* 1970 November;23(5):460–474.

13. Swank RL, Dugan BB. Effect of low saturated fat diet in early and late cases of multiple sclerosis. *Lancet* 1990 July 7;336(8706):37–39.

14. Marrie RA, Rudick R, Horwitz R, Cutter G, Tyry T, Campagnolo D et al. Vascular comorbidity is associated with more rapid disability progression in multiple sclerosis. *Neurology* 2010 March 30;74(13):1041–1047.

15. Weinstock-Guttman B, Zivadinov R, Mahfooz N, Carl E, Drake A, Schneider J et al. Serum lipid profiles are associated with disability and MRI outcomes in multiple sclerosis. *J. Neuroinflam.* 2011;8:127.

16. Simopoulos AP. The importance of the omega-6/omega-3 fatty acid ratio in cardiovascular disease and other chronic diseases. *Exp. Biol. Med. (Maywood)* 2008 June;233(6):674–688.

17. Das UN. Essential fatty acids: Biochemistry, physiology and pathology. *Biotechnol. J.* 2006;1(4):420–439.

18. Willctt WC. Dietary fats and coronary heart disease. *J. Intern. Med.* 2012 May;272(1):13–24.

19. Willett WC. The role of dietary n-6 fatty acids in the prevention of cardiovascular disease. *J. Cardiovasc. Med. (Hagerstown).* 2007 September;8 Suppl 1:S42–S45.

20. Lauer K. The risk of multiple sclerosis in the U.S.A. in relation to sociogeographic features: A factor-analytic study. *J. Clin. Epidemiol.* 1994 January;47(1):43–48.

21. Kampman MT, Wilsgaard T, Mellgren SI. Outdoor activities and diet in childhood and adolescence relate to MS risk above the Arctic Circle. *J. Neurol.* 2007 April;254(4):471–477.

22. Gallai V, Sarchielli P, Trequattrini A, Franceschini M, Floridi A, Firenze C et al. Cytokine secretion and eicosanoid production in the peripheral blood mononuclear cells of MS patients undergoing dietary supplementation with n-3 polyunsaturated fatty acids. *J. Neuroimmunol.* 1995 February;56(2):143–153.

23. Shinto L, Marracci G, Baldauf-Wagner S, Strehlow A, Yadav V, Stuber L et al. Omega-3 fatty acid supplementation decreases matrix metalloproteinase-9 production in relapsing-remitting multiple sclerosis. *Prostaglandins Leukot. Essent. Fatty Acids* 2009 March;80(2–3):131–136.

24. Bates D, Cartlidge NE, French JM, Jackson MJ, Nightingale S, Shaw DA et al. A double-blind controlled trial of long chain n-3 polyunsaturated fatty acids in the treatment of multiple sclerosis. *J. Neurol. Neurosurg. Psychiatr.* 1989 January;52(1):18–22.

25. Torkildsen Ø, Wergeland S, Bakke S, Beiske A, Bjerve K, Hovdal H et al. ω-3 fatty acid treatment in multiple sclerosis (OFAMS Study): A randomized, double-blind, placebo-controlled trial. *Arch. Neurol.* 2012 April 16;1.

26. Hughes D, Keith AB, Mertin J, Caspary EA. Linoleic acid therapy in severe experimental allergic encephalomyelitis in the guinea-pig: Suppression by continuous treatment. *Clin. Exp. Immunol.* 1980 June;40(3):523.

27. Meade CJ, Mertin J, Sheena J, Hunt R. Reduction by linoleic acid of the severity of experimental allergic encephalomyelitis in the guinea pig. *J. Neurol. Sci.* 1978 February;35(2–3):291–308.

28. Millar JH, Zilkha KJ, Langman MJ, Wright HP, Smith AD, Belin J et al. Double-blind trial of linoleate supplementation of the diet in multiple sclerosis. *Br. Med. J.* 1973 March 31;1(5856):765–768.

29. Bates D, Fawcett PR, Shaw DA, Weightman D. Trial of polyunsaturated fatty acids in non-relapsing multiple sclerosis. *Br. Med. J.* 1977 October 8;2(6092):932–933.

30. Bates D, Fawcett PR, Shaw DA, Weightman D. Polyunsaturated fatty acids in treatment of acute remitting multiple sclerosis. *Br. Med. J.* 1978 November 18;2(6149):1390–1391.

31. Paty DW. Double-blind trial of linoleic acid in multiple sclerosis. *Arch. Neurol.* 1983 October 21;40(11):693–694.

32. Martín R, Carvalho-Tavares J, Hernández M, Arnés M, Ruiz-Gutiérrez V, Nieto ML. Beneficial actions of oleanolic acid in an experimental model of multiple sclerosis: A potential therapeutic role. *Biochem. Pharmacol.* 2010 January 15;79(2):198–208.

33. Martín R, Hernández M, Córdova C, Nieto M. Natural triterpenes modulate immune-inflammatory markers of experimental autoimmune encephalomyelitis: Therapeutic implications for multiple sclerosis. *Br. J. Pharmacol.* 2012 July;166(5):1708–1723.

34. Reynolds EH, Bottiglieri T, Laundy M, Crellin RF, Kirker SG. Vitamin B12 metabolism in multiple sclerosis. *Arch. Neurol.* 1992 June;49(6):649–652.

35. Zhu Y, He Z-Y, Liu H-N. Meta-analysis of the relationship between homocysteine, vitamin $B_{12}$, folate, and multiple sclerosis. *J. Clin. Neurosci.* 2011 July;18(7):933–938.

36. Goodkin DE, Jacobsen DW, Galvez N, Daughtry M, Secic M, Green R. Serum cobalamin deficiency is uncommon in multiple sclerosis. *Arch. Neurol.* 1994 November;51(11):1110–1114.

37. Najafi MR, Shaygannajad V, Mirpourian M, Gholamrezaei A. Vitamin B(12) deficiency and multiple sclerosis; Is there any association? *Int. J. Prev. Med.* 2012 April;3(4):286–289.

38. Miller A, Korem M, Almog R, Galboiz Y. Vitamin B12, demyelination, remyelination and repair in multiple sclerosis. *J. Neurol. Sci.* 2005 June 15;233(1–2):93–97.

39. Peracchi M, Bamonti Catena F, Pomati M, De Franceschi M, Scalabrino G. Human cobalamin deficiency: Alterations in serum tumour necrosis factor-alpha and epidermal growth factor. *Eur. J. Haematol.* 2001 August;67(2):123–127.

40. Mastronardi FG, Min W, Wang H, Winer S, Dosch M, Boggs JM et al. Attenuation of experimental autoimmune encephalomyelitis and nonimmune demyelination by IFN-beta plus vitamin B12: Treatment to modify notch-1/sonic hedgehog balance. *J. Immunol.* 2004 May 15;172(10):6418–6426.

41. Wade DT, Young CA, Chaudhuri KR, Davidson DLW. A randomised placebo controlled exploratory study of vitamin B-12, lofepramine, and L-phenylalanine (the "Cari Loder regime") in the treatment of multiple sclerosis. *J. Neurol. Neurosurg. Psychiatr.* 2002 Sepember;73(3):246–249.

42. Tavazzi B, Batocchi AP, Amorini AM, Nociti V, D'Urso S, Longo S et al. Serum metabolic profile in multiple sclerosis patients. *Mult. Scler. Int.* 2011;2011:167156.

43. Mangas A, Coveñas R, Bodet D, de León M, Duleu S, Geffard M. Evaluation of the effects of a new drug candidate (GEMSP) in a chronic EAE model. *Int. J. Biol. Sci.* 2008;4(3):150–160.

44. Mai J, Sørensen PS, Hansen JC. High dose antioxidant supplementation to MS patients. Effects on glutathione peroxidase, clinical safety, and absorption of selenium. *Biol. Trace Elem. Res.* 1990 February;24(2):109–117.

45. Holick MF. Vitamin D deficiency. *N. Engl. J. Med.* 2007 July 19;357(3):266–281.

46. Pierrot-Deseilligny C. Clinical implications of a possible role of vitamin D in multiple sclerosis. *J. Neurol.* 2009 September;256(9):1468–1479.

47. Munger KL, Levin LI, Hollis BW, Howard NS, Ascherio A. Serum 25-hydroxyvitamin D levels and risk of multiple sclerosis. *JAMA* 2006 December 20;296(23):2832–2838.

48. Mowry EM, Krupp LB, Milazzo M, Chabas D, Strober JB, Belman AL et al. Vitamin D status is associated with relapse rate in pediatric-onset multiple sclerosis. *Ann. Neurol.* 2010 May;67(5):618–624.

49. Mirzaei F, Michels KB, Munger K, O'Reilly E, Chitnis T, Forman MR et al. Gestational vitamin D and the risk of multiple sclerosis in offspring. *Ann. Neurol.* 2011 July;70(1):30–40.

50. Beretich BD, Beretich TM. Explaining multiple sclerosis prevalence by ultraviolet exposure: A geospatial analysis. *Mult. Scler.* 2009 August;15(8):891–898.

51. Taylor BV, Lucas RM, Dear K, Kilpatrick TJ, Pender MP, van der Mei IAF et al. Latitudinal variation in incidence and type of first central nervous system demyelinating events. *Mult. Scler.* 2010 April;16(4):398–405.

52. Ramagopalan SV, Dyment DA, Cader MZ, Morrison KM, Disanto G, Morahan JM et al. Rare variants in the CYP27B1 gene are associated with multiple sclerosis. *Ann. Neurol.* 2011 December;70(6):881–886.

53. Runia TF, Hop WCJ, de Rijke YB, Buljevac D, Hintzen RQ. Lower serum vitamin D levels are associated with a higher relapse risk in multiple sclerosis. *Neurology* 2012 July 17;79(3):261–266.

54. Smolders J, Menheere P, Kessels A, Damoiseaux J, Hupperts R. Association of vitamin D metabolite levels with relapse rate and disability in multiple sclerosis. *Mult. Scler.* 2008 November;14(9):1220–1224.

55. Pedersen LB, Nashold FE, Spach KM, Hayes CE. 1,25-Dihydroxyvitamin D3 reverses experimental autoimmune encephalomyelitis by inhibiting chemokine synthesis and monocyte trafficking. *J. Neurosci. Res.* 2007 August 15;85(11):2480–2490.

56. DeLuca HF, Plum LA. Vitamin D deficiency diminishes the severity and delays onset of experimental autoimmune encephalomyelitis. *Arch. Biochem. Biophys.* 2011 September 15;513(2):140–143.

57. Soilu-Hänninen M, Aivo J, Lindström B-M, Elovaara I, Sumelahti M-L, Färkkilä M et al. A randomised, double blind, placebo controlled trial with vitamin D3 as an add on treatment to interferon β-1b in patients with multiple sclerosis. *J. Neurol. Neurosurg. Psychiatr.* 2012 May;83(5):565–571.

58. Stein MS, Liu Y, Gray OM, Baker JE, Kolbe SC, Ditchfield MR et al. A randomized trial of high-dose vitamin D2 in relapsing-remitting multiple sclerosis. *Neurology* 2011 October 25;77(17):1611–1618.

59. Romagnoli E, Mascia ML, Cipriani C, Fassino V, Mazzei F, D'Erasmo E et al. Short and long-term variations in serum calciotropic hormones after a single very large dose of ergocalciferol (vitamin D2) or cholecalciferol (vitamin D3) in the elderly. *J. Clin. Endocrinol. Metab.* 2008 August;93(8):3015–3020.

60. Ross ACAC, Taylor CLCL, Yaktine ALAL, Del Valle HBHB, eds. *Dietary Reference Intakes for Calcium and Vitamin D* (Internet). Washington, DC: National Academies Press; 2011. (Cited 2012 July 26). Available from: http://www.ncbi.nlm.nih.gov/pubmed/21796828

61. Burton JM, Kimball S, Vieth R, Bar-Or A, Dosch H-M, Cheung R et al. A phase I/II dose-escalation trial of vitamin D3 and calcium in multiple sclerosis. *Neurology* 2010 June 8;74(23):1852–1859.

62. Marcus JF, Shalev SM, Harris CA, Goodin DS, Josephson SA. Severe hypercalcemia following vitamin d supplementation in a patient with multiple sclerosis: A note of caution. *Arch. Neurol.* 2012 January;69(1):129–132.

63. Goudarzvand M, Javan M, Mirnajafi-Zadeh J, Mozafari S, Tiraihi T. Vitamins E and D3 attenuate demyelination and potentiate remyelination processes of hippocampal formation of rats following local injection of ethidium bromide. *Cell. Mol. Neurobiol.* 2010 March;30(2):289–299.

64. Ghadirian P, Jain M, Ducic S, Shatenstein B, Morisset R. Nutritional factors in the aetiology of multiple sclerosis: A case-control study in Montreal, Canada. *Int. J. Epidemiol.* 1998 October;27(5):845–852.

65. Gusev E, Boiko A, Lauer K, Riise T, Deomina T. Environmental risk factors in MS: A case-control study in Moscow. *Acta Neurol. Scand.* 1996 December;94(6):386–394.

66. Dyckner T, Ek B, Nyhlin H, Wester PO. Aggravation of thiamine deficiency by magnesium depletion. A case report. *Acta Med. Scand.* 1985;218(1):129–131.

67. Johnson S. The multifaceted and widespread pathology of magnesium deficiency. *Medical Hypotheses* 2001 February;56(2):163–170.
68. Yasui M, Yase Y, Ando K, Adachi K, Mukoyama M, Ohsugi K. Magnesium concentration in brains from multiple sclerosis patients. *Acta Neurol. Scand.* 1990 March;81(3):197–200.
69. Ramsaransing GSM, Mellema SA, De Keyser J. Dietary patterns in clinical subtypes of multiple sclerosis: An exploratory study. *Nutr. J.* 2009;8:36.
70. Goldberg P, Fleming MC, Picard EH. Multiple sclerosis: Decreased relapse rate through dietary supplementation with calcium, magnesium and vitamin D. *Medical Hypotheses* 1986;21(2):193–200.
71. Rossier P, van Erven S, Wade DT. The effect of magnesium oral therapy on spasticity in a patient with multiple sclerosis. *Eur. J. Neurol.* 2000;7(6):741–744.
72. Biewenga GP, Dorstijn MA, Verhagen JV, Haenen GR, Bast A. Reduction of lipoic acid by lipoamide dehydrogenase. *Biochem. Pharmacol.* 1996 February 9;51(3):233–238.
73. Marracci GH, Jones RE, McKeon GP, Bourdette DN. Alpha lipoic acid inhibits T cell migration into the spinal cord and suppresses and treats experimental autoimmune encephalomyelitis. *J. Neuroimmunol.* 2002 October;131(1–2):104–114.
74. Biewenga GP, Haenen GR, Bast A. The pharmacology of the antioxidant lipoic acid. *Gen. Pharmacol.* 1997 September;29(3):315–331.
75. Chaudhary P, Marracci GH, Bourdette DN. Lipoic acid inhibits expression of ICAM-1 and VCAM-1 by CNS endothelial cells and T cell migration into the spinal cord in experimental autoimmune encephalomyelitis. *J. Neuroimmunol.* 2006 June;175(1–2):87–96.
76. Salinthone S, Schillace RV, Tsang C, Regan JW, Bourdette DN, Carr DW. Lipoic acid stimulates cAMP production via G protein coupled receptor dependent and independent mechanisms. *J. Nutr. Biochem.* 2011 July;22(7):681–690.
77. Wong A, Dukic-Stefanovic S, Gasic-Milenkovic J, Schinzel R, Wiesinger H, Riederer P et al. Anti-inflammatory antioxidants attenuate the expression of inducible nitric oxide synthase mediated by advanced glycation endproducts in murine microglia. *Eur. J. Neurosci.* 2001 December;14(12):1961–1967.
78. Miller RL, James-Kracke M, Sun GY, Sun AY. Oxidative and inflammatory pathways in Parkinson's disease. *Neurochem. Res.* 2009 January;34(1):55–65.
79. Chaudhary P, Marracci G, Yu X, Galipeau D, Morris B, Bourdette D. Lipoic acid decreases inflammation and confers neuroprotection in experimental autoimmune optic neuritis. *J. Neuroimmunol.* 2011 April;233(1–2):90–96.
80. Yadav V, Marracci G, Lovera J, Woodward W, Bogardus K, Marquardt W et al. Lipoic acid in multiple sclerosis: A pilot study. *Mult. Scler.* 2005 April 1;11(2):159–165.
81. Reljanovic M, Reichel G, Rett K, Lobisch M, Schuette K, Möller W et al. Treatment of diabetic polyneuropathy with the antioxidant thioctic acid (alpha-lipoic acid): A two year multicenter randomized double-blind placebo-controlled trial (ALADIN II). Alpha Lipoic Acid in Diabetic Neuropathy. *Free Radic. Res.* 1999 September;31(3):171–179.
82. Ziegler D, Hanefeld M, Ruhnau KJ, Hasche H, Lobisch M, Schütte K et al. Treatment of symptomatic diabetic polyneuropathy with the antioxidant alpha-lipoic acid: A 7-month multicenter randomized controlled trial (ALADIN III Study). ALADIN III Study Group. Alpha-Lipoic Acid in Diabetic Neuropathy. *Diabetes Care* 1999 August;22(8):1296–1301.
83. Izzo AA, Borrelli F, Capasso R, Di Marzo V, Mechoulam R. Non-psychotropic plant cannabinoids: New therapeutic opportunities from an ancient herb. *Trends Pharmacol. Sci.* 2009 October;30(10):515–527.
84. Corey-Bloom J, Wolfson T, Gamst A, Jin S, Marcotte TD, Bentley H et al. Smoked cannabis for spasticity in multiple sclerosis: A randomized, placebo-controlled trial. *CMAJ* 2012 July 10;184(10):1143–1150.
85. Collin C, Davies P, Mutiboko IK, Ratcliffe S, Group for the SS in MS. Randomized controlled trial of cannabis-based medicine in spasticity caused by multiple sclerosis. *Eur. J. Neurol.* 2007;14(3):290–296.

86. Wade DT, Makela PM, House H, Bateman C, Robson P. Long-term use of a cannabis-based medicine in the treatment of spasticity and other symptoms in multiple sclerosis. *Mult. Scler.* 2006 September 1;12(5):639–645.
87. Zajicek J, Fox P, Sanders H, Wright D, Vickery J, Nunn A et al. Cannabinoids for treatment of spasticity and other symptoms related to multiple sclerosis (CAMS study): Multicentre randomised placebo-controlled trial. *The Lancet* 2003 November 8;362(9395):1517–1526.
88. Zajicek JP, Sanders HP, Wright DE, Vickery PJ, Ingram WM, Reilly SM et al. Cannabinoids in multiple sclerosis (CAMS) study: Safety and efficacy data for 12 months follow up. *J. Neurol. Neurosurg. Psychiatr.* 2005 December;76(12):1664–1669.
89. Zajicek JP, Hobart JC, Slade A, Barnes D, Mattison PG. Multiple sclerosis and extract of cannabis: Results of the MUSEC trial. *J. Neurol. Neurosurg. Psychiatr.* (Internet). 2012 July 12. (Cited 2012 August 7); Available from: http://jnnp.bmj.com/content/early/2012/07/11/jnnp-2012-302468
90. Robson P. Abuse potential and psychoactive effects of δ-9-tetrahydrocannabinol and cannabidiol oromucosal spray (Sativex), a new cannabinoid medicine. *Expert Opin. Drug. Saf.* 2011 September;10(5):675–685.
91. Honarmand K, Tierney MC, O'Connor P, Feinstein A. Effects of cannabis on cognitive function in patients with multiple sclerosis. *Neurology* 2011 March 29;76(13):1153–1160.
92. Johnson SK, Diamond BJ, Rausch S, Kaufman M, Shiflett SC, Graves L. The effect of Ginkgo biloba on functional measures in multiple sclerosis: A pilot randomized controlled trial. *Explore* (NY). 2006 January;2(1):19–24.
93. Lovera J, Bagert B, Smoot K, Morris CD, Frank R, Bogardus K et al. Ginkgo biloba for the improvement of cognitive performance in multiple sclerosis: A randomized, placebo controlled trial. *Mult. Scler.* 2007 April;13(3):376–385.
94. Lovera JF, Kim E, Heriza E, Fitzpatrick M, Hunziker J, Turner AP et al. Ginkgo biloba does not improve cognitive function in MS: A randomized placebo-controlled trial. *Neurology* 2012 September 18;79(12):1278–1284.
95. Lim W, Mudge KW, Vermeylen F. Effects of population, age, and cultivation methods on ginsenoside content of wild American ginseng (Panax quinquefolius). *J. Agric. Food Chem.* 2005 November 2;53(22):8498–8505.
96. Bowie LE, Roscoe WA, Lui EMK, Smith R, Karlik SJ. Effects of an aqueous extract of North American ginseng on MOG((35–55))-induced EAE in mice. *Can. J. Physiol. Pharmacol.* 2012 July;90(7):933–939.
97. Hwang I, Ahn G, Park E, Ha D, Song J-Y, Jee Y. An acidic polysaccharide of Panax ginseng ameliorates experimental autoimmune encephalomyelitis and induces regulatory T cells. *Immunol. Lett.* 2011 August 30;138(2):169–178.
98. Yuan C-S, Wei G, Dey L, Karrison T, Nahlik L, Maleckar S et al. Brief communication: American ginseng reduces warfarin's effect in healthy patients: A randomized, controlled Trial. *Ann. Intern. Med.* 2004 July 6;141(1):23–27.
99. Siegel RK. Ginseng abuse syndrome. Problems with the panacea. *JAMA* 1979 April 13;241(15):1614–1615.
100. Ritchie MR, Gertsch J, Klein P, Schoop R. Effects of Echinaforce® treatment on ex vivo-stimulated blood cells. *Phytomedicine* 2011 July 15;18(10):826–831.
101. Bowling AC. *Alternative Medicine and Multiple Sclerosis.* New York: Demos Medical Publishing; 2001.
102. Wu D, Wang J, Pae M, Meydani SN. Green tea EGCG, T cells, and T cell-mediated autoimmune diseases. *Mol. Aspects Med.* 2012 February;33(1):107–118.
103. Liuzzi GM, Latronico T, Branà MT, Gramegna P, Coniglio MG, Rossano R et al. Structure-dependent inhibition of gelatinases by dietary antioxidants in rat astrocytes and sera of multiple sclerosis patients. *Neurochem. Res.* 2011 March;36(3):518–527.
104. Herges K, Millward JM, Hentschel N, Infante-Duarte C, Aktas O, Zipp F. Neuroprotective effect of combination therapy of glatiramer acetate and epigallocatechin-3-gallate in neuroinflammation. *PLoS One* (Internet). 2011 October 13. (Cited 2012 August 31);6(10). Available from: http://www.ncbi.nlm.nih.gov/pmc/articles/PMC3192751/

105. Wang J, Ren Z, Xu Y, Xiao S, Meydani SN, Wu D. Epigallocatechin-3-gallate ameliorates experimental autoimmune encephalomyelitis by altering balance among CD4+ T-cell subsets. *Am. J. Pathol.* 2012 January;180(1):221–234.

106. Aktas O, Prozorovski T, Smorodchenko A, Savaskan NE, Lauster R, Kloetzel P-M et al. Green tea epigallocatechin-3-gallate mediates T cellular NF-κB inhibition and exerts neuroprotection in autoimmune encephalomyelitis. *J. Immunol.* 2004 November 1;173(9):5794–5800.

107. Safety and Neuroprotective Effects of Polyphenon E in MS; Phase II—Full Text View—ClinicalTrials.gov (Internet). (Cited 2012 September 8). Available from: http:// www.clinicaltrials.gov/ct2/show/NCT01451772?term = green+tea+multiple+sclerosis& rank = 2

108. Gold MS, Dackis CA, Pottash AL, Sternbach HH, Annitto WJ, Martin D et al. Naltrexone, opiate addiction, and endorphins. *Med. Res. Rev.* 1982 September;2(3):211–246.

109. Gironi M, Martinelli-Boneschi F, Sacerdote P, Solaro C, Zaffaroni M, Cavarretta R et al. A pilot trial of low-dose naltrexone in primary progressive multiple sclerosis. *Mult. Scler.* 2008 September;14(8):1076–1083.

110. Zagon IS, McLaughlin PJ. Naltrexone modulates tumor response in mice with neuro-blastoma. *Science* 1983 August 12;221(4611):671–673.

111. Bodnar RJ. Endogenous opiates and behavior: 2010. *Peptides* 2011 December; 32(12):2522–2552.

112. Zagon IS, Rahn KA, Turel AP, McLaughlin PJ. Endogenous opioids regulate expression of experimental autoimmune encephalomyelitis: A new paradigm for the treatment of multiple sclerosis. *Exp. Biol. Med. (Maywood)* 2009 November;234(11):1383–1392.

113. Cree BAC, Kornyeyeva E, Goodin DS. Pilot trial of low-dose naltrexone and quality of life in multiple sclerosis. *Ann. Neurol.* 2010 August;68(2):145–150.

114. Sharafaddinzadeh N, Moghtaderi A, Kashipazha D, Majdinasab N, Shalbafan B. The effect of low-dose naltrexone on quality of life of patients with multiple sclerosis: A randomized placebo-controlled trial. *Mult. Scler.* 2010 August;16(8):964–969.

115. Gillson G, Wright JV, DeLack E, Ballasiotes G. Transdermal histamine in multiple sclerosis: Part one—clinical experience. *Altern. Med. Rev.* 1999 December;4(6):424–428.

116. Gillson G, Wright JV, DeLack E, Ballasiotes G. Transdermal histamine in multiple sclerosis, part two: A proposed theoretical basis for its use. *Altern. Med. Rev.* 2000 June;5(3):224–248.

117. Gillson G, Richard TL, Smith RB, Wright JV. A double-blind pilot study of the effect of Prokarin on fatigue in multiple sclerosis. *Mult. Scler.* 2002 February;8(1):30–35.

118. Fleming JO, Cook TD. Multiple sclerosis and the hygiene hypothesis. *Neurology* 2006 December 12;67(11):2085–2086.

119. Correale J, Farez M. Association between parasite infection and immune responses in multiple sclerosis. *Ann. Neurol.* 2007 February;61(2):97–108.

120. Correale J, Farez MF. The impact of parasite infections on the course of multiple sclerosis. *J. Neuroimmunol.* 2011 April;233(1–2):6–11.

121. Fleming JO, Isaak A, Lee JE, Luzzio CC, Carrithers MD, Cook TD et al. Probiotic helminth administration in relapsing-remitting multiple sclerosis: A phase 1 study. *Mult. Scler.* 2011 June;17(6):743–754.

122. Helminth-induced Immunomodulation Therapy (HINT) in Relapsing-remitting Multiple Sclerosis—Full Text View—ClinicalTrials.gov (Internet). (Cited 2012 July 26). Available from: http://clinicaltrials.gov/ct2/show/NCT00645749?term = multiple+sclerosis+AND +helminth&rank = 1

123. Summers RW, Elliott DE, Urban JF Jr, Thompson RA, Weinstock JV. Trichuris suis therapy for active ulcerative colitis: A randomized controlled trial. *Gastroenterology* 2005 April;128(4):825–832.

124. Summers RW, Elliott DE, Urban JF Jr, Thompson R, Weinstock JV. Trichuris suis therapy in Crohn's disease. *Gut* 2005 January;54(1):87–90.

125. Bager P, Arnved J, Rønborg S, Wohlfahrt J, Poulsen LK, Westergaard T et al. Trichuris suis ova therapy for allergic rhinitis: A randomized, double-blind, placebo-controlled clinical trial. *J. Allergy Clin. Immunol.* 2010 January;125(1):123–130, e1–e3.
126. Castro HJ, Mendez-Lnocencio JI, Omidvar B, Omidvar J, Santilli J, Nielsen HS Jr et al. A phase I study of the safety of honeybee venom extract as a possible treatment for patients with progressive forms of multiple sclerosis. *Allergy Asthma Proc.* 2005 December;26(6):470–476.
127. Wesselius T, Heersema DJ, Mostert JP, Heerings M, Admiraal-Behloul F, Talebian A et al. A randomized crossover study of bee sting therapy for multiple sclerosis. *Neurology* 2005 December 13;65(11):1764–1768.
128. Song HS, Wray SH. Bee sting optic neuritis. A case report with visual evoked potentials. *J. Clin. Neuroophthalmol.* 1991 March;11(1):45–49.
129. Alqutub AN, Masoodi I, Alsayari K, Alomair A. Bee sting therapy-induced hepatotoxicity: A case report. *World J. Hepatol.* 2011 October 27;3(10):268–270.

# VIII
## Methodological Issues

# 18

# Methodological Issues in Conducting Epidemiological Research on Nutraceuticals

Somdat Mahabir

## Contents

## Introduction

The term nutraceutical is used to imply the use of nutrition (food and food components) in a manner parallel to pharmaceuticals. The assertion is that nutraceuticals enhance health and well-being and provide disease prevention, treatment, and survival benefits. Nutraceuticals cover a vast classification of products that include diet, specific nutrients, specific trace minerals, other isolated dietary components, specific parts of plants, vitamin/mineral supplements, herbal supplements, other alternative supplements, and specially formulated foods such as cereals, yogurt, soups, and beverages. Genetically modified foods would also be included in the categorization of nutraceuticals. In general, nutraceuticals are widely used in the United States. For example, national surveys have shown that the use of herbs and dietary supplements rose in the United States during the 1990s and 2000s and a report in 2011 indicated that the number of adults in the United States who used herbs or supplements grew slightly in 2007 to 55.1 million compared to 50.6 million users in 2002 [1].

There is substantial interest in the public health potential of nutraceuticals in health enhancement, disease prevention, and disease treatment. Commercialization of nutraceuticals, based mostly on speculative health claims, has skyrocketed. There is also a belief that many nutraceuticals would benefit by designing delivery systems for consumption. This could be risky as the evidence is consistent that, for example, vitamin/mineral supplements increase risk for cancer (see Chapter 2). Epidemiological research on nutraceuticals has lagged behind for a multitude of reasons including funding and methodological challenges. Except for diet and dietary patterns and vitamin/mineral supplements, very limited epidemiological studies have been conducted to address the role of nutraceuticals in disease prevention, risk, treatment, and survival.

## Epidemiological considerations to diet–disease research
### Calibration of diet assessment tools

When studying human populations, measurement of dietary or nutraceutical intakes and associations with disease outcome of interest is very important. Most of the existing tools were not designed to adequately capture the great diversity of nutraceuticals that exist today. In terms of more classic dietary assessment, researchers traditionally compared two dietary tools, such as a food frequency questionnaire (FFQ) or diet history, to 24 h food recalls or records as the reference method; either self-reported or interview-administered data are collected over less than or up to a year. Reported dietary data are subject to substantial correlated error [2–5]. The correlation coefficients derived from such relative validity studies may be biased, and in the absence of information on true intake, the magnitude of the bias cannot be evaluated [3,6,7]. Thus, it is not appropriate to simply correlate FFQ with food recalls or records (usually used as reference instruments) because they measure different things—FFQ measures habitual intake, and the food recalls and records measure short-term intake. Dietary calibration research, which examines the comparative performance of dietary tools, suffers from systematic respondent-related reporting bias, and thus, this type of research does not compare reported intake with a biomarker that is independent of response bias [8].

## Biomarker validation of diet assessment tools

Because of the errors associated with diet calibration assessment tool development, the criteria for diet assessment validation over the past 30 years shifted from an analysis of the performance of the FFQ or diet history questionnaire compared to a referent method such as food diaries to an examination of the performance of multiple dietary tools concurrent with biomarker collections. Biomarkers do not rely on self-reports of food intake, and thus, random measurement errors of the biomarker are not likely to be correlated with those of the dietary assessment method [3,5,7]. There are different types of biomarkers used in nutritional assessment. For example, we [9] and others [5,7] have used doubly labeled water (DLW), which reflects a balance between intake and output over a specific time period, and have errors unrelated to true intakes and unrelated to errors in reported dietary tools, and can be used to assess under- and overreporting of food intake and therefore assess validity. Using the DLW biomarker, we [9] and others [5,7] have reported substantial dietary underreporting that has implications for diet–disease association studies that exclusively rely on the FFQ.

## Reproducibility of diet assessment tools

Several investigators [10–15] have published studies designed to examine whether dietary intake as reported in an FFQ is reliable. Dietary data are usually reported in either interviewer or self-administered FFQs at $\geq 2$ points in time. Reproducibility is assessed by kappa statistics, Spearman rank correlation coefficients, and concordance, that is, the percentage jointly classified in the same category of intake over time. Results are reported as a range of correlations or kappa statistics, with delineation of food items that have poor agreement over time due to high intra- or interindividual variability in intake. A study has shown that the season of completion of the FFQ may affect reproducibility among disadvantaged populations [16]. Reproducibility research has demonstrated a range in correlation coefficients, for which investigators have concluded that the FFQ of interest can "adequately" rank participants by intake of food and nutrients over time [17]. An important issue in reproducibility research is whether differences in intake reflect real changes in diet and food sources over time or reflect poor repeatability.

## Epidemiological consideration of specific nutraceuticals
### Database issues

In nutritional epidemiology research, nutrient levels are calculated by tapping into the U.S. Department of Agriculture (USDA) National Nutrient Database. However, for many nutraceuticals, there are no analytical values in the database. For example, in an analysis, we conducted on the joint effects of dietary boron intake and hormone replacement therapy (HRT) for lung cancer risk in women [18]; we had to use a separate created database for boron because it was not available in the USDA National Nutrient Database. The values for boron were derived from analytical values available in the literature for foods consumed in the United States, but these values may not represent an adequate variety of foods by regions, or they may be grown under different conditions that may have different levels of boron. Boron is a trace metal that is ubiquitous in foods commonly consumed in the United States, such as fruits, vegetables, nuts, legumes, wine, coffee, milk, and other beverages. There is no established recommended dietary intake for boron, because it is not established as an essential trace metal.

Today, the USDA National Nutrient Database for Standard Reference contains nutrient information for approximately 8000 foods. The database provides valuable information on macronutrients, vitamins/minerals, and phytonutrients, but numerous nutraceuticals including bioactive compounds, ingredients, and products are not represented.

## Example of dietary trace metals

Americans do not consume adequate amounts of several essential minerals such as zinc, copper, magnesium, and calcium [19–22]. There is also a concern that minerals/trace metals may interact with cigarette smoking or alone may be influencing the current trends for lung cancer incidence and mortality and other diseases.

The need for accurate exposure assessment of minerals/trace metals is underscored by the fact that none of the few published epidemiological studies of essential minerals and lung cancer risk have used a questionnaire validated for dietary mineral/trace metal intake. For studies of toxicological metals and lung cancer risk, the exposures were primarily based on surrogate estimates such as levels in drinking water, job histories, and air quality data. While accurate biomarker assessment of metals would be a more objective measurement of metal status, validated dietary assessments to reflect habitual intakes are still required because of the need to adjust for several dietary exposures in lung cancer risk assessment, such as alcohol, fruit, and vegetables. This is important because dietary factors may attenuate some of the risk due to toxicological metals. In addition, for example, studies can assess the association between heme iron (from red meat, poultry, and seafood) and nonheme iron on lung cancer risk. Biomarkers for iron do not distinguish between heme iron and nonheme iron, which have differential lung cancer risk [23].

The FFQ for dietary assessment is a mainstay of epidemiological studies due to cost-effectiveness, less participant burden, and ease of administration. For the same reasons, dietary intervention research has relied on self-administered dietary tools to measure dietary change as, for example, in the recently completed Women's Health Initiative [24]. Because essential mineral/

trace metals include elements that are ubiquitous in several foods, the first challenge to research is the development of a valid user-friendly assessment tool to estimate intake. The lack of diet assessment tools focused on essential minerals/trace metal intakes may be one reason why few data are available on the role of dietary essential metals and disease risk. For example, most studies of lung cancer have used a generic FFQ to assess diet–lung cancer associations. Further, although dietary supplements are widely used in the United States, we are not aware of any validated FFQ to capture this exposure. Thus, the first phase of a population-based research would involve the development and validation of an epidemiology-friendly FFQ for trace metal and macro-mineral intakes with appropriate corresponding biological markers. Since the FFQ is used to assess habitual intakes, the biomarkers for the essential metals should not reflect short-term intake, such as indicated by serum/plasma measurements.

## Trace metal-specific biomarkers

An important challenge to this area of research is the selection of the appropriate biomarkers for the essential and toxicological metals. The biomarker must take into consideration the homeostatic mechanisms that control levels in the blood circulation. For the essential nutritional trace metals, the biomarker must also reflect homeostatic mechanisms when dietary intake is marginal or inadequate.

*Plasma or serum*: Despite the penchant for using plasma and serum samples in clinical and epidemiological research, these samples are unsatisfactory for most essential and toxicological assessments because levels in the plasma and serum reflect short-term intake/exposures or fluctuation from intracellular pools and storage sites [25]. For example, in epidemiology studies, serum ferritin, which reflects iron stores in the liver and bone marrow, has been widely used, but it is well known that serum ferritin is an acute-phase reactant that is elevated in various conditions such as acute and chronic infections and inflammation [26]. In the case of zinc, which has attracted a lot of attention because of zinc deficiency, plasma concentration has been a widely used biomarker, but no correlation was observed between either intake or absorption and plasma zinc [27].

*Cellular components*: In both clinical and epidemiological research, cellular blood components are rarely used for trace metal assessments. Isolation of cells for trace metal assessment is very demanding in terms of sample preparation, and thus, it has restricted enthusiasm for these samples in trace metal research. As an example, erythrocyte-membrane zinc has been reported to be sensitive to dietary zinc restriction in some [28], but not all, studies [29]. For essential metals such as iron, screening practice includes hemoglobin or hematocrit, but these only reflect severe iron deficiency [26]. Other measures include signs of restricted iron supply in circulating red cells by measuring the mean corpuscular volume (MCV), but this is one of the last parameters to change in iron deficiency [26].

*Whole blood*: Whole-blood levels reflect exposures from one or more sources, depending on how much of the metal entered the body through all routes of exposure, including ingestion, inhalation, or dermal absorption, and how the metal is distributed in the blood cells. For whole blood, the procedure measures total concentration of the metal, regardless of the biochemical form and regardless of partitioning of the metals in blood fractions such as cells, plasma, or serum. Many essential metals have important functions inside cells and membranes, and therefore, whole-blood levels would also capture the cellular and extracellular compartments. There is already some evidence that whole-blood metal measurements may provide a better alternative to the commonly used serum/plasma measurements. For examples, whole-blood selenium has been found to be sensitive to dietary intake of this trace metal and provides a longer-term biomarker [25].

*Urine samples*: In order to use urine samples, the potential value of urinary excretion data must be understood in terms of the physiological role or the organs and biological systems in maintaining trace metal homeostasis. For example, for those metals for which the kidneys are not involved or have a minor role in maintaining homeostasis, measurement of urinary metal levels is not very useful. The kidney does not play a major role in the homeostasis of iron and copper and only play a minor role in zinc homeostasis [25]. It plays an important major role in selenium homeostasis [25].

*Nail, hair, and other tissue samples*: For several of the trace metals, nail and hair samples have provided a tantalizing potential biomarker for the long-term status of selected metals. However, nail samples are prone to contamination with soil that contains trace metals, and although washing has been employed to remove contamination, ultrasensitive measures such as the inductively coupled plasma mass spectroscopy (ICP-MS) method can detect even minute contamination. Hair samples based on our experience have been very difficult to acquire in the amount needed for analysis because participants are not usually interested in providing samples of their hair. For some metals such as iron, although hematologists still rely on the assessment of stainable iron or aspirated marrow smears or biopsy for the diagnosis of iron-deficiency anemia, studies have shown that the reliability of the iron stain is often suboptimal [26,30,31].

## Measurement technique

The ICP-MS method is an innovative, powerful analytical technique to measure metal concentrations in biological fluids. This machine is the most versatile atomizer and element ionizer available today for metal analysis. It has outstanding properties, such as a high sensitivity, the highest quantification accuracy, and the speed of analysis compared to other analytical methods for metal measurements. Today, the ICP-MS technique is the method of choice for detecting and reliably quantifying metals [32,33]. The ICP-MS method has been used by the FDA to analyze dietary supplements for arsenic, cadmium, mercury, and lead [34] and the CDC for toxicological metal measurements in urine samples [35].

In general, the ICP-MS method has primarily been used in the earth sciences. Epidemiology studies have generally not used the ICP-MS method because it was not well developed for measurement of trace metals in biological samples. Now, with innovative advancement in ICP-MS sensitivity and accuracy of metal measurements, due to technological advancements in detectors, nebulizers, better instrument design, and high-resolution mass spectrometers, the technique is expanding in the life sciences research because it requires a small amount of biological sample and it is a rapid quantification method. The major advantage of this method is that it allows for the simultaneous quantification of many elements in small samples of whole blood, plasma, and serum at concentration equal to one-hundredth of the lower limits of the normal ranges [36].

## Study design considerations
### Cell culture studies

A vast number of published studies of nutraceuticals and disease inference in humans have utilized the cell culture or in vitro design. Many of these studies use single-lineage epithelial cell monoculture, but others utilize coculture of cells with other host cell types such as neutrophils, eosinophils, monocytes, and lymphocytes. An important limitation for cell culture experiments using nutraceuticals is that the in vitro method is limited by the loss of critically important anatomical and biochemical characteristics. Based on preexisting evidence, especially from drug testing, many of the nutraceuticals that have shown promise in cell culture models will not be effective in humans for several reasons including the knowledge that isolated cells lack the complexity of the human biological network.

### Animal studies

These are very useful experimental model systems, but ethical concerns about the usefulness of animals to scientific research are a concern. Some animals may better reflect the human situation than others. Animal studies of nutraceuticals and disease do not reliably predict links in humans. Similarly, using preexisting evidence, a vast majority of drugs that appear promising in animal experimentation do not pan out in humans.

### Case–control studies

Case–control studies are retrospective studies. In terms of nutraceutical–disease association, case–control studies are conducted to make inference about disease based on the frequency of consumption/exposure among cases and controls. The assumption is that all the cases have the disease and all the controls are free of the disease. However, even though investigators set criteria for inclusion of cases and controls, this assumption is prone to error. Most importantly, in case–control studies, nutraceutical intake suffers from selection bias and recall bias. For example, cases and controls are known to recall their intake of diet, dietary supplement, and other nutraceuticals differently, a problem known as differential recall.

## Prospective cohort studies

Prospective cohort studies start with a large group of subjects who are free of the disease of interest and are followed over several years for disease outcome. Usually, criteria are established for the inclusion of subjects in the cohort for follow-up. The advantage of prospective cohort studies is that the exposure or nutraceutical intake is known prior to the occurrence of the disease. Because of long follow-up periods prior to disease development, nutraceutical intake can be measured several times in periodic intervals to increase accuracy and to get at more habitual or usual intake patterns.

## Intervention studies

Intervention studies of nutraceutical–disease relationship test the efficacy of a defined dose of the agent over a define duration usually in a well-defined population. There are different types of intervention studies, but the randomized, double-blind, placebo-controlled, crossover trials are the most advantageous. Even though the randomized controlled trials (RCTs) represent a superior design, strong scientific justification must be accounted for in selecting the dose and duration of the nutraceutical intervention, and study population must be homogenous in characteristics. In addition, there should be clearly defined accounting for diet and nutritional factors and other confounders.

## Obesity considerations

In the United States, more than 65% of adults are overweight or obese, with nearly 31% of adults—over 61 million people meeting the criteria for obesity (BMI > 30) [37]. Overweight and obesity disproportionately affect racial and ethnic minority populations and those of lower socioeconomic status [38]. In the United States, Mexicans and Blacks have the highest rates of obesity [38]. During 1999–2002 period, approximately 71% Blacks and 72% Mexican Americans were overweight, and 39% Blacks and 33% Mexican Americans were obese [39]. The diets of Mexican Americans may be different compared to other Hispanic and non-Hispanic groups. In addition, Mexican Americans born in Mexico may have differences in dietary intake patterns compared to those born in the United States [40]. Obesity is a strong risk factor for several diseases. For example, the incidence of type 2 diabetes mirrors the obesity epidemic [41]. Besides diabetes and cardiovascular disease, overweight and obesity may also explain about 15%–20% of all cancer deaths in the United States [42]. Among children, the rising levels of pediatric obesity have been accompanied by an increase prevalence of type 2 diabetes in adolescents and preadolescents over the past 20 years [43,44].

For some nutraceuticals, obesity itself may be an important effect modifier. A good example is folate metabolism. As would be expected for most nutraceuticals, research on the relationship between obesity and folate levels is only now evolving and includes only a handful of studies to date. Among U.S. women of childbearing age, increasing BMI was associated with lower serum folate levels both before and after folic acid fortification [45]. Obese Thai subjects were found to have lower serum folate than normal-weight subjects [46]. A study of obese adolescents in Austria reported that serum folate correlated inversely with BMI [47]. However, no significant differences in serum folate in overweight and normal-weight Brazilian adolescents [48] or obese and normal-weight adults in Israel [49] were seen. None of these studies had strict control for the confounding influence of diet. Using a tightly controlled dietary feeding study of postmenopausal women, we hypothesized that increased adiposity measured by BMI and DEXA would result in decreased serum folate concentrations. We investigated the relationship of serum folate, vitamin B12, homocysteine, and methylmalonic acid to obesity as assessed by BMI and other measures of adiposity among healthy postmenopausal women who did not smoke or take HRT. In multivariate analysis, we found that women who were overweight had 12% lower folate concentrations and obese women had 22% lower serum folate concentrations compared to normal-weight women (p trend=0.02). Increased BMI, percent body fat, and absolute amounts of central and peripheral fat were all significantly associated with decreased serum folate concentrations [50]. Because all the women in our study consumed the same diet, the lower serum folate concentrations found in women with higher BMI and fat mass may reflect a perturbation of whole-body steady-state folate concentrations with increased fat mass. Three possibilities could explain lower steady-state serum folate concentrations with increased fat mass: increased cellular uptake, increased intracellular retention, and/or increased renal excretion. Similar findings have been observed for vitamin D concentrations in the blood of obesity subjects. Therefore, in designing studies to investigate nutraceuticals in health and disease, an important consideration is the body composition of the subjects.

## Summary

The nutraceutical industry is large and growing, and nutraceutical products are sold on the basis of health claims without supporting epidemiological research evidence. From several large-scale nutrition intervention studies around the world (see, e.g., Chapter 2), there is clear evidence that dietary supplementation to adequately nourished populations does not provide additional health benefits. In the case of cancer, vitamin/mineral supplements to well-nourished populations increase the risk. This scenario is biologically plausible for other nutraceuticals and extracted bioactive compounds to other disease risk as well. There is no denying the value of proper nutrition for good health and disease prevention, but most of the western populations are adequately nourished, and it is uncertain whether this segment of the population, if supplemented with nutraceuticals in the forms of pills and capsules, will get health benefits or suffer harm. In some experimental model systems (nonhumans), there is preliminary evidence showing that nutraceuticals have broad-ranging physiological activities, and some of these effects could be harnessed for pharmaceutical interventions. However, it is very important that research using well-defined epidemiological methods be conducted to investigate their effects on health outcomes.

# References

1. Wu C-H, Wang C-C, Kennedy J. Changes in herb and dietary supplement use in the US adult population: A comparison of the 2002 and 2007 national health interview surveys. *Clin Therapeut* 2011;33(11):1749–1758.
2. Kipnis V, Carroll R, Freedman L et al. Implications of a new dietary measurement error model for estimation of relative risk: Application to four calibration studies. *Am J Epidemiol* 1999;150:642–651.
3. Kipnis V, Midthune D, Freedman L et al. Bias in dietary-report instruments and its implications for nutritional epidemiology. *Public Health Nutr* 2002;5:915–923.
4. Kipnis V, Midthune D, Freedman L et al. Empirical evidence of correlated biases in dietary assessment instruments and its implications. *Am J Epidemiol* 2001;153:394–403.
5. Kipnis V, Subar A, Midthune D et al. Structure of dietary measurement error: Results of the OPEN biomarker study. *Am J Epidemiol* 2003;158:14–21.
6. Carroll R, Midthune D, Freedman L et al. Seemingly unrelated measurement error models, with application to nutritional epidemiology. *Biometrics* 2006;62:75–84.
7. Subar A, Kipnis V, Troiano R et al. Using intake biomarkers to evaluate the extent of dietary misreporting in a large sample of adults: The OPEN study. *Am J Epidemiol* 2003;158:1–13.
8. Clevidence B, Reichman M, Judd J et al. Effects of alcohol consumption on lipoproteins of premenopausal women. *Arterioscler Thromb Vasc Biol* 1995;15:179–184.
9. Mahabir S, Baer D, Giffen C et al. Calorie intake misreporting by diet record and food frequency questionnaire compared to doubly labeled water among postmenopausal women. *Eur J Clin Nutr* 2006;60:561–565.
10. Roddam A, Spencer E, Banks E et al. Reproducibility of a short semi-quantitative food group questionnaire and its performance in estimating nutrient intake compared with a 7-day diet diary in the Million Women Study. *Public Health Nutr* 2005;8:201–213.
11. Sebring N, Denkinger B, Menzie C et al. Validation of three food frequency questionnaires to assess dietary calcium intake in adults. *J Am Diet Assoc* 2007;107:752–759.
12. Nagel G, Zoller D, Ruf T et al. Long-term reproducibility of a food-frequency questionnaire and dietary changes in the European Prospective Investigation into Cancer and Nutrition (EPIC)-Heidelberg cohort. *Br J Nutr* 2007;98:194–200.
13. Khani B, Ye W, Terry P et al. Reproducibility and validity of major dietary patterns among Swedish women assessed with a food-frequency questionnaire. *J Nutr* 2004;134:1541–1545.

14. Riboli E, Toniolo P, Kaaks R et al. Reproducibility of a food frequency questionnaire used in the New York University Women's Health Study: Effect of self-selection by study subjects. *Eur J Clin Nutr* 1997;51:437–442.
15. Smith-Warner S, Elmer P, Fosdick L et al. Reliability and comparability of three dietary assessment methods for estimating fruit and vegetable intakes. *Epidemiology* 1997;8:196–201.
16. Forman M, Zhang J, Nebeling L et al. Relative validity of a food frequency questionnaire among tin miners in China: 1992/93 and 1995/96 diet validation studies. *Public Health Nutr* 1999;2:301–315.
17. Goldbohm R, van den Brandt P, Brants H et al. Validation of a dietary questionnaire used in a large-scale prospective cohort study on diet and cancer. *Eur J Clin Nutr* 1994;48:253–265.
18. Mahabir S, Spitz M, Barrera S et al. Dietary boron and hormone replacement therapy as risk factors for lung cancer in women. *Am J Epidemiol* 2008;167(9):1070–1080.
19. Ma J, Betts N. Zinc and copper and their major food sources for older adults in the 1994–96 continuing survey of food intakes by individuals (CSFII). *J Nutr* 2000;130:2838–2843.
20. Karppanen H, Karppanen P, Mervaala E. Why and how to implement sodium, potassium, calcium, and magnesium changes in food items and diets? *J Hum Hypertens* 2005;19:S10–S19.
21. Fulgoni III V, Nicholls J, Reed A et al. Dairy consumption and related nutrient intake in African-American adults and children in the United States: Continuing survey of food intakes by individuals 1994–1996, 1998, and the National Health and Nutrition Examination Survey 1999–2000. *J Am Diet Assoc* 2007;107:256–264.
22. Ervin R, Kennedy-Stephenson J. Mineral intakes of elderly adult supplement and non-supplement users in the third national health and nutrition examination survey. *J Nutr* 2002;132:3422–3427.
23. Zhou W, Park S, Liu G et al. Dietary iron, zinc, and calcium and the risk of lung cancer. *Epidemiology* 2005;16:772–779.
24. Jackson RD, LaCroix AZ, Cauley JA et al. The Women's Health Initiative calcium-vitamin D trial: Overview and baseline characteristics of participants. *Ann Epidemiol* 2003 October;13(9 Suppl):S98–S106.
25. Hambdige M. Biomarkers of trace mineral intake and status. *J Nutr* 2003;133: 945S–948S.
26. Beutler E, Hoffbrand A, Cook J. Iron deficiency and overload. *Hematology Am Soc Hematol Educ Program* 2003:40–62.
27. Sian L, Mingyan X, Miller L et al. Zinc absorption and intestinal losses of endogenous zinc in young Chinese women with marginal zinc intakes. *Am J Clin Nutr* 1996;63:348–353.
28. Bettger W, Taylor C. Effects of copper and zinc status of rats on the concentration of cooper and zinc in erythrocyte membrane. *Nutr Res* 1986;6:451–457.
29. Ruz M, Cavan K, Bettger W et al. Erythrocytes, erythrocyte membranes, neutrophils and platelets as biopsy materials for the assessment of zinc status in humans. *Br J Nutr* 1992;68:515–527.
30. Muretto P, Angelucci E, Lucarelli G. Reversibility of cirrhosis in patients cured of thalassemia by bone marrow transplantation. *Ann Intern Med* 2002;136:667–672.
31. Barron BH, Hoyer JD, Tefferi A. A bone marrow report of absent stainable iron is not diagnostic of iron deficiency. *Ann Hematol* 2001;80:166–169.
32. Ammann A. Inductively coupled plasma mass spectrometry (ICP MS): A versatile tool. *J Mass Spectrom* 2007;42:419–427.
33. Heumann K. Isotope-dilution ICP-MS for trace element determination and speciation: From a reference method to a routine method? *Anal Bioanal Chem* 2004;378:318–329.
34. Dolan S, Nortrup D, Bolger M et al. Analysis of dietary supplements for arsenic, cadmium, mercury, and lead using inductively coupled plasma mass spectrometry. *J Agric Food Chem* 2003;51:1307–1312.

35. CDC. *Third National Report on Human Exposure to Environmental Chemicals.* Department of Health and Human Services, Centers for Disease Control and Prevention (CDC), Atlanta, GA. pp. 1–467, 2005.
36. Krachler M, Irgolic K. The potential of inductively coupled plasma mass spectrometry (ICP-MS) for the simultaneous determination of trace elements in whole blood, plasma and serum. *J Trace Elem Med Biol* 1999;13:157–169.
37. USDHHS. U.S. Department of Health and Human Services, National Institutes of Health. Strategic Plan for NIH Obesity Research. A Report of the NIH Obesity Research Task Force. NIH Publication No. 04-5493. 2004.
38. NCHS. *Prevalence of Overweight among Children and Adolescent: United States,* 1999. Hyattsville, MD: National Center for Health Statistics. 2001.
39. Hedley A, Ogden C, Johnson C et al. Prevalence of overweight and obesity among US children, adolescents, and adults, 1999–2002. *JAMA* 2004;291:2847–2850.
40. Dixon L, Sundquist J, Winkleby M. Differences in energy, nutrient, and food intakes in a sample of Mexican Women and Men: Findings from the Third National Health and Nutrition Examination Survey, 1988–1994. *Am J Epidemiol* 2000;152(6):548–557.
41. Calle E, Kaaks R. Overweight, obesity and cancer: Epidemiological evidence and proposed mechanisms. *Nat Rev* 2004;4:579–591.
42. Calle E, Rodriguez C, Walker-Thurmond K et al. Overweight, obesity, and mortality from cancer in a prospectively studied cohort of U.S. adults. *N Engl Med* 2003;348:1625–1638.
43. WHO. *Obesity: Preventing and Managing the Global Epidemic.* Geneva, Switzerland: WHO, pp. 1–276. (Report of a WHO consultation on obesity, Geneva, 3–5 June 1997), 1998.
44. Mokdad A, Ford E, Bowman B et al. Prevalence of obesity, diabetes, and obesity-related health risk factors, 2001. *JAMA* 2003;289:76–79.
45. Mojtabai R. Body mass index and serum folate in childbearing age women. *Eur J Epidemiol* 2004;19:1029–1036.
46. Tungtrongchitr R, Pongpaew P, Tongboonchoo C et al. Serum homocysteine, B12 and folic acid concentration in Thai overweight and obese subjects. *Int J Vitam Nutr Res* 2003;73:8–14.
47. Gallistl S, Sudi K, Mangge H et al. Insulin and independent correlate of plasma homocysteine levels in obese children and adolescents. *Diabetes Care* 2000;23: 1348–1352.
48. Hirsch S, Poniachick J, Avendano M et al. Serum folate and homocysteine levels in obese females with non-alcoholic fatty liver. *Nutrition* 2005;21:137–141.
49. Reitman A, Friedrich I, Ben-Amotz A et al. Low plasma antioxidants and normal plasma B-vitamins and homocysteine in patients with severe obesity. *Isr Med Assoc J* 2002;4:590–593.
50. Mahabir S, Ettinger S, Johnson L et al. Measures of adiposity and body fat distribution in relation to serum folate levels in postmenopausal women in a feeding study. *Eur J Clin Nutr* 2008;62:644–650.

# Index

Printed and bound by CPI Group (UK) Ltd, Croydon, CR0 4YY

21/10/2024

01777105-0012